An Introduction to String Algorithms

An Introduction to String Algorithms

Carl Kingsford

Princeton University Press

Princeton and Oxford

Published by Princeton University Press
41 William Street, Princeton, New Jersey 08540
99 Banbury Road, Oxford OX2 6JX

press.princeton.edu

GPSR Authorized Representative: Easy Access System Europe - Mustamäe tee 50, 10621 Tallinn, Estonia, gpsr.requests@easproject.com

ISBN 9780691274539
ISBN (e-book) 9780691274546

Library of Congress Control Number: 2025948270

British Library Cataloging-in-Publication Data is available

Editorial: Hallie Stebbins and Chloe Coy
Production Editorial: Mark Bellis
Text and Jacket Design: Wanda España
Production: Erin Suydam
Publicity: William Pagdatoon
Copyeditor: Lor Campbell Gehret

This book has been composed in Minion Pro

10 9 8 7 6 5 4 3 2 1

To Jessica, Henry, and Teddy

Contents

Preface

About this book

This book grew out of the "String Algorithms" course taught at Carnegie Mellon University. That class is aimed at senior undergraduates, master's students, and early Ph.D. students. It is designed as a fast-pace introduction to the area, covering the topics any string algorithm practitioner or researcher should know. The chapters are designed to cover the important aspects, the high-level ideas, and induce an understanding of the description of the algorithms and why they work as they do. The chapters are not intended to cover every detail, variant, or implementation challenge associated with the algorithms. We focus on the underlying algorithms (generally with proofs of correctness, running time, and space usage). We do not focus on particular applications, though we point out some tools that use some of these ideas occasionally as motivation.

Who should read this book?

You should read this book if you are interested in string algorithms but either have no experience with them or have familiarity with only a few of them. The book is aimed at people who want to get a solid foundation in string algorithms on which to build a later, deeper understanding. It is an on-ramp to more advanced texts (of which there are several good ones) including Gusfield [1997]; Crochemore and Rytter [2003]; Mäkinen et al. [2015]; Crochemore et al. [2007]; Durbin et al. [1998].

We assume you understand the language of algorithms: loops, if statements, functions, arrays, etc. We also assume some mathematical ability—big-O notation and some basic mathematical manipulations are used. By design, no seriously "advanced" mathematics is used in the text.

How to read this book

The course that this book is based on has used the text in two ways. It is possible and reasonable to cover the material in the order presented here (skipping a few chapters as needed). An alternative ordering jumps around, covering some exact matching algorithms, then some inexact matching algorithms, then some data structures, and then jumping back to more advanced exact algorithms, etc. In a typical one-semester course, I cover all the chapters except Chapters 10, 11, 22, 27, 33, and 34, though each of these could be added at the expense of removing some other material.

If you are reading this book on your own, you should feel free to similarly jump around. While it was not possible to make each chapter self-contained, links to other chapters are included where needed, and it's possible to pick any chapter that is of interest to you and start there. A few chapters use concepts (such as suffix trees and Huffman coding) that are only introduced later on—skimming the relevant later chapter should make understanding the earlier chapter easier. There are a few sections with titles notated with a * that are more challenging or that more heavily rely on later concepts.

Blind spots

We are not aiming to cover every area of string algorithms—the area is too vast for a book such as this, and besides such an endeavor would go counter to our goal of being an "on ramp" to more in-depth exploration. One area, where there has been enough work to fill another book of this size and which is extremely important in today's world, is parallel string algorithms that make use of more than one processor in clever ways to achieve speed ups. Unfortunately, we will not cover such algorithms. Online string algorithms (which must make decisions while streaming various inputs) are another interesting area that we omit due to our focus on the foundation. Nor will we cover the very new and emerging area of quantum string algorithms.

Acknowledgements

I would like to thank the students of my *String Algorithms* course over the years for their questions and perseverance working with earlier versions of this text, particularly, Atto Allas, Benjamin Kleyner, Hong-Sheng Lai, Kieran Carroll, Litian Liang, Lydia Yang, Molly Borowiak, Shijie Tang, Aresh Pourkavoos, Linqi Zhang, Mercury Liu, and Yaoyuan Gan, who used and commented on a nearly final draft of the book.

I also thank Guillaume Marçais for writing an early draft of the chapter on minimizers, a topic for which he has been leading the charge to put on a solid foundation. Much of his text for this chapter remains unchanged. I'd also like to thank Mingfu Shao, Ke Chen, and Paul Medvedev for careful comments on an early version of this manuscript. I must also happily thank Danny Sleator for comments on some chapters and the other instructors of 15-451 at CMU where some of these topics were presented.

Nearly none of the results in this book (excluding some work on minimizers, Section 23.5, and genome graphs) are my own. Instead, the interesting algorithms and data structures presented here have been developed over decades by insightful and clever researchers, from whom I have learned a lot. I have tried to cite the primary and secondary literature that is relevant, but I am sure I have missed many good citations in this vast field— please let me know if you think an additional citation is needed, and I will be happy to update future versions of the book. Whenever we say "we" or "our," this should be read in the royal sense, referring to the collaboration between the writer and reader in understanding the material.

I have to acknowledge that I learned about most of these topics first through secondary sources like Gusfield's excellent book and others. This has greatly informed my presentation of these topics, and in some cases I have adopted the notation used in those sources. Each chapter ends with a "Presentation Notes" box that describes those sources and gives a (incomplete) list of other explanations of the material of the chapter to both acknowledge that influence and to direct people to other non-primary sources that cover the topic. This material was written over the course of nearly a decade, initially without the intention of assembling it into a single collection, so I likely have missed some of these sources.

Although this work was not directly supported by any grants (and was written in support of my teaching responsibilities and during my free time), I would like to acknowledge several grants that made my research possible while writing this book. These include grants from the US National Science Foundation (1937540, 2232121), the US National Institutes of Health (R01HG012470, R01GM122935), and the Gordon and Betty Moore Foundation (4554). These organizations played no part in deciding on the existence or the content of this book.

Finally, I would like to thank my family, Henry, Teddy, and Jessica, for putting up with the long hours of work it took to write this book.

How the book was typeset

The book was written in LaTeX on top of the TeX system. Nearly all (except for a handful) of the figures were created using Ti*k*Z. The pseudocode was typeset using the `algorithmic` package. The images at the start of each part were drawn using custom Processing 4 code.

The figures and displayed equations are numbered jointly and sequentially within each chapter, and every displayed equation and figure is numbered (to facilitate interactive questioning by students) even if it is not referenced in the text. Theorems, lemmas, definitions, etc. are numbered sequentially within each chapter (separately from equations and figures). Algorithms are numbered separately from equations, figures, theorems, and other statements (but still sequentially within each chapter) to facilitate, finding them in the list of algorithms at the end of the book.

An Introduction to String Algorithms

CHAPTER 1

Introduction

1.1   What are strings?

In this book we focus on algorithms that operate on strings. What do we mean by strings?

Definition 1.1 (String). A *string* is a one-dimensional, ordered list of symbols drawn from some finite alphabet Σ. ∎

Typically, we assume $|\Sigma|$ is "small" compared with the length of the string. Usually, in fact, we will assume $|\Sigma|$ is a constant. For example, Σ may be the ASCII character set, or the Unicode character set, or the set of stock ticker symbols. On the other hand, choosing $\Sigma = \mathbb{R}$ (the real numbers) would produce something that is *not* a string.

A string is typically encoded into the memory of a computer using a sequence of contiguous, fixed-length fields, where each field (of say 8 bits) codes for some character of Σ. In the C language, for example, `char s[10]` represents a string of length 9 where Σ equals the set of symbols represented by the type `char`. In C, and some other languages, rather than encode the length of a string, a terminal special character (`\0`) is placed at the end of the representation. In other representations, an explicit length is stored.

None of these representations are essential for our definition of "string." They are implementation details. A string stored with each character in successive nodes of a linked list is also a string. Although in that case accessing the ith character incurs a cost of $O(i)$ instead of the cost $O(1)$ in the more typical encoding. While storage of a string in linked list or other more complex data structures still encodes something we will call a string, unless otherwise noted, we will assume that we are using a representation where accessing the ith character takes $O(1)$ time.

To encode a symbol from Σ using a simple encoding requires

$$\lceil \log_2 |\Sigma| \rceil \qquad (1.1)$$

bits. The notation $\lceil \dots \rceil$ indicates rounding up to the next integer; $\lfloor \dots \rfloor$ indicates rounding down.

1.2   Why study string algorithms?

The ability to process, store, search, and manipulate strings with computational efficiency has changed the world in many ways. Web search engines regularly process terabytes of strings, internet shopping sites deal with many product descriptions, and social networks handle large collections of comments and posts. Version control systems and compilers must process text that forms the source code of programs. Expert systems (like IBM Watson [Ferrucci et al., 2010] or NELL [Mitchell et al., 2015; Carlson et al., 2010]) must process strings to form databases of knowledge. Large Language Models (LLMs) [Radford et al., 2018; Devlin et al., 2018] train on massive amounts of text to generate human-like answers to queries.

The field of genomics has been a fruitful and motivating area for string algorithms, where, to a good first approximation, the genome of an organism can be encoded as a long string of letters representing the nucleotides that make up the DNA (or RNA) molecules. These genome strings can be huge—megabytes for a bacterium, gigabytes for an organism like human. Storing, searching, comparing, and analyzing these strings requires efficient algorithms. Reconstructing genomes or parts of them from fragmented measurements requires reconstructing strings from shorter strings. Collections of protein sequences are also well represented by strings. Genome graphs in Chapter 29 are motivated by applications in pan-genomics, and minimizers in Chapter 25 have their primary current application in genomics, though they were originally proposed in other contexts.

String algorithms are also fundamental from a computer science perspective. One can view the sequence of bits in a computer's RAM as a long string of 1s and 0s. Questions about what problems on strings can be efficiently solved have spurred innovation in computational techniques that can be widely applied and adapted to other problems that do not at first appear to be string-related. Solving a one-dimensional version of a problem is often a good first step toward understanding pattern matching problems in higher dimensions. Much of the theory of computational complexity is based around sets of strings. Finally, string algorithms are a fun and interesting type of algorithm that have given rise to interesting algorithmic techniques and neat data structures that are elegant and useful.

1.3   What are our goals?

This book aims to describe algorithmic solutions to string problems. We will generally take a "theoretical" viewpoint, meaning that we will be concerned with proving correctness and runtime for our various approaches rather than empirical evaluation, though we will discuss some practical considerations. While the emphasis is on provable properties, we aim to not be unnecessarily mathematical, but instead aim to draw out the key ideas that may be more generally useful. We are not trying to treat you as a "mathematics compiler" nor to provide the definitive proof of every result. Instead, we present the material in such a way that we hope you will understand why the algorithms work the way they do. We strive to cover everything you need to believe that the presented algorithms have the properties we claim they have, though, in the interest of focusing on the key ideas, some details are omitted from proofs and other material.

If you master the topics in this book, you will be well positioned to use and adapt string algorithms to solve new problems. You will also have the necessary background to begin making contributions to the design of new string algorithms and applications.

The field of string algorithms is vast, and we cannot hope to cover everything. There are many more advanced or esoteric topics that are of great interest—both theoretically and for particular applications—that we must omit. An entire second volume could be written covering advanced topics that are only hinted at here. We also don't aim to cover the complete history of the presented algorithms. Once you grasp what is included here, you will have the tools to learn about these other topics on your own.

1.4   A roadmap

We start (Part I) with the classical problem of finding occurrences of a short string P exactly within a longer text T. This is, of course, the string problem that is most used and most useful, on its own or as a building block for bigger string processing systems.

In Part II, we move to the edit distance problem that supports finding matches with some deviation from the pattern (extra characters, mismatching characters, etc.); this allows us to introduce the classical dynamic programming approach for this problem and also some tricks that enable faster or more space-efficient realizations of the underlying dynamic programming technique. This collection of problems is particularly important in genomic analyses, where sequences have mutations.

Part III introduces data structures that have been developed to store strings and solve a variety of problems associated with them. These include the classical (and very important) suffix trees and suffix arrays, as well as newer (though still decades old) data structures such as wavelet trees and RRR compressed bit vectors. The building blocks introduced in this part are foundational to a number of string processing algorithms, and often useful outside of string problems.

Part IV expands on the techniques for creating small, lossy representations of strings or sets of strings allowing for some loss of accuracy but with great gains in space or time efficiency. Classical approaches such as locality sensitive hashing and Bloom filters are introduced, as are techniques that are more modern such as sketching for edit distances and the minimizers technique. These sketching techniques are frequently used to deal with extremely large strings or collections of strings.

Part V turns to looking at, broadly, generative models for strings. By encoding a family (perhaps of infinite size) of strings in a compact way, we can answer questions about that family of strings and relate other strings to the family. Techniques introduced in this part include the classical hidden Markov models, but also newer approaches such as genome graphs.

Finally, Part VI collects miscellaneous topics that do not fit in other parts.

1.5   A Primer

Here, we give a primer on computational and discrete mathematical concepts that will be useful in the text.

1.5.1   String notations and definitions

$|S|$ is the length of string S. ($|A|$ is also the size of set A.) Σ as a variable will always be an alphabet over which strings are formed. Typically, we start strings at index 1, which is convenient for explaining the algorithms, but not the norm for actually programming the algorithm, where indexing starting at 0 is more common. $T[i]$ is the character of string T at position i.

A *substring* of a string S is a contiguous sequence of characters from S. $T[i \ldots j]$ denotes a substring of string T starting at index i and continuing to index j, inclusive. We will sometimes use other notation for substrings that will be introduced when used.

A *prefix* is a substring of characters starting from the start of the string to some higher index in the string. A *suffix* is a substring of characters starting from some position in the string and continuing until the end of the string. A *proper* prefix (suffix) is a prefix (suffix) that is neither empty nor the entire string. The *empty string*, denoted ϵ, consists of 0 characters. It is a (non-proper) suffix and prefix of any string.

Concatenation of two strings α and β is denoted $\alpha\beta$ or $\alpha \circ \beta$ when we want to emphasize the operation.

1.5.2   Graph theory

An *undirected graph* is a pair of sets (V, E), where each element of V is a *vertex* (also called a *node*), and E is a subset of $\{\{u, v\} : u \in V, v \in V\}$, each element of which is an *edge* that connects two vertices. If E is instead a subset of $\{(u, v) : u, v \in V\}$, the resulting pair (V, E) is a *directed graph*.

Examples of undirected (left) and directed (right) graphs are:

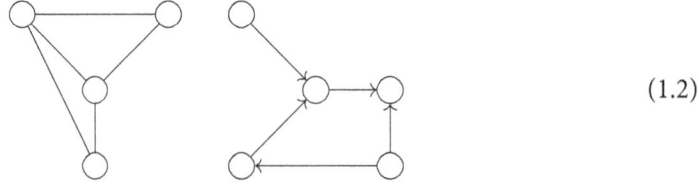

(1.2)

A *path* through a graph is a sequence of vertices $v_1, \ldots, v_k$ such that each successive pair in the sequence is connected by an edge (if the graph is directed, the edge must be directed from a vertex to the successive one, i.e., $(v_i, v_{i+1}) \in E$ for edge set E). A *cycle* is a path where $v_1 = v_k$. A *simple* path or cycle is one in which no two vertices (except v_1, v_k in the case of a cycle) are equal. An undirected graph is *connected* if there is a path between every pair of vertices.

A graph is *acyclic* if it does not contain any cycles. If the graph is directed and acyclic, it is called a *DAG* (for directed acyclic graph). If an undirected graph is connected and contains no cycles it is a *tree*. If it is not connected and contains no cycles, it is a *forest*.

A *rooted tree* is a directed graph for which the corresponding undirected graph is a tree and for which there is some node r (the *root*) such that every edge goes from a vertex closer to the root to a vertex that is farther from the root. For an edge (u, v) in a rooted tree, u is called the *parent* of v and v is a *child* of u. In a rooted tree, there is a unique path from the root to any node. If v is contained on that path to a node u then v is an *ancestor* of u and u is a *descendent* of v. If a node has no descendants, it is a *leaf*. An example of a rooted tree:

$$(1.3)$$

A *traversal* of a tree is an algorithm to visit each vertex of a (rooted) tree. A *breadth-first search* (BFS) of a tree visits each node in increasing distance from a starting node r; that is a vertex at distance i from the root is visited before any vertices at distance $i+1$. A *depth-first search* of a tree starting from node r recursively visits children until a leaf is reached, then returns to the most recently visited node with unvisited children, and continues by visiting one of its unvisited children.

1.5.3   Running times

Big-O notation $O(f(n))$, where f is a non-negative function of an integer n, represents an upper bound on the asymptotic behavior of a resource usage as the problem instance size, n, grows. For example, with $f(n) = n$, $O(f(n)) = O(n)$ indicates linear growth; $O(n^2)$ indicates quadratic growth, and $O(1)$ indicates growth that is a constant independent of the size of the problem.

More formally, let $f(n)$ be a function that gives the worst-case runtime (in terms of number of fundamental computational steps) of an algorithm on an instance of size n. We say an algorithm with running time $f(n)$ is in $O(g(n))$ if there exists constants c, n_0 such that for all $n \geq n_0, f(n) \leq cg(n)$. A consequence of this asymptotic definition is that only the "leading" term in $f(n)$ matters: $f(n) = 10n^3 + 5n$ is in $O(n^3)$ and $f(n) = \log n + 2\log\log n$ is in $O(\log n)$. An important special case is when $f(n) \leq k$ for some constant (meaning not depending on n) k. Then $f(n)$ is in $O(1)$ since with $g(n) = 1, f(n) \leq kg(n)$. The $O(\cdot)$ notation can be used for resources besides runtime (like memory usage).

If a computational problem has an algorithm that runs in time $O(n^k)$ for some fixed k, we say that the problem is solvable in *polynomial time*. The set of problems that are solvable in polynomial time is denoted P. While it is an imperfect definition, P is viewed as the set of problems with *efficient* algorithms. See Chapter 34 for more discussion of the set P and related sets of problems. If an algorithm is in $O(n)$ for problem size n, we say the algorithm runs in *linear time*, which is usually the ideal time (not always achievable) for string algorithms.

A relative of Big-O is Ω notation, where $\Omega(f(n))$ indicates a *lower bound* on the growth of f for increasing n. An algorithm said to take $\Omega(n^2)$ time will take at least quadratic time asymptotically. If $f(n)$ is both $O(f(n))$ and $\Omega(f(n))$ then we say it is $\Theta(f(n))$.

Sometimes it is convenient to express the running time of an algorithm on instances of size n as a *recurrence relation*, where the time $T(n)$ is expressed as a function of the time for a smaller problem:

$$T(n) \leq f(n, T(n')), \qquad (1.4)$$

where $n' < n$ and f is some function of the instance size and the running time of smaller instances. "Solving" such a recurrence means finding a non-recursive expression that is an upper bound on $T(n)$. This is especially relevant when the algorithm itself is recursive or can be analyzed inductively.

1.5.4   Sets and combinatorics

A *set* is a collection of unique items. If A and B are sets, $A \cup B$ is the set of items that are in one or both of B, called the *union*. $A \cap B$ is the set of items that are in both A and B, called the *intersection*. A *multiset* is a set that allows duplicated items.

Given a set A of size n, the number of ways to choose k distinct items from that set is denoted $\binom{n}{k}$, pronounced *n choose k*. We have $\binom{n}{k} = \frac{n!}{(n-k)!k!}$, where $x!$ is the *factorial* function $x \cdot x - 1 \cdot x - 2 \cdot \ldots \cdot 1$.

An important special case of the *binomial theorem* is:

$$2^n = \sum_{k=0}^{n} \binom{n}{k}. \tag{1.5}$$

1.5.5   Numbers

A base-2 (aka *binary*) representation of a number is a sequence of n bits $b_{n-1}, b_{n-2}, \ldots b_0$, where each $b_i \in \{0, 1\}$ and that represents the number $\sum_{i=0}^{n-1} 2^i b_i$. Adding an additional bit doubles the range of numbers that can be represented. Hence, representing a number M requires $\lceil \log_2 M \rceil$ bits.

Base-16 is *hexadecimal*, where a number is represented with digits $\{0, \ldots, 15\}$, denoted $\{0, \ldots, 9, A, B, C, D, E, F\}$. In hexadecimal, a number $d_{n-1} d_{n-2} \ldots d_0$ represents $x = \sum_{i=0}^{n-1} d_i 16^i$. To distinguish hexadecimal from other bases, the number is typically preceded by the symbol 0x.

The bitwise "and" function is defined as $x \& y = 1$ if and only if bits $x = 1$ and $y = 1$ (otherwise 0). The bitwise "or" function is defined as $x | y = 1$ if and only if $x = 1$ or $y = 1$ (otherwise 0). When x and y are equal-length bit vectors, the $\&$ and $|$ operations are applied bitwise to each bit.

If x and y are positive integers, $x \pmod y$ is the remainder after x is divided by y.

1.5.6   Data structures

An array A is a contiguous sequence of values, indexed by an integer i. $A[i]$ is the value stored at position i. Arrays, like strings, may be indexed starting at 0 or 1 as needed.

A *hash table* is a data structure that stores key-value pairs (k, v). There are many implementations of this data structure. We assume that the average time to access the value v associated with key k is $O(1)$. This is a non-trivial assumption, but there are many data structures where this is true. A *hash function* is a function $h(k)$ that maps a key to a location in a hash table. Both hash functions and hash tables have a rich theory surrounding specific implementations, which we assume the reader can find if they are interested.

1.5.7 Probability

Let U be a universe of things that could happen. An *event A* is a subset of U meaning that one of the things in A happened. $\Pr[A]$ is a function that assigns a *probability* to one of the things in A happening. This probability is a number between 0 and 1. The notation $\Pr_X[A]$ means that probability that A happens over random variable X.

The conditional probability $\Pr[A \mid X]$ is the probability that A happens given that X happens. $\Pr[A \mid X] = \frac{\Pr[A \cap X]}{\Pr[X]}$.

Part I

Exact Matching

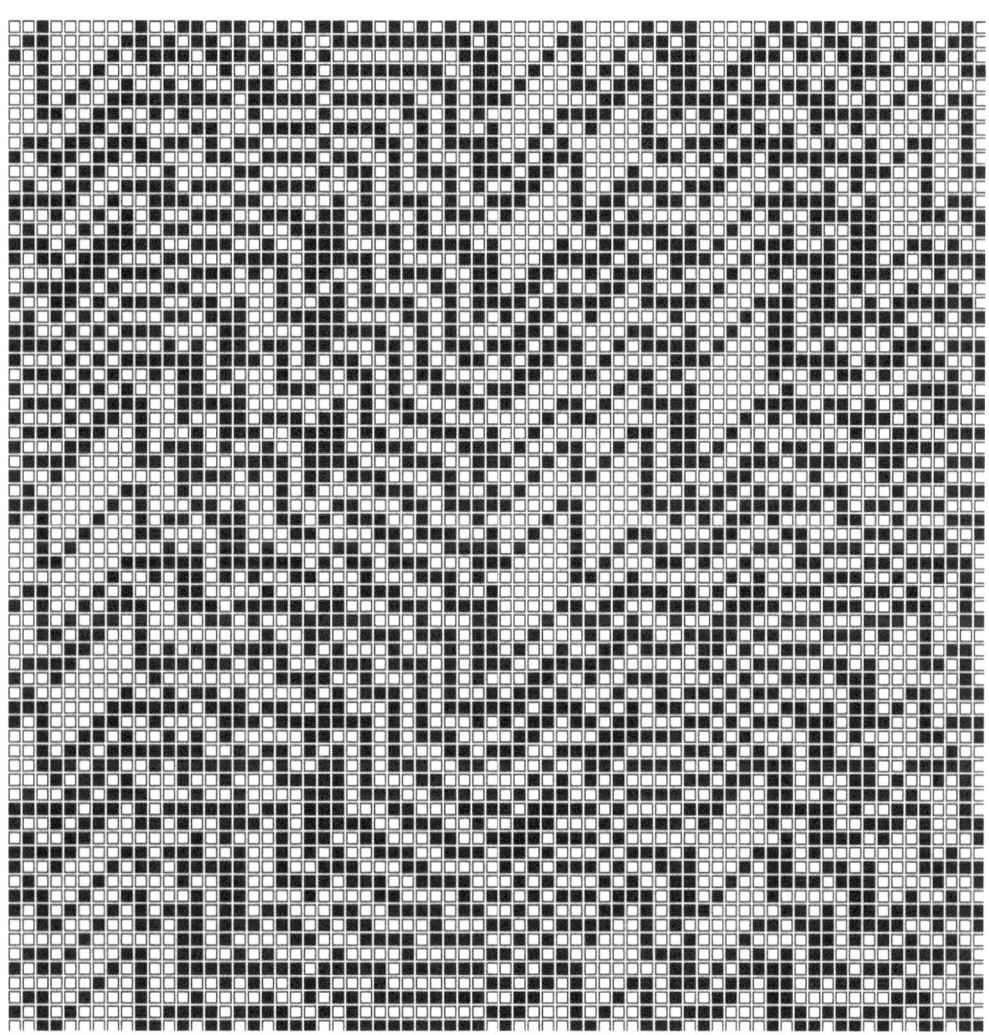

CHAPTER 2

The Z Algorithm

2.1 The exact matching problem

We start with the most obvious, and most widely used, problem on strings:

Problem 2.1 (Exact String Matching). *Given a (long) string T and a shorter string P, find all occurrences of P in T. Occurrences of P are allowed to overlap.* ◆

Here, T is the "text" (our common name for a big, relatively unchanging string) and P is the "pattern" (the name for a shorter string, usually specified by a user or an application).

The motivation for this problem is clear. It is used to implement "Ctrl-F" in your web browser or word processor. It needs to be solved to find occurrences of functional substrings (like transcription factor binding sites) in a genome sequence. Your source code editor may use it to find places where a variable is defined or used. A web search engine needs to find whether your search term occurs on a web page. Even in cases where inexact matching is the most natural, practical solutions often are built from exact matches. The problem is foundational.

2.2 Simple (slow) solution

The naïve algorithm solves Problem 2.1 via two loops.

Algorithm 2.1. Simple exact match search.

> **function** SIMPLEMATCH(T, P)
>   **for** $i \leftarrow 1, \ldots, |T| - |P| + 1$ **do**
> $j \leftarrow 1$
>   **while** $j \leq |P|$ **and** $T[i + j - 1] = P[j]$ **do**
> $j \leftarrow j + 1$
>   **if** $j = |P| + 1$ **then**
>   **output** "Occurs at i"

This algorithm checks whether an instance of P occurs at each location in T. Because of the two loops, it runs in time $O(|T| \times |P|)$. When P is short, this might be fine, but as P grows, this algorithm becomes impractical.

The key idea behind improving over the runtime of the naïve algorithm is recognizing that iteration $i+1$ of the outer `for` loop operates entirely without consideration of what happened during iteration i. But the algorithm may have "learned" a lot about the part of T that is relevant for iteration $i+1$ during iteration i because iteration i may have looked at characters $i \ldots i + |P| - 1$ already.

For example:

$$
\begin{array}{l}
\quad\quad i \\
\quad\quad \downarrow \\
\texttt{All this happened, more or less} \\
\quad\quad\quad\quad\quad \texttt{happy} \\
\quad\quad\quad\quad\quad\quad \texttt{happy.}
\end{array}
\tag{2.1}
$$

After comparing `happy` to "`happe`" at iteration i:

- we know that $T[i \ldots i+3] = $ "`happ`" $= P[1 \ldots 4]$;
- we can deduce that there can be no match at $i+1$ because $T[i+1] = P[2] = $ "`a`" but $P[1] = $ "`h`";
- in fact, since "`h`" does not appear in $T[i \ldots i+3] = P[1 \ldots 4]$, we could set $i \leftarrow i + 4$.

Because it is the parts of P that matched T that we have learned about, it is the similarities between various parts of P that allow us to make these deductions. Therefore, this motivates the idea of preprocessing P to speed things up to enable incrementing i by more than 1 when we can in an execution of the outer loop.

What speed should we hope for? Any algorithm must take $\Omega(|P| + |T|)$ time, since that is the size of the input. The maximum output size is $O(|T|)$. So it would be great if we could find an algorithm that takes $O(|P| + |T|)$. (Since $|P|$ must be smaller than $|T|$, $O(|T| + |P|) = O(2|T|) = O(|T|)$.) Finding such a linear-time algorithm was a big open question for a long time, but now there are several such algorithms known. We will see three in this and the next two chapters.

2.3   The Z algorithm

The "Z" algorithm was introduced by Gusfield [1997] to synthesize and unify several exact matching algorithms. It is based on what Gusfield calls the "fundamental preprocessing" that will compute values $Z_i(P)$ defined as follows:

Definition 2.2 (Z values). $Z_i(P)$ is the length of the longest substring of P that starts at i and matches a prefix of P. The Z_i are defined for $1 < i \le n$ for strings starting at index 1. We omit the (P) when it's clear. ∎

Here is the definition in a picture.

$$\tag{2.2}$$

The narrow bar represents a string P and each thicker bar represents an equal string of length $Z_i(P)$. The two characters following each bar must not be equal (or the bars would be longer).

Some examples of Z_i values are:

- $P = $ "aardvark": $Z_2 = 1, Z_6 = 1$,
- $P = $ "alfalfa": $Z_4 = 4$,
- $P = $ "photophosphorescent": $Z_6 = Z_{10} = 3$.

For any string, $Z_1(P) = |P|$, so a Z_1 value would generally be uninteresting, and that is why Z_i is only defined for $i > 1$.

Using Z values for exact matching. If the Z values have been computed, then there is a linear-time algorithm, ZMatch, to find all the locations of the pattern P in T.

Algorithm 2.2. Using Z values for exact match.

    function ZMATCH(T, P)
       $S \leftarrow P\$T$
       Compute all $Z_i(S)$
       return all $i - |P| - 1$ such that $Z_i(S) = |P|$

Here, \$ is some character that is not in P or T and $P\$T$ denotes concatenating the two strings separated by \$. The expression $i - |P| - 1$ simply converts indices in S into the corresponding index of T.

Why does this work? Since P doesn't contain \$, no index $i < |P| + 1$ in S will be returned, so all returned indices will be in the T region of S. Similarly, no $Z_i(S)$ can be greater than $|P|$ since \$ doesn't occur in T. So, $Z_i(S) = |P|$ if and only if the string starting at i in S matches P.

The running time is $O(|P| + |T| + z_S)$ where z_S is the time to compute the Z values for S. The concatenation to create S can be done in $O(|P| + |T|)$ time, and we can check every Z_i value in the same amount of time. The challenge is to create a fast algorithm to compute the Z_i values for a given string. We will now see a linear-time algorithm to do that.

2.4   Computing the Z values

The algorithm to compute the Z values uses the concept of a "Z box", defined as follows:

Definition 2.3 (Z-box). The Z-box at i is the non-empty substring starting at i and continuing to $i + Z_i - 1$. This is the substring that matches the prefix. There is no Z-box at i if $Z_i = 0$. ∎

The Z-box at i is the string that is the witness to the value of Z_i. The algorithm to compute the Z values will iteratively compute Z_k given:

- $Z_2 \ldots Z_{k-1}$
- the boundaries ℓ and r of the *rightmost* Z-box found so far starting someplace in $2 \ldots k - 1$. Here, rightmost means the box with its right end farthest down the string.

The algorithm will also return new values for ℓ and r. The algorithm computes Z_i in increasing order of i, and maintains the ℓ and r values at each iteration. Suppose we've computed up to Z_{k-1}.

To compute Z_k. The input is $Z_2, \ldots, Z_{k-1}$ and ℓ and r, where ℓ and r are the left and right boundaries of the rightmost Z-box starting someplace in $2 \ldots k-1$. The output is Z_k and updated values for ℓ and r. There are two main cases:

Case $k > r$: Explicitly compute Z_k by comparing the prefix with the substring starting at k. If $Z_k > 0$, then set $\ell = k$ and $r = k + Z_k - 1$ since this is a new farther-right Z-box. This is also the base case for computing Z_2, where we can assume $\ell = r = 0$ to start.

Case $k \leq r$: This is the more complex case. The situation looks like (2.3).

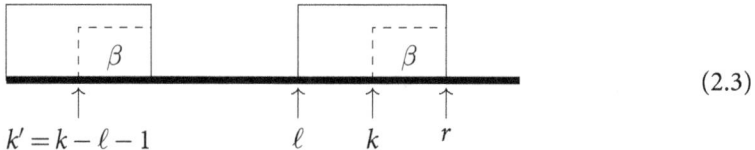

$$(2.3)$$

k must be between ℓ and r: it's before r by assumption of the case, and ℓ must precede it because we're computing the Z_i in increasing order so we couldn't have found a Z-box starting at a location after k yet.

In (2.3), the leftmost solid box represents the prefix that matches the rightmost Z-box. The strings are the same inside the solid boxes by the definition of Z boxes. k' is the position corresponding to k inside this prefix. Note that $Z_{k'}$ has already been computed.

Define, as in figure (2.3), β to be the length of the string from k to r. There are now two subcases:

Subcase $Z_{k'} < \beta$: Set $Z_k = Z_{k'}$ and leave ℓ, r unchanged. The reason this is correct is that the situation looks like (2.4)

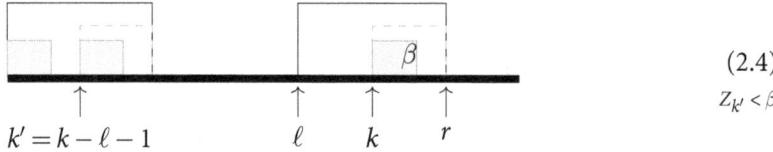

$$(2.4)$$
$$Z_{k'} < \beta$$

where the length of the three small shaded boxes are all $Z_{k'}$. The strings under each shaded box are equal. The first two are equal by the definition of $Z_{k'}$, and the second two are equal because they are both (by the assumption of the case) under the equal solid boxes. Since $Z_{k'} < \beta$, we know that the character following the first shaded box in (2.4) must differ from the character following the second shaded box (otherwise $Z_{k'}$ would have been larger). Since the two length-β strings are the same, we know that the character following the third shaded box must also differ from the character following the first shaded box. Hence, Z_k must equal $Z_{k'}$.

Subcase $Z_{k'} \geq \beta$: Here, we have the following situation:

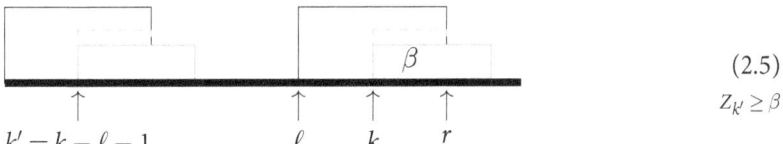

$$(2.5)$$
$$Z_{k'} \geq \beta$$

where again the shaded boxes are of length $Z_{k'}$. The strings under the shaded boxes *within the solid-line boxes* are the same, and they both equal a prefix of P by the definition of the Z values. But we know nothing about the part of the second shaded box that follows r: we've never looked at these characters. In this case, we explicitly compare *after* r to find the end of the string corresponding to Z_k. We set $\ell = k$ and r to the point just before this comparison failed. This case lets us skip comparing the characters inside the length-β string.

Correctness of the algorithm follows by induction and the previous arguments.

Runtime. For a string S, this runs in $O(|S|)$ time. To see this, we can count the number of comparisons that are done, classifying them as either *matches* or *mismatches*.

No comparisons happen in the $Z_{k'} < \beta$ subcase. Comparisons can happen either in the $k > r$ case or in the subcase $Z_{k'} \geq \beta$. In both of those cases, characters that are compared are only those that are not already contained in a Z box.

Characters that are matched will be placed into a new Z box (and therefore not compared in any future step). Therefore, characters covered by a Z box will be matched at most once, and once matched they will be placed in a Z box. So there are $O(|S|)$ matches.

Every iteration contains at most one mismatch (since a mismatch ends an iteration). Hence, there are $O(|S|)$ mismatches.

So, there are at most $O(|S| + |S|)$ comparisons during the algorithm. Finally, the $Z_{k'} < \beta$ case takes $O(1)$ time and can happen at most $O(|S|)$ times. The total runtime is therefore $O(|S|)$.

2.5   Summary and notes

The previous text immediately gives the $O(|P| + |T|)$-time algorithm for the exact match problem: compute the Z_i values for $P\$T$, which takes $O(|P| + |T|)$, and then apply the ZMATCH algorithm, which also takes $O(|P| + |T|)$ time.

This is optimal in the worst case, since you might have to look at the whole input. However, better algorithms exist in practice, that for many real strings, avoid looking at some input characters. We'll see one of these in the next chapter. Another practical issue with ZMATCH as we have described it is memory usage: the algorithm keeps around $O(|T| + |P|)$ Z_i values. See Exercise 2.8 to reduce this.

The Z algorithm is *alphabet independent* meaning that that algorithm views the string only in terms of character comparisons: the algorithm would still work if you could only access strings via a function EQUAL(a, b) that returns whether two characters a, b are equal.

The Z algorithm [Gusfield, 1997] neatly exposes the central preprocessing ideas used in several algorithms. We will reuse the Z values, and values related to them, in future algorithms.

Presentation Notes

Our presentation follows that of Gusfield [1997], Chapter 1, which also introduces the figures in the style of (2.3).

2.6   Exercises

2.1   Give a pattern of length n and a text of length m where the naïve algorithm takes $O(n+m)$ time.

2.2   Give an instance where the naïve algorithm takes $\Omega(|T| \times |P|)$ time.

2.3   Suppose $Z_2 = k > 0$. What are $Z_3, \ldots, Z_{k+2}$? Why?

2.4   Consider the case in the algorithm to compute the Z_i values when $k \leq r$ and $Z_{k'} \geq \beta$ (this is the last subcase in the algorithm description). Explain why if $Z_{k'} > \beta$ then $Z_k = \beta$ and no character comparisons are needed.

2.5   Let x and y be strings of length n and m, respectively. Give an $O(n+m)$-time algorithm to find the longest suffix of x that exactly matches a prefix of y.

2.6   (a) Give a string of length 10 where all the Z_i values are 0. (b) Can all the Z_i values equal 1 for an arbitrarily long string? If so, give an example; if not explain why not. (c) Can $Z_i = Z_{i+1} = 1$ for some value of i in an arbitrarily long string? If so, give an example; if not, why not?

2.7   Suppose for some string $Z_i = k \geq i$. What is Z_{2i-1}?

2.8   Explain how to implement the idea behind the ZMATCH algorithm (including creating any required Z_i values) in $O(|P|)$ space.

2.9   Given only the Z_i values for a string T (and not the string itself), give an algorithm to find the positions in T where the unknown character $T[0]$ occurs.

CHAPTER 3

Boyer-Moore

Another algorithm for the exact matching problem (Problem 2.1) that is often very good in practice was given by Boyer and Moore [1977]. We'll go over this algorithm at a high level so we can see some of its unique features. The version we are going to cover in this chapter runs in $O(|T| \times |P|)$ time for a text T and pattern P. This means that, theoretically in the worst case, this version of Boyer-Moore is not as good as the Z algorithm from Chapter 2. But even this version often does much better than the worst case running time on typical strings, and extensions, which we will not cover, exist to make it run in $O(|P| + |T|)$ time in the worst case.

There are two big ideas in this algorithm. Idea #1 is that we will move the pattern P along text T from the left-to-right as before, but at each location where we put the pattern, we will compare P and T *right-to-left* instead of left-to-right.

$$
\begin{array}{l}
\texttt{t h e q u i c k b r o w n f o x} \\
\qquad\qquad\quad \texttt{x | | | |} \\
\qquad\qquad\quad \texttt{c r o w n}
\end{array}
\tag{3.1}
$$

Idea #2 is that we will use some rules to shift the pattern by more than one position after a mismatch (or complete match) when we can. The algorithm has two different rules for this. These rules will guarantee that no match will be missed, but they may not always apply, so in the version we give here, the algorithm could do many single-position shifts.

3.1   High-level description

3.1.1   First rule: Next Matching Character

We start by defining some values that encode properties of the pattern.

Definition 3.1 (R_i). $R_i(x)$ is the position of the rightmost occurrence of character x before position i in the pattern P. For example, if $P =$ "abaaab", then $R_3(\texttt{b}) = 2$ and $R_3(\texttt{a}) = 1$. ■

Next Matching Character Rule: When a mismatch occurs at position i in the pattern against position k in the text, like so:

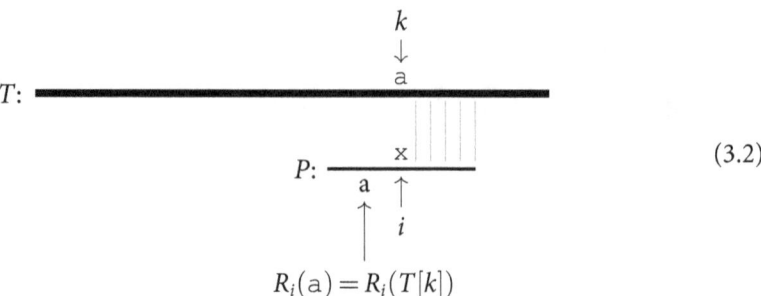

$$(3.2)$$

shift by $i - R_i(T[k])$ so that the next occurrence of $T[k]$ in the pattern is aligned with position k in T.

Since we mismatched, we know we have to shift at least 1 character, but we also know the current character must match if the pattern is going to match. The rightmost instance of the character before the current position is the next chance for the current character to match.

3.1.2   Second rule: Good Shift Rule

When a mismatch occurs, we will have matched some suffix α of P.

$$(3.3)$$

Apply the following cases in order, doing the first one that shifts by a non-zero amount.

Case (A): Shift so that the rightmost, non-suffix occurrence in P of the matched suffix α is aligned to the matched part of T.

$$(3.4)$$

We additionally require that the character in (3.4) at which we mismatched (x) does not equal the character directly before the earlier occurrence of α (y). The two occurrences of α might overlap.

Case (B): Let β be the longest prefix of P that matches a suffix of α. Shift so that the prefix β is matched to the region of T that last matched the suffix β.

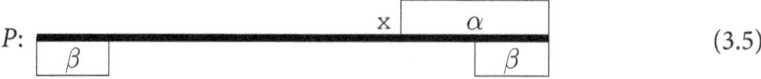

$$(3.5)$$

Case (C): If neither (A) or (B) apply, shift by $|P|$ characters.

Why will we never miss a match? If there is a match that overlaps the part of T we have compared during this iteration, then either case (A) or (B) must apply: that is the matched suffix of P, which we now know occurs in T either matches someplace else earlier in the pattern (case A) or the end of this matched part matches the beginning of the pattern (case B). You can think of case (B) as really the same as case (A) conceptually, except it handles the case where the earlier match "falls off of the start of P." Since we take the rightmost instance of these, we won't ever miss a match.

Another way to view the good shift rule is as a generalization of the idea of the next matching character rule to strings that are longer than one character. We know the end of α must match someplace in T, so we shift to the first (rightmost) place where that happens.

3.1.3   Complete algorithm

Putting these together, we have the Boyer-Moore algorithm (given in pseudocode with some edge cases missing in order to highlight the main ideas).

Algorithm 3.1. High-level Boyer-Moore.

function BOYERMOORE-HIGHLEVEL(P, T)
    $k \leftarrow 1$
    while $k \le |T| - |P| + 1$ **do**   ▷ *until we have looked at the whole text*
        Compare P to $T[k \ldots k + |P| - 1]$ from right to left. Stop at mismatch or complete match.
        if complete match **then**
            output report match
            $k = k +$ "match shift"   ▷ *we'll define this soon*
        else
            $k = k + \max\{$ next matching character rule, good shift rule, $1\}$

Since this may shift by only 1 each time through the outer loop, it could take $O(|P| \times |T|)$ time plus the time it takes to implement the good shift and next matching character rules. By the previous arguments, this algorithm is correct: it will find only matches and won't miss any matches. The remaining question is how to implement those shifting rules by precomputing various features of the pattern.

3.2   Computing the R_i values for the next matching character rule

We could implement the R_i values by creating a two-dimensional array $R[x, i]$ where $R[x, i]$ is the position of the rightmost occurrence of character x before position i. This allows us to compute $R_i(x)$ in constant time. However, the size of such an array depends on the alphabet size, which is often not desirable (i.e., suppose your alphabet is Unicode—that's a "constant" size, yes, but it's a big constant, and the 2-D array could be huge). For each character, this simple matrix implementation would use $O(|P|)$ space for that row.

A better way that also highlights a new idea is to use an array `Occur` of lists where `Occur[x]` is a list of positions in decreasing order where x occurs in P. To compute $R_i(x)$ using this list, we scan down list `Occur[x]` until we find the first index that is $< i$. The total space for the list in this structure is $O(|P|)$ since the sum of the lengths of the lists equals $|P|$. In a simple implementation, the array would still have length equal to $|\Sigma|$, but we could also replace the array with a hash table to only store rows for characters that occur in P.

Runtime to compute $R_i(x)$ using this multiple list implementation is at most $O(|P| - i)$ time, since there can be at most $|P| - i$ items (positions) on the list that are $\geq i$, and those are the only ones that we will have to skip over when traversing the list. This is worse than constant, but—and here's the neat idea—we only call this routine after matching $O(|P| - i)$ characters. So, this at most doubles the running time; that is, it only adds a constant factor overhead.

3.3   Formalizing the Good Shift Rule

To implement the good shift rule efficiently, we need to do some preprocessing on the pattern, just as we did for the R_i values and the Z_i values for the Z algorithm. In fact, some of our preprocessing will use the Z_i values from Chapter 2. To get started, we need some notation to formalize the good shift rule.

Definition 3.2 $(L(i))$. Let n be the length of P. $L(i)$ is the largest index such that $P[i \dots n]$ matches a suffix of $P[1 \dots L(i)]$ and $P[i-1] \neq$ the character preceding that suffix. $(L_P(i) = 0$ if no such index exists.)

$$P: \qquad \begin{array}{ccc} \boxed{\text{y} \quad \alpha} & \boxed{\text{x} \quad \alpha} \\ \quad\uparrow \qquad \uparrow & \qquad\qquad\uparrow \\ \quad L(i) \qquad i & \qquad\qquad n \end{array} \tag{3.6}$$

∎

In other words, $L(i)$ is the end of the rightmost place that matches the suffix of P starting at i (and also has a different preceding character). The reason this is a useful value is that one of the cases of the good shift rule corresponds to shifting the pattern so that $L(i)$ is moved to the position in T where the end of the pattern currently is.

Definition 3.3 $(\ell(i))$. Let n be the length of P. $\ell(i)$, for $i \geq 2$, is the length of the largest suffix of $P[i \dots n]$ that equals some prefix of P (0 if none exists).

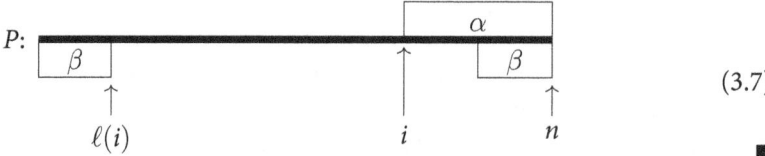

$$\tag{3.7}$$

∎

In other words, $\ell(i)$ is the rightmost end of the longest prefix that matches some suffix of the matched suffix.

These two definitions formalize cases (A) and (B) of the good shift rule. Now, we can state that rule as:

Mismatch Case (A): shift by $|P| - L(i)$ if $L(i) > 0$.
Mismatch Case (B): if $L(i) = 0$ and $\ell(i) > 0$: shift by $|P| - \ell(i)$.
Mismatch Case (C): if not (A) or (B), shift by $|P|$.

We can also use the definition of ℓ to formalize the "match shift" amount by which we shift when we find a complete match:

If match: shift by $|P| - \ell(2)$ places.

Why can we shift by this amount when we match? Where does $\ell(2)$ come from? On a match, the situation looks like (3.8).

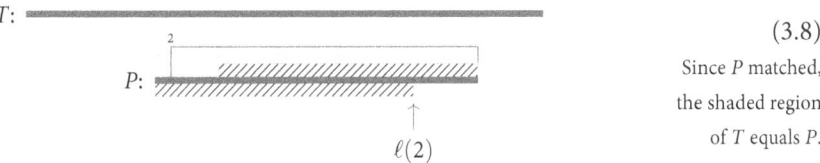

$$(3.8)$$

Since P matched, the shaded region of T equals P.

We need to shift so that the bottom hatched bar covers the region of the top hatched bar. The $\ell(2)$ chooses the longest such bar. We can't use $\ell(1)$—if that even existed—because then we wouldn't shift at all.

3.4   Implementing the Good Shift Rule

To implement the good shift rule, we need to preprocess P to compute $L(i)$ and $\ell(i)$. To do this, we will use a set of values that are related to the Z_i values.

Definition 3.4. $N_j(P)$ is the length of longest suffix of $P[1 \ldots j]$ that is also a suffix of P. In a picture, this is (3.9).

$$\vdash N_j(P) \dashv$$

$$P: \quad \boxed{\alpha} \qquad \boxed{\alpha} \qquad (3.9)$$

$$\uparrow$$
$$j$$

$$\blacksquare$$

Recall the definition of the Z_i values from Chapter 2.

Definition 3.5 (Z_i). $Z_i(P)$ is the length of the longest substring of P that starts at i and matches a prefix of P. In a picture, this is (3.10).

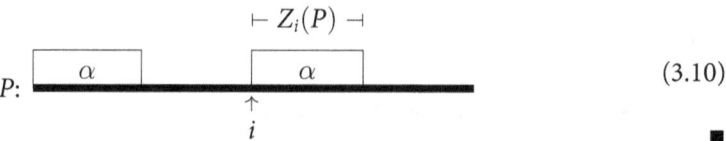

$$P: \qquad\qquad\qquad\qquad \tag{3.10}$$

You can see that $N_j(P)$ and $Z_i(P)$ are reverses of each other: the Z_i deal with prefixes, while the N_j deal with suffixes. It makes sense that we would need the reverse of the Z values because we're matching right-to-left instead of left-to-right. This allows us to easily compute the N_j values using the Z values algorithm:

$$N_j(P) = Z_{n-j+1}(P^r), \tag{3.11}$$

where P^r is the string P reversed, and n is the length of P (here and for the rest of the chapter). (The notation can help you remember this: an N looks like rotated Z.)

3.4.1   Computing the $L(i)$ values

We use the N_j values to compute $L(i)$ in a way that is similar to how we used the Z_i values in the ZMATCH algorithm. Recall that $L(i)$ is the largest j such that $P[i \ldots n]$ a suffix of $P[1 \ldots j]$ and $P[i-1]$ is not equal to the character preceding that suffix, as shown in (3.12).

$$P: \qquad\qquad\qquad\qquad \tag{3.12}$$

The candidate locations for $L(i)$ are the places where the substring α (the suffix $P[i \ldots n]$) occurs. We want the rightmost one of those that has a different preceding letter. Similar to the ZMATCH algorithm, the places where the suffix occurs are the places where the N_j values equal the length of the suffix:

$$N_j(P) = \text{length of longest suffix of } P[1 \ldots j] \text{ that is also a suffix of } P;$$
$$\implies \quad L(i) = \text{largest index } j \text{ such that } N_j(P) = |P[i \ldots n]| = n - i + 1.$$

When this is true, we also know that the preceding characters (x and y in the previous figure) are different. If they had been the same, N_j would have been larger.

Based on this, we can rewrite our definition of $L(i)$:

$$L(i) = \max\{j : N_j(P) = n - i + 1\} \tag{3.13}$$
$$= \max\{j : i = n - N_j(P) + 1\}, \tag{3.14}$$

where (3.14) is just rearranging terms. This leads to this algorithm to compute the $L(i)$:

Algorithm 3.2. Efficient procedure to compute L values.

> **function** COMPUTEL(P)
>> Compute $N_j(P)$ via the Z-algorithm and the relationship (3.11) for all j.
>> Initialize $L[i] \leftarrow 0$ for all i.
>> **for** $j = 1, \ldots, n-1$ **do**
>>> $i \leftarrow n - N_j[P] + 1$ ▷ *using the relation in Eqn. 3.14*
>>> $L[i] \leftarrow j$

The expression $i = n - N_j(P) + 1$ in (3.14) forms a relationship between pairs of (i, j): for any i, the j that satisfy that equation are the possible candidates. The algorithm (3.2) computes each of these (i, j) pairs, but saves, for any i, only the one with the largest j (since we compute with j in increasing order).

The running time is $O(|P|)$ since we can compute the Z values in $O(|P|)$ time (Section 2.4) and the loop does a constant amount of work for each of its $O(|P|)$ iterations.

3.4.2   Computing the $\ell(i)$ values

Here's the situation with the $\ell(i)$ values. We're looking for the size of largest suffix of $P[i \ldots n]$ that equals some prefix of P. That is, in (3.15) we are looking for the largest solid box.

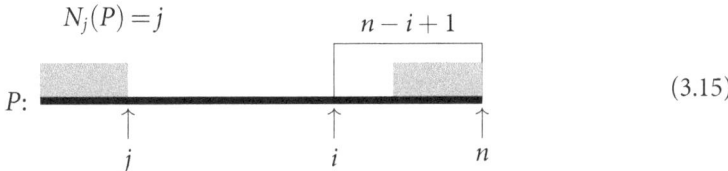

$$(3.15)$$

Candidate values for j are those j for which $N_j(P) = j$. When $N_j(P) = j$ that means that the string ending at j of length j matches a suffix of P. A string "ending at j of length j" is just another description of a prefix of P, which is what we need for the ℓ values.

We need to only consider suffixes of $P[i \ldots n]$. This leads to the following theorem.

Theorem 3.6 (Gusfield 2.2.4). $\ell(i) = \max\{j \leq |P[i \ldots n]| : N_j(P) = j\}$. *In other words,* $\ell(i)$ *is the largest position j where the prefix ending at j matches a suffix of P (that is "$N_j(P) = j$") and that suffix is shorter than* $|P[i \ldots n]|$.

Computing these $\ell(i)$ values given this definition and the precomputed N_j values is straightforward.

We've seen how to compute the L, ℓ, and R values, and that completes the description of this version of Boyer-Moore.

3.5   Summary and notes

The version of Boyer-Moore that we have described here does not run in worst-case $O(|P| + |T|)$ time. Several extensions exist that transform this algorithm into one that runs in linear time [Galil, 1979; Cole, 1991]. Many proofs of the linear running time of the faster version are quite complicated.

The key ideas are the various examples of how to preprocess the pattern P into a number of arrays Z_j, N_j, $L(i)$, $\ell(i)$, and R_i to make all of the various shifting steps run quickly.

> ### Presentation Notes
>
> Our presentation of Boyer-Moore follows that of Gusfield [1997], showing the connection to his Z values; we have mostly adopted his notation. Gusfield [1997] contains many more details about the algorithm, including the analysis to prove it can be made to run in linear time.

3.6   Exercises

3.1   Give an example where the Next Matching Character rule lets you shift by more than the Good Shift rule.

3.2   Prove that the total length of the lists of positions in the list-based storage of the R values is $O(|P|)$.

3.3   Give the details of an efficient algorithm to precompute the $\ell(i)$ values.

3.4   Give a data structure to store the $R_i(x)$ values that is similar to the hash-table-and-linked-lists and that (a) stores the $R_i(a)$ values in $O(|P|)$ total space; (b) enables accessing $R_i(a)$ in time proportional to $|P| - i$ when applied in the context of the "next matching character rule"; and that (c) does not use a hash table.

3.5   Describe a situation where the 1 term is the largest in the max in Algorithm 3.1.

3.6   Let n be the length of a pattern and m be the length of the text. Give an example where the version of Boyer-Moore from this chapter takes $O(nm)$ time.

3.7   *Bad good shift rule.* Give an example of a text and pattern where fewer comparisons are made when the "next matching character rule" is used alone instead of including the "good shift rule" in the algorithm.

CHAPTER 4

Knuth-Morris-Pratt

Let's continue with the exact matching problem, where we are given strings T and P and want to find all places where P occurs in T. One of the most well known algorithms for exact string matching is Knuth-Morris-Pratt (KMP; [Knuth et al., 1977]). This was in fact the first algorithm given for this problem that runs in $O(|P| + |T|)$ time. KMP will shift the pattern by more than 1 position each time based on what it has learned so far about T. We will see two different views of this algorithm, one based on deterministic finite automata, and one based on the Z values from Chapter 2.

4.1   KMP via deterministic finite automata

KMP will build a *deterministic finite automaton* (DFA), that depends on the pattern. A DFA is a directed graph where the nodes are *states*, edges represent possible transitions between states, and each edge is labeled with some character or condition that says when the edge can be used. Before we see KMP's DFA, let's look at a slightly more intuitive DFA.

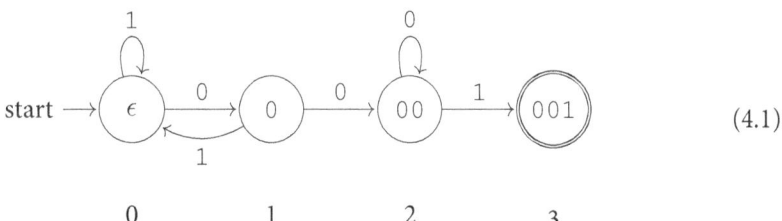

$$(4.1)$$

The state labeled by the "start" arrow is the state we are initially in. The state with a double circle is the special "accepting state." This DFA was constructed from the pattern $P = 001$ over the alphabet $\Sigma = \{0, 1\}$. It allows us to find the first occurrence of P in a longer string T by "feeding" the characters of T through the DFA. As we see each new character in T, we follow the corresponding edge. If we ever reach the accepting state, then we have found the first occurrence of P in T. The DFA is "deterministic" because each state has at most one edge labeled with any given character leaving it, so we never have any choice about what is the next state. It's "finite" because there are a finite number of states.

Why these edges? The left-to-right edges are the "match" edges: if we're in state j, we've matched the first j characters of the pattern, and if we match the next character, we should move to the next state $j + 1$. The right-to-left and stay-in-same-state edges are more interesting. Let's look at the first such edge: the edge labeled 1 that goes from state 0 to state 0. If we haven't matched any of the pattern, and we see a character different than the first character, then we still haven't matched any of the pattern and should stay in the first state. Look at the right-to-left edge labeled 1: if we're in the state 1, we've matched the 0 prefix of the pattern. After seeing a 1, since the pattern doesn't start with 01, we have to go back to the start. Now, look at the loop edge labeled 0. In this case, we've matched the 00 prefix of P. If we see another 0 that means we've seen 000 in T; this still means we've matched the 00 prefix of P, so we can stay here.

We reach the accepting state exactly when we have seen the complete pattern. If we want to continue to find additional occurrences of P, we need to add some additional edges that tell us what state to go to after a match. We add edges out of the final state to the earlier states that correspond to how much of P we know we have matched, given that we've seen the whole pattern and observed the character after that pattern match.

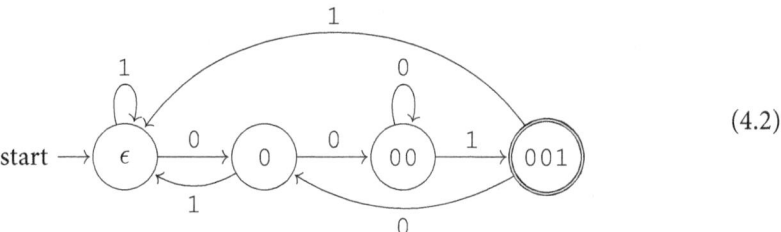

(4.2)

The edges we've added in (4.2) go from the accepting state to the number of characters we would have matched after observing the character following an occurrence of P.

Let's look at another example of a similar DFA. The DFA for $P = $ "01101" is illustrated in (4.3).

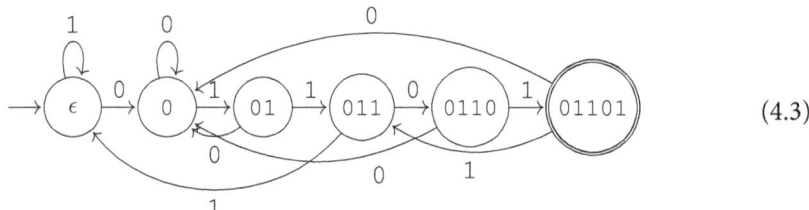

(4.3)

Once such a DFA is created, we can find all occurrences of P in $O(|T|)$ time by using the characters of T to determine a path through the DFA, recording the locations of P each time we reach the accepting state.

4.1.1   The KMP DFA

The KMP algorithm actually creates a slightly different DFA with only two edge types: "match" and "mismatch," rather than labeling each edge with a specific character of Σ. This has the advantage that the DFA size does not depend on $|\Sigma|$: every state will have 2 edges leaving it. For a pattern P, the KMP DFA will have $|P| + 1$ states $q_0, q_1, \ldots, q_{|P|}$, with the interpretation that being in state q_i means we have currently matched i characters of P (where i ranges from 0 to $|P|$). The state $q_{|P|}$ is the accepting state. There is a "match" edge between state q_{i-1} and q_i for $i = 1, \ldots, |P|$.

The "mismatch" edge leaving each state will tell us where to go when the next character of T doesn't match the next character of P. For a pattern $P = p_1 p_2 \ldots p_{|P|}$, let P_k denote the prefix $p_1 \ldots p_k$ and P_0 denote the empty prefix. Each state q_k means we've matched up through P_k.

Suppose we are at state q_k. We want to see where to go when we see the character c after this. If $c = p_{k+1}$, then we go to the state q_{k+1} corresponding to P_{k+1} via a "match" edge. If $c \neq p_{k+1}$, where could the next match begin? Here in (4.4) is a picture of the situation after a mismatch at $k + 1$.

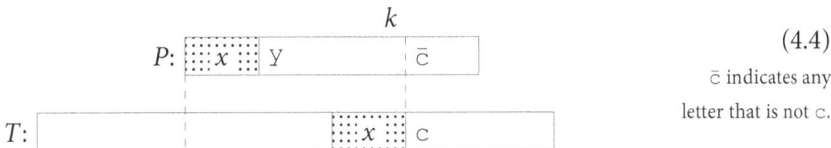

$$(4.4)$$

$\bar{c}$ indicates any
letter that is not c.

If a match were to start someplace in the region currently matched to P then:

- that match would start with a prefix of P (since all matches start that way), and
- that prefix would have to match in T up to the characters we've seen so far.

Such a string is marked as the shaded x region in the figure (4.4). We want the *longest* such region so that we don't skip over a match in T. This means we need to find the *prefix* of P that corresponds to the longest *suffix* of P_k. We record these mismatch edges in a precomputed array *memo*.

Definition 4.1 (*memo*). Let *memo*$[k]$ be the length of the longest *proper* suffix of P_k which is also a prefix of P. ∎

We require a proper suffix because we already know that the full suffix of P_k doesn't match. The *memo* array gives the "mismatch" edges for the KMP DFA. Since we require a proper suffix, *memo*$[j] < j$.

Suppose P is the string "1110111101." Here is *memo* in (4.5).

$i =$	0	1	2	3	4	5	6	7	8	9	10
$P =$		1	1	1	0	1	1	1	1	0	1
memo $=$	0	0	1	2	0	1	2	3	3	4	5

$$(4.5)$$

4.1.2   Using the *memo* array for string search

Here's the pseudocode for the complete algorithm, assuming *memo* has been computed.

Algorithm 4.1. Knuth-Morris-Pratt.

> **function** KMP(P, T)
> $\quad j \leftarrow 0$ ▷ *our current state*
> $\quad$ **for** $i \leftarrow 1, \ldots, |T|$ **do**
> $\quad\quad$ **while** $j > 0$ **and** $T[i] \neq P[j+1]$ **do**
> $\quad\quad\quad j \leftarrow memo[j]$ ▷ *mismatch edge*
> $\quad\quad$ **if** $T[i] = P[j+1]$ **then** $j \leftarrow j+1$ ▷ *match edge*
> $\quad\quad$ **if** $j = |P|$ **then**   ▷ *in final state*
> $\quad\quad\quad$ **output** "match found at", $i - |P|$
> $\quad\quad\quad j \leftarrow memo[j]$

The **while** loop is the hardest part to understand. The loop keeps hopping backward in P until it finds one of the suffix-prefix mismatches where the character in P matches the current character in T (or it gets to the start of the pattern in state q_0). It hops back, using *memo* (the body of the **while** loop) and checks to see if the next character would be a match (the condition in the **while** loop). If not, it continues hoping back. Here's a picture to help understand the **while** loop.

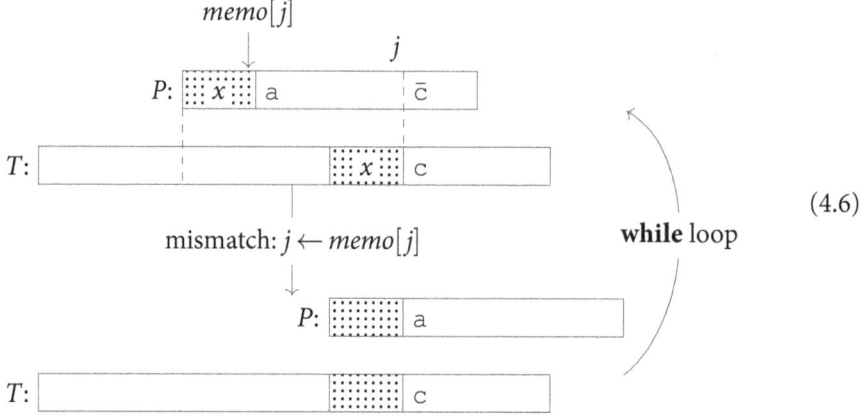

$$(4.6)$$

If a $\neq$ c, then we're in the same situation we started in. The **while** loop repeats the check until a = c. When $j = 0$ and we continue to mismatch, nothing in the body of the **for** loop will happen, and we will just be scanning T by incrementing i.

4.1.3   Correctness and running time

The algorithm is simulating a simplified DFA for P. Each time there is a mismatch, P is shifted by the *least* amount for which a match could continue at i. No matches will be missed.

Theorem 4.2. *The running time of the algorithm is linear in* $|T|$.

Proof: Each iteration of the inner **while** loop decreases j (all the mismatch DFA edges point backward since we require a *proper* suffix of P). Each iteration of the outer loop increases i and increases j by at most 1. Consider the quantity $q = 2i - j$. This increases for every bit of work the algorithm does. For example,

- in the **while** loop, i is constant while j decreases $\implies q$ increases for each bit of work the while loop does.
- in the **for** loop if both i and j are incremented, q increases by 1. (This is why we use $2i$ instead of i in q.) If i is incremented, and j stays the same or decreases, then q increases by at least 2.

Therefore, q never decreases.

Finally, q is bounded between 0 and $2|T|$. q is $\leq 2|T|$ because $|T| \geq i$ and $j \geq 0$. j will never get ahead of i since j is only incremented when i is (so $q \geq 0$). Therefore, q can be increased at most $2|T|$ times; since q is increased with every unit of work, the running time is $O(|T|)$. $\qquad\square$

4.1.4   Computing *memo*

Last step: how do we compute *memo* quickly? Here's the code.

Algorithm 4.2. Computing the memo array.

function Cation ComputeMemo(P)
    ▷ *memo[i] will store the length of the longest prefix of P that matches a suffix of $P[2 \ldots i]$.*
    $memo \leftarrow$ new integer array of length $|P| + 1$
    $j \leftarrow 0$
    for $i \leftarrow 2, \ldots, |P|$ **do**
        while $j > 0$ **and** $P[i] \neq P[j+1]$ **do**
            $j \leftarrow memo[j]$
        if $P[i] = P[j+1]$ **then** $j \leftarrow j+1$
        $memo[i] \leftarrow j$ ▷ (*)
    return *memo*

This should look very familiar! It is essentially Algorithm 4.1 but with $T = P$. Why should this make sense? When we are matching in T, we're always looking to extend the prefix of P that matches the suffix of our current place in T. Using P in place of T does the same thing—we just have to record how long a prefix we matched at the line marked with (*), whereas in the match code, we only cared about full-length matches.

In Algorithm 4.2, we are both creating and using the *memo* array at the same time. We again use the fact that j will never be ahead of i, so we will have already computed all the *memo[x]* that we need in the **while** loop for i.

The running time of computing *memo* is $O(|P|)$ by the same argument as for the KMP code. Therefore, the total runtime of KMP is $O(|P| + |T|)$.

4.2   KMP via the Z-values

We can see a similar algorithm to KMP that is recast in the language of the Z values of Chapter 2. To do this, we start by preprocessing the pattern P into a spm_i array.

Definition 4.3. $spm_i(P)$ is the length of the longest substring of P that ends at $i > 1$ and matches a prefix of P and such that $P[i+1] \neq P[spm_i+1]$ as shown in (4.7). ("spm" stands for suffix, prefix, mismatch.)

$$(4.7)$$

As we are moving a pattern down T, on a mismatch, we are in this situation shown in (4.8).

$$(4.8)$$

We can shift P so that the spm_i regions coincide. That is, we can shift by $i - spm_i$ characters. This leads to the following algorithm due to Gusfield [1997].

Algorithm 4.3. KMP via spm_i.

function KMPviaSPM(T, P)
    $c \leftarrow p \leftarrow 1$ ▷ *pointers into T and P, respectively*
    while $c + (|P| - p) \leq |T|$ **do**   ▷ *until pattern falls off end of T*
        ▷ *compare P and T, advancing c and p in parallel*
        while $P[p] = T[c]$ **and** $p \leq |P|$ **do**
            $p \leftarrow p + 1$
            $c \leftarrow c + 1$
        if $p = |P| + 1$ **then**   ▷ *we got all the way through P*
            output "Found at", $c - |P|$.
        ▷ *failure at start of P means stay at the start of P and try the next character in T*
        if $p = 1$ **then**
            $c \leftarrow c + 1$
        else
            ▷ *"shift" by $p - spm[p-1]$ (even if $p = |P| + 1$)*
            $p \leftarrow spm[p-1] + 1$

In the last line of algorithm 4.3, p is set as shown here in (4.9).

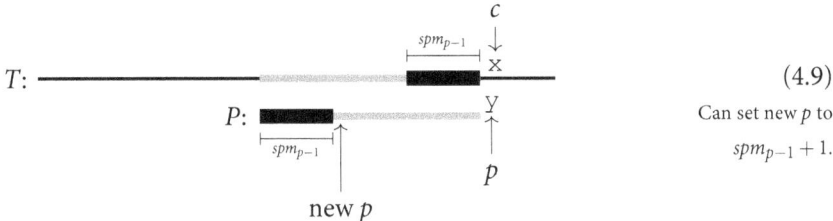

$$(4.9)$$

Can set new p to

$$spm_{p-1} + 1.$$

While c is always increasing, p decreases on an occurrence of $|P|$ or a mismatch.

There is a connection between our spm_i values and the edges implied by the *memo* array. In both, when we're at state i, on a mismatch, we jump back to the longest prefix of P that matches a suffix of the string ending at i. For *memo*, we dropped the "mismatch" requirement and just looked for the longest prefix matching a proper suffix and dealt with the mismatches through the **while** loop of Algorithm 4.1, which looped until we found a match. The spm_i array instead requires that we skip back to someplace at least that the next possible match in P doesn't repeat the same mismatch we have already tried.

Why won't Algorithm 4.3 miss any matches? If there is to be a match that overlaps the part of T we've matched against P in the current iteration, it must start with a prefix of P and that prefix must match a suffix of that matched part, and it must not put the same mismatching character aligned with the current position in T. Since we take the longest such suffix, we shift by the least possible amount.

4.2.1   Running time of the spm_i version of KMP

The argument for the running time of this version of KMP relates the number of comparisons to the number of shifts.

Theorem 4.4. *Once the spm_i values have been determined, Algorithm 4.3 runs in $O(|T|)$ time to search a string T.*

Proof: After every shift, we start comparing either at the character of T where we mismatched or just after that character in T. Hence, at most one character of T is re-examined per shift of the pattern. Therefore, the total number of comparisons is $\leq |T| + \#\text{shifts}$. Since each shift is at least by 1, $\#\text{shifts} \leq |T|$, and therefore the number of comparisons is $\leq 2|T|$. $\qquad\square$

4.2.2   Computing the spm_i array

Recall the Z values.

Definition 4.5 (Z_i). $Z_i(P) =$ the length of the longest substring of P that starts at i and matches a prefix of P.

$$P: \qquad \qquad i \qquad i + Z_i - 1 \tag{4.10}$$

∎

Recall also that a "Z-box" is a region of P corresponding to some Z value. The Z-box starting a position j is the witness of the Z_j value.

Definition 4.6 (f). Let $f(j) =$ the right end of the Z-box (if any) that starts at position j.

$$P: \qquad j' \qquad j \qquad f(j) = f(j') \tag{4.11}$$

∎

This is well-defined because at most one Z-box can *start* at any position j. As in (4.11), many Z boxes could *end* at the same position. We therefore define a kind of "inverse" of f, taking this into account.

Definition 4.7 (g). Let $g(i) = \min\{j : f(j) = i\}$ or 0 if this is the empty set. ∎

For example, in (4.11), $g(f(j)) = j'$. As you might expect by comparing the pictures there is a very close relationship between the spm_i array and the Z_i array. The $g(i)$ function lets us specify that relationship.

Theorem 4.8. *For each i, $spm_i = Z_{g(i)}$ if $g(i) > 0$; otherwise spm_i is 0.*

Proof: Here's the situation illustrated for some i.

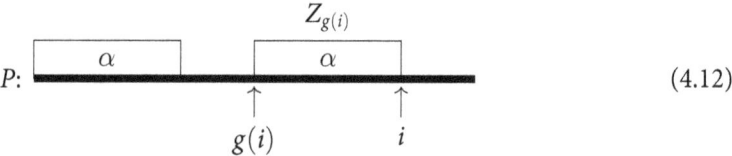

$$\tag{4.12}$$

Suppose $g(i) > 0$. We have $\alpha = P[g(i) \ldots i] = P[1 \ldots Z_{g(i)}]$ by definition of the Z values. So we have a suffix of $P[1 \ldots i]$ matching a prefix of P.

Also, $P[i+1] \neq P[Z_{g(i)} + 1]$, otherwise $Z_{g(i)}$ would have been larger. This means we have a mismatch following the ends of these two regions.

Therefore, $spm_i \geq Z_{g(i)}$ since (4.12) provides an α of length $Z_{g(i)}$ matching the requirements of spm_i.

But spm_i can't be bigger than this, otherwise $g(i)$ would be smaller because there would be an earlier index where a Z-box ended at i.

This shows that $spm_i = Z_{g(i)}$ when $g(i) > 0$. If $g(i) = 0$, then there is no proper suffix that matches a prefix ending at this point (because there was no Z-box that ended at i) and spm_i is correctly 0. $\qquad\square$

To compute spm_i we need only to compute the Z_i for P, which we know how to do (Section 2.4), and to compute the function $g(i)$. See Exercise 4.2.

4.3   Summary and notes

We've seen now two versions of the KMP algorithm. These two versions are essentially the same: they both shift the pattern by the least amount that could match some suffix of the already matched part. They differ in the "language" they are presented in (DFA vs. spm_i derived from Z values). They also differ in how they handle the requirement that we know the next character can't match the same pattern character. For the spm_i version, this is built into the requirements of the spm_i values. For the DFA version, we only require a proper suffix and rely on the tests in the **while** loop to find the right state where we can again start matching.

While the DFA we started with is the inspiration for the KMP approach, KMP actually creates a simpler DFA with only two types of edges. Since we only record a single edge for "mismatch," we can't have self-loops. (Because we don't record *what character* the mismatch was; that might require us to jump to an earlier state when, if we took into account the mismatching character, we wouldn't have to.)

Presentation Notes

Our discussion of the version of KMP implemented using the Z values follows Gusfield [1997], who introduced the Z values partially to explain exact matching algorithms in a common framework. Our discussion of the DFA-based KMP is based on a lecture that I and others have given in CMU's 15-451 Algorithms class.

4.4   Exercises

4.1   True or false: when a character is matched in KMP, it is never matched again.

4.2   Describe how to compute $g(i)$ in $O(|P|)$ time.

4.3   For each position i in pattern P, $memo[i]$ is the length of the longest proper suffix of $P[1 \ldots i]$ that matches a prefix of P.

        For a position i, let r_i be the rightmost end of a Z-box that begins at or before i and ends after i, and let ℓ_i be the corresponding left end of this Z-box. If no such Z-box exists, r_i and ℓ_i are undefined.

(a) True or false: $memo[i] = i - \ell_i$ whenever r_i and ℓ_i are defined?

(b) If true, prove it. If false, give a counterexample.

4.4   Let T be a rooted tree with each edge labeled by 1 character of the alphabet Σ (assume $|\Sigma|$ is a constant). T represents a collection of strings spelled out by following edges from the root to each non-root node. That is T represents strings $\{S_u \mid u$ is a node in $T\}$. Let $|T|$ be the number of edges in T (which is equal to the number of characters in T).

Sketch an algorithm that runs in time $O(|T| + |P|)$ to find the strings represented by T that contain a string P.

4.5   Give an example where the KMP DFA will make more comparisons than the DFA in the style of (4.2).

4.6   Implement KMP, Boyer-Moore, and the Z-algorithm and compare the number of comparisons each makes on various strings.

4.7   Explain each of the edges in (4.3).

4.8   Give an example where the DFA and spm_i versions of KMP will make different numbers of comparisons.

4.9   Give an example where the KMP DFA can be improved to reduce the number of comparisons, while still only allowing match and mismatch edges and $|P| + 1$ states.

4.10   In KMP (either version), we are concerned about a suffix of the current match compared with a prefix of P. That is similar to case (B) of Boyer-Moore (Chapter 3). But in Boyer-Moore, we also had to consider a case (A): where the matched suffix matches an internal (non-prefix) substring of P. Why don't we need such a case here in KMP?

CHAPTER 5

Seminumerical String Matching

We will again consider the exact match problem: given a long text T and a shorter query string P, find all the occurrences (possibly overlapping) of P in T. Our previous algorithms for this problem had the property that they only accessed the string via comparison operations. This is nice because you don't have to assume anything about the alphabet or how it is encoded. But of course the strings are encoded as bits in memory, and we can just as easily treat strings as numbers. That's the idea behind seminumerical string matching.

We'll see two techniques for doing this, one based on hashing and one based on bitwise operations. Neither has a worst-case guarantee as good as KMP or the Z algorithm, but again, they can be quite fast since they better exploit the types of arithmetic operations a computer is so good at.

5.1 Rabin-Karp fingerprinting

We start with the well-known Rabin-Karp approach [Karp and Rabin, 1987]. In addition to being a good way to solve the exact match problem, this approach introduces some ideas that can be adapted to other settings.

To make the explanation of Rabin-Karp simpler, assume for a moment that $\Sigma = \{0, \dots, 9\}$. A string then can be thought of as the decimal representation of a number. For example, we could think of the string "427328" as the integer 427328. In general, if $|\Sigma| = d$, then a string represents a number in base d. Let p be the number represented by the query P using this idea.

Now suppose t_s is the number represented by the $|P|$ "digits" of T that start at position s. Then:

- P occurs at position s of T if and only if $p = t_s$ (in the numeric equality sense).

This leads to a nice algorithm that is similar in spirit to the ZMATCH algorithm. If we can compute t_s for every s, then the positions we want to output are exactly those for which $t_s = p$. The question becomes how to compute p and t_s efficiently.

5.1.1 Computing p

We can use Horner's rule to compute p for a pattern P of length m in time $O(|P| = m)$. We have defined p and t_s so that the most significant digit is to the "left" (earlier) in the string.

This leads to the following expression for p, assuming still a "base-10" alphabet:

$$p = P[m] + 10\left(P[m-1] + 10(P[m-2] + \cdots + 10\left(P[2] + 10P[1]\right)\right)\right). \tag{5.1}$$

For example:

$$427328 = (8 + 10(2 + 10(3 + 10(7 + 10(2 + 10 \times 4))))). \tag{5.2}$$

The idea is that we can "build" the number by first building a two-digit number from the rightmost two digits ($10 \times 4 + 2$, then "shifting" it to the left by multiplying by 10 and adding in the next number, and so on. This uses approximately 1 multiplication and 1 addition per digit. We assume we can multiply and add numbers in $O(1)$ time, so the whole process to compute p takes $O(|P|)$ time. We'll come back to the assumption that the addition and multiplication take $O(1)$ time soon.

We're using base 10 for convenience, but typically in software, your base is chosen to be a power of 2 (i.e., if characters take up 1 byte, then $|\Sigma| = 256 = 2^8$). When the base is a power of 2, the "shift" operation (which was "multiply by 10" in (5.2)) becomes "multiply by a power of 2." This is usually a very fast operation because it can be implemented directly with a "left shift" computer instruction (<< in C/C++ for example).

5.1.2   Computing t_s for every position s

Definition 5.1 (t_s, first definition). t_s is the number represented by the $|P|$ characters starting at s in T. ■

We can use the same idea that we used to compute p to compute t_1 in $O(|P|)$ time. But if we naïvely computed t_j for $j > 1$ in the same way, we'd end up with a $\Theta(|P| \times |T|)$ algorithm, which is too slow.

The fix is to exploit the relationship between t_s and t_{s-1}. To get t_s from t_{s-1}, we need to subtract the contribution that character $T[s-1]$ made to t_{s-1} (since character $T[s-1]$ is not involved in t_s), shift the result to make room for the new character in the least-significant place and add that new character. That reasoning gives the following expression for t_s given t_{s-1}:

$$t_s = \underbrace{10\times}_{\text{shift}} \underbrace{(t_{s-1} - 10^{m-1}T[s-1])}_{\text{remove most significant digit}} + \underbrace{T[s+m-1]}_{\text{add next digit}}. \tag{5.3}$$

The second part subtracts out the no-longer-relevant part of t_{s-1}. The value 10^{m-1} is a constant that we can precompute — we don't have to find the $(m-1)$ power of the alphabet size each time. This uses a product, a subtraction, and an addition to compute t_s. Assuming each of those operations run in constant time, this gives us a $O(|T|)$ algorithm to compute all the t_s values.

Overall, we then have an $O(|P| + |T|)$ algorithm for string matching assuming each of the arithmetic operations and number comparisons take constant time. We do not need to store all of the t_s values. We can compute a t_s, compare it with p, output s if they are equal, use t_s to compute t_{s+1} and then discard t_s.

5.1.3   Time for addition, product, and comparison

The sizes of the numbers involved depend on $|P|$, which is not constant. In many instances, the pattern might be short enough so that the operations take $O(1)$ time. However, in many other cases, the pattern is long enough that it is not realistic to assume that addition, product, and comparison take $O(1)$ time. For a large pattern, each of these numbers may be *much* larger than a computer word that the CPU can process in constant time with a single `ADD` instruction. Somehow, we have to handle arbitrarily large numbers.

The way we do this is by computing everything (p and t_s) modulo some prime number q. That is we redefine t_s to be:

Definition 5.2 (new t_s). t_s is the number represented by the $|P|$ digits starting at s (mod q). ∎

We redefine p the same way. (We'll continue to sometimes write the explicit "(mod q)" when we want to emphasize that these numbers are modulo q.) Now, we have:

- if P occurs at position s then $p \equiv t_s$ (mod q).

The largest number we will need to deal with when computing t_s or p is $10q$ (assuming a base-10 alphabet). So if we choose q small enough (i.e., so that $10q$ fits in a constant number of computer words) then our algorithm will take $O(|P| + |T|)$ time as desired to compute the t_s values. It will also only use a constant amount of space (in addition to the strings).

But alas, we have introduced a new problem. That problem is false positives. Yes, when there is a match at position s, then $p \equiv t_s$ (mod q), but this can also happen when there isn't a match. It is not the case that if $p \equiv T_s$ (mod q) then P definitely occurs at position s.

One solution to this is to check every potential match explicitly. That is when $p = t_s$ (mod q), we double check whether there is a match by checking the relevant characters explicitly. This works but leads to a worst-case runtime of $O(|P| \times |T|)$ if every position is a match or a false positive.

There's no way to eliminate false positives entirely. So, we'll always end up with a worst-case $O(|P| \times |T|)$ algorithm. But we can try to make the probability of a false positive very low. That's the next idea.

5.1.4   Quantifying and reducing false positives

If our strategy is to "make the probability of a false positive low" then we have to introduce some randomness. As a first step, let's choose q to be a random prime less than some number M. We pick M (deterministically) to be small enough that we can get constant-time operations, but $q \leq M$ is chosen randomly each time we run the algorithm.

Consider pattern P and the substring S of length $|P|$ starting at position s in T. If $P = S$, then $p = t_s$, and we don't have any false negatives. Assuming $P \neq S$, what is the chance $p = t_s$ for the q we picked? The following theorem provides the answer:

Theorem 5.3. *If we chose M large enough, then if $P \neq S$, then $\Pr_q[p = t_s \pmod{q}] \leq 0.2$.*

Proof: Reimagine P and S as bit strings of length N, and let p' and t'_s be the numbers represented by these bit strings *before* the modulo q is applied. We can do this no matter what our alphabet is. This means that p' and t'_s are $< 2^N$.

We have a false positive when $p' = t'_s \pmod{q}$. This happens when $(p' - t'_s) = 0 \pmod{q}$. In other words, when the chosen q exactly divides the difference $D = |p' - t'_s|$. Given a number D, what is the chance that a randomly chosen prime q divides it?

D is also $< 2^N$ since it's the difference between two non-negative numbers $< 2^N$. D can be written as the product of primes $q_1 q_2 \ldots q_k$ (allowing primes to repeat) since any integer can be written that way. Since $D < 2^N$ and each prime is ≥ 2, we can't have more than N primes in this list. That is $k \leq N$. So D has at most N prime divisors. Those prime divisors are the "bad" values of q for this D: any of these prime divisors, if chosen as q, would divide D exactly, leading to $D = 0 \pmod{q}$.

The probability that the q that we picked is one of the divisors of D is:

$$\frac{\text{bad choices}}{\text{possible choices}} \leq \frac{N}{\text{number of primes in } \{1, \ldots, M\}}. \tag{5.4}$$

If we choose M large enough so that there are at least $5N$ primes among $\{1, \ldots, M\}$, then we have:

$$\Pr_q[p = t_s \pmod{q}] \leq \frac{N}{\text{number of primes in } \{1, \ldots, M\}} \tag{5.5}$$

$$\leq \frac{N}{5N} \leq 0.2. \tag{5.6}$$

$\square$

How big does M have to be to ensure there are $5N$ primes?

Fact 5.4. *If we want at least $n \geq 4$ primes between 1 and M, it suffices to have $M \geq 2n \log_2 n$.*

This fact follows from the prime number theorem. We will not prove it. We want at least $5N$ primes, so we choose M to be $\geq 10N \times \log_2 5N$.

Is this small enough? To have constant-time addition, multiplication, etc. we wanted M to be small; to get a small probability of a false positive, we want M to be big. Is our lower bound requirement on M still small enough to make it reasonable that the operations can be done in constant time? To deal with numbers up to M, we need $O(\log M)$ bits. When $M = 10N \log_2(5N)$,

$$O(\log(10N \log 5N)) = O(\log(10N) + \log \log 5N) = O(\log N). \tag{5.7}$$

So we need only a number of bits for our prime that is logarithmic in the number of bits to represent our pattern. This is the same number of bits that we'd need just for a pointer into a location into our pattern (i.e., less than needed to represent a position). It is reasonable to assume we can operate on such numbers in $O(1)$ time.

5.1.5   Reducing the error probability

A false positive rate of 0.2 for any potential position is okay, but if we have a lot of positions (i.e., a long T) a 20% false positive rate is pretty high. We can reduce it even further by repeating the algorithm with independent choices of q. Any real match will have $p = t_s$ no matter the choice of q, but a false positive must have this be true for every chosen q. Hence, if we repeat the algorithm R times, our probability of a false positive is $\leq 0.2^R$.

5.2   The Shift-And algorithm

We will now see a completely different algorithm that works well because it uses only simple processor instructions. It's practical mostly for small patterns. It was devised by Baeza-Yates and Gonnet [1992], and our presentation follows Gusfield [1997].

The algorithm will conceptually construct a 0/1 matrix M, where the rows correspond to prefixes of P and the columns correspond to substrings of T.

Definition 5.5. Let $M[i, j]$ be 1 if prefix i of P matches the substring of T ending at j, and 0 otherwise. ∎

Schematically, M looks like (5.8).

$$M = \begin{array}{c} \\ p_1 \\ p_2 \\ p_3 \\ p_4 \\ \vdots \\ p_n \end{array} \quad \begin{array}{ccccccccc} t_1 & t_2 & t_3 & t_4 & t_5 & t_6 & t_7 & t_8 & \cdots & & t_m \\ \hline & & & & & & & & & & \\ & & & \boxed{1} \leftarrow 1 \text{ if } P[1\ldots i] = T[j - i + 1 \ldots j] & & & & \\ & & & & & & & & & & \\ & & & & & & & & & & \\ \hline \end{array} \tag{5.8}$$

Ones in the last row indicate where P occurs. When will $M[i, j]$ be 1? We can break it down into two parts: (a) the i character of P and jth character of T match, and (b) the rest of P matches.

$$\begin{array}{l} T: \rule{3cm}{0.4pt} \\ \qquad P: \end{array} \quad \begin{array}{c} j \\ \downarrow \\ \\ \uparrow \\ i \end{array} \tag{5.9}$$

We can define this in terms of $M[i-1, j-1]$:

$$M[i,j] = 1 \text{ if } (P[i] = T[j]) \text{ and } (P[1 \ldots i-1] = T[\text{ending @ } j-1]) \qquad (5.10)$$
$$\Longleftrightarrow \text{ if } (P[i] = T[j]) \text{ and } (M[i-1, j-1] = 1). \qquad (5.11)$$

Why is this? Clearly, we need $P[i] = T[j]$ (the first condition). The second condition of (5.10) is simply the definition of M: we also need the previous $i-1$ characters of P to match the string ending at $j-1$ in T. But this is exactly what the entry $M[i-1, j-1]$ means, and so (5.11) follows.

Filling out the matrix M each entry at a time would be a slow $O(|P| \times |T|)$ algorithm. Instead, the idea of the Shift-And algorithm is to compute M column-by-column, computing each column from the previous one in a single manipulation of those bit vectors. To do this, we preprocess P to compute the following set of vectors (one per character in Σ):

Definition 5.6 (U_P). $U_P(x) = |P|$-bit vector where the ith entry is 1 if $P[i] = x$, and 0 otherwise. ∎

$U_P(T[j])$ is a bit vector indicating where in P the character $T[j]$ occurs. In other words

$$U_P(T[j])[i] = 1 \Longleftrightarrow P[i] = T[j], \qquad (5.12)$$

which is exactly the first condition we had in (5.11).

Let $M[\cdot, j]$ denote column j of matrix M. We can compute column j of M via:

$$M[\cdot, j] = U_P(T[j]) \text{ \& } v, \qquad (5.13)$$

where & indicates bitwise "and" and v is a vector for which $v[i] = 1$ if and only if $M[i-1, j-1] = 1$ (which is the second condition in (5.11)). Note that $M[\cdot, j-1]$ is a very good candidate for such a vector:

Want: $v[i] = 1$ when $P[1 \ldots i-1] = T[\text{ending @ } j-1]$
Have: $M[\cdot, j-1][i-1] = 1$ when $P[1 \ldots i-1] = T[\text{ending @ } j-1]$.

Comparing these two, we see that $u = M[\cdot, j-1]$ is exactly what we want except that the entry we want at $v[i]$ is at location $u[i-1]$. We can fix this just by shifting everything down. Let $rs(u)$ be the right shift of vector u that prepends a 1 to u and drops its final dimension. That is $rs(u) = \langle 1; u[1 \ldots m-1] \rangle$. This leads to our final expression to compute column j of M:

$$\underbrace{M[\cdot, j]}_{j\text{th column}} = \underbrace{U_P(T[j])}_{1 \text{ when } P[i] = T[j]} \text{ \& } \underbrace{rs(M[\cdot, j-1])}_{\text{prev. col prepended with 1}}. \qquad (5.14)$$

This completes the algorithm: repeatedly compute $M[\cdot, j]$ for increasing j using (5.14). Why do we prepend with 1 instead of 0? This is because the second condition is always trivially true for the first row since there is no shorter prefix to consider.

5.2.1 Space and runtime

Only the current and previous columns of M are needed, so space is $O(|P|)$. If each bit has to be manipulated independently, this has worst case running time $O(|P| \times |T|)$. However, if the operations can be done fast in hardware, we end up with a good running time. In particular, if $|P|$ in bits is $\leq$ the length of computer word, each column of M can be computed in constant time, leading to an $O(|T|)$ algorithm to do the actual matching. The preprocessing takes $O(|P| \times |\Sigma|)$, which is $O(|P|)$ for a constant-sized alphabet, so for patterns that are small enough, we have an $O(|P| + |T|)$ algorithm.

5.2.2 Extension to approximate matching

With a little more space and a slightly more complicated expression, Shift-And can be extended to support mismatches. Consider the problem:

Problem 5.7 (ℓ mismatch string search). *Given strings T, P and integer L, find all occurrences of P in T with $\leq L$ mismatching characters.* ◆

To solve this, we create $L + 1$ matrices M^ℓ (for $\ell = 0 \ldots L$) that are defined similarly to M in the original Shift-And algorithm.

Definition 5.8. $M^\ell[i, j] = 1$ if the ith prefix of P matches the suffix of T ending at j with $\leq \ell$ mismatches. ∎

When should $M^\ell[i, j] = 1$? The following gives both the conditions when it should be 1 and a bitwise expression to compute it. (For ease of notation, we've left off the "i," part of the indexing expressions since we're always dealing with the entire column of M. We've also renamed U_P to U):

$$M^\ell[j] = \underbrace{M^{\ell-1}[j]}_{(a)} \text{ or } \underbrace{\left(rs(M^\ell[j-1]) \text{ and } U(T[j])\right)}_{(b)} \text{ or } \underbrace{rs(M^{\ell-1}[j-1])}_{(c)}. \tag{5.15}$$

The rationales for these three parts are:

(a) ith prefix of P matches the string ending at j with $\leq \ell - 1$ mismatches. This says that if we've matched using $\leq \ell - 1$ mismatches, then clearly the string also matches with $\leq \ell$ mismatches.

(b) $i - 1$ characters of P match with $\leq \ell$ mismatches and the ith and jth characters match. This middle term is analogous to the Shift-And algorithm: if this character matches and the previous column of M^ℓ indicates we can match using $\leq \ell$ mismatches, then we match with $\leq \ell$ mismatches.

(c) $(i - 1)$st prefix of P matches the string ending at $j - 1$ with $\leq \ell - 1$ mismatches. The last condition checks whether we matched the first $i - 1$ characters of P with strictly fewer than ℓ mismatches. If that is the case, then it doesn't matter whether we match or mismatch the current character; either way, we'll use fewer than ℓ mismatches.

The extension to mismatches (and additional extensions to handle small insertions and deletions) is due to Manber and Wu [1992].

5.3   Summary and notes

The two algorithms in this chapter present more "numerical" ways to compare strings, using two very different approaches. Rabin-Karp is a probabilistic algorithm based on a simple hash function that indicates where matches might occur. It can wrongly output some positions where there is no match, but we can control this probability.

The Shift-And algorithm takes a different view and does character comparisons "in parallel" using operations on bit strings. The plain Shift-And version is probably mostly useful as a way to introduce some of these ideas. After all, if we assume P can fit into a computer word, and computer words can be tested for equality in constant time (reasonable), then even the naïve string matching algorithm we saw in Chapter 2 runs in $O(|T|)$ time. But Shift-And supports several extensions that are harder to do directly, such as finding locations with a small number of mismatches and supporting wildcard "don't care" locations in the pattern (see Exercise 5.3). In addition, these kinds of algorithms provide a different perspective on the string matching problem.

Presentation Notes

Our discussion of Rabin-Karp is based on lecture notes from 15-451 that were used in several of the semesters I taught that class. It, and our discussion of Shift-And, is also based on the discussion of this material in Gusfield [1997].

5.4   Exercises

5.1   Explain the $n \geq 4$ condition in Fact 5.4.

5.2   Suppose we want the probability of a false positive to be ≤ 0.1. Describe how to modify Rabin-Karp to achieve this without running it multiple times.

5.3   Describe how to support wildcards "?" in the Shift-And algorithm that can match any letter in the pattern or text.

5.4   The Karp-Rabin fingerprinting string search algorithm has an advantage that we did not discuss. Suppose you have a text T over the alphabet $\{0, 1\}$, and you need to search for many patterns $P_1, \ldots, P_k$, where all of the patterns are the same length m. With Karp-Rabin, you need to compute the hash values for each m-long window of T only once and you can reuse them for each pattern.

The big downside with this approach is that all the patterns must be the same length, which this problem asks you to fix. Let $\hat{T}[i, j]$ be the number formed by taking the bits from index i to index j (inclusive) of T. Let p be a prime and let $h_p(x) = x \bmod p$. Give a way to preprocess T in $O(|T|)$ time that uses $O(|T|)$ space so that you can compute $h_p(\hat{T}[i,j])$ for any $1 \leq i \leq j \leq |T|$ in constant time. (Assume arithmetic operations mod p take constant time.)

5.5   Let $h_p(A)$ be a Rabin-Karp hash function applied to string A (for some prime p). Let $S[i \ldots j]$ represent the substring of S from positions i to j inclusive. Assume $\Sigma = \{0, 1\}$. This problem asks you to use Rabin-Karp ideas to sort substrings of a string more quickly. Let S_i and S_j be substrings of some length k starting at positions i and j of a string S. Normally, it would take $O(k)$ time to compare these strings to decide which is alphabetically first.

Assuming h_p has no hash collisions (i.e., no false positives), sketch an $O(\log k)$ time algorithm to decide whether $S_i < S_j$ alphabetically.

5.6   Suppose you have an efficient algorithm (e.g., [Agrawal et al., 2004]) for testing if a number $\leq x$ is prime that runs in time $f(x)$ for some function f. Give a randomized algorithm that runs in $O(f(m) \log M)$ time and finds a prime $\leq M$ with probability $\geq 1/2$.

5.7   Give the base cases for filling out matrix M (5.10).

5.8   If $|P| > 1$, then clearly it cannot match starting at position m of T. Which entries of M (5.10) must therefore be 0?

CHAPTER 6

Searching for Multiple Patterns

Suppose we have a fixed text T and a set of query strings $\{P_1, \ldots, P_k\}$ that we want to search for in T. That is, we have the following multi-pattern search problem:

Problem 6.1 (Multi-pattern exact match). *Find every occurrence of each $P_1, \ldots, P_k$ in string T.* ◆

This generalizes the search problem we saw solved by the Z algorithm, Boyer-Moore, and KMP to the case of multiple patterns. An instance of this problem consists of $k + 1$ strings $(T, P_1, \ldots, P_k)$. While we can preprocess T or P_i, that preprocessing counts against our running time.

We are looking for a runtime that is $O(z + |T| + \sum_i |P_i|)$ where z is the number of locations to output. Such a runtime would be optimal in the worst case, since we have to read T and the patterns, and do at least some work for any match. But the running time of repeatedly applying KMP (say), once for each pattern, is $O(k|T| + \sum_i |P_i|)$ since the ith KMP execution takes $O(|T| + |P_i|)$ time. Later (Chapter 13), we will see how a data structure called a suffix tree can solve this problem in $O(|T| + \sum_i |P_i| + z)$ time, which matches what we want. But here, we will consider two other approaches that also work well for this problem—one with the desired worst case runtime and one that doesn't have such a bound. Seeing multiple approaches is useful because practical considerations may cause one approach to work better than others in a given situation, and different algorithmic ideas to solve the same problem can be adapted in different ways. In addition, our first approach to multiple pattern search, the Aho-Corasick algorithm, predates suffix trees and is therefore of historical interest.

One might ask, why didn't we have a term similar to $O(z)$ in our KMP or Z algorithm analysis? Didn't we also care about output time there? In the single-pattern case, there can be at most $O(|T|)$ positions output, so the $O(z)$ term was "included" in the $O(|T|)$ term. For the multi-pattern case, we could have a match for each pattern at each location, which is in worst case $O(k|T|)$, but we don't want this $O(k|T|)$ term to show up in our running time, otherwise we could just run KMP k times!

The first approach that we will see for this problem, Aho-Corasick, is based on keyword trees and is really a generalization of KMP. We'll have a nice runtime guarantee for it of the form just discussed. The second approach we will see is based on hashing, and it uses ideas similar to Rabin-Karp. We won't have any theoretical argument for its performance, but it can work well in practice.

6.1   Aho-Corasick—a prefix-based approach

The Aho-Corasick algorithm [Aho and Corasick, 1975] is based on putting the patterns into a *keyword tree*.

Definition 6.2 (keyword tree). A keyword tree $K(P)$ of a set of patterns $P = \{P_1, \ldots, P_k\}$ is the smallest, rooted tree where:

1. each edge is labeled with a letter from the alphabet Σ;
2. edges leading from any node to its children all have different labels;
3. and there is a function $n(i)$ that gives the node such that pattern i is spelled out on the unique path from the root to $n(i)$.   ■

Each letter in the patterns is represented at most once, so this means that the size of the keyword tree is $O(\sum_{s \in P} |s|)$. Here in (6.1) is an example with $P = \{\text{"abandon"}, \text{"abduct"}, \text{"abacus"}\}$.

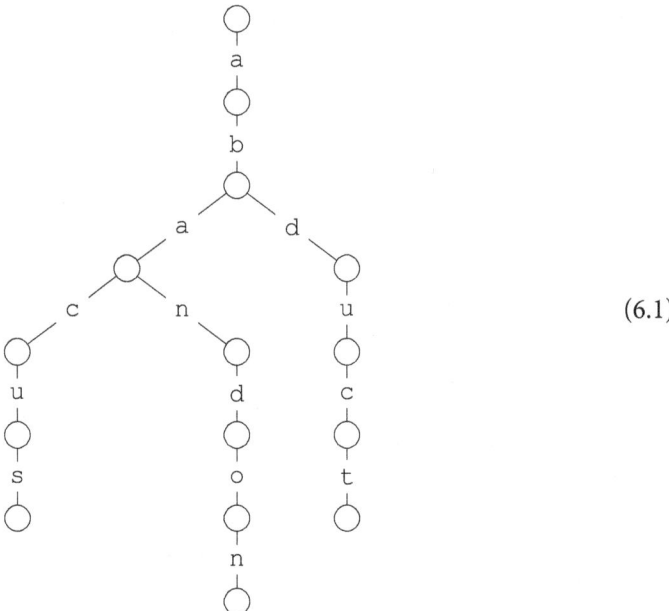

$$(6.1)$$

Definition 6.3 (keyword tree node properties). Define the following notation for a keyword tree $K(P)$ of a set P:

- $L(v) :=$ the string spelled out by the path from the root to v.
- $lp(v) :=$ the longest proper suffix of $L(v)$ that is also a prefix of some pattern in the set P.
- If $|lp(v)| > 0$, define $f(v)$ to be the node in $K(P)$ representing the string $lp(v)$; otherwise $f(v)$ is the root of the keyword tree.   ■

Pictorially, these definitions are:

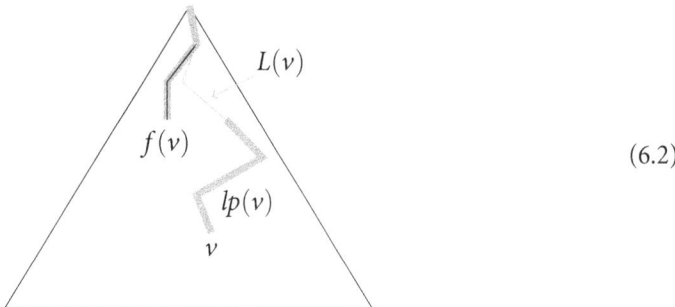

$$(6.2)$$

where the gray paths are equal.

Note the similarity of $lp(v)$ to the ideas we used in KMP: the *memo* and spm_i arrays. $lp(v)$ gives which "prefix" of the tree we should move to after a mismatch.

Clearly $L(v)$ and $lp(v)$ are well defined—allowing for $lp(v)$ to possibly be the empty string. What about $f(v)$? Does such a node always exist? Yes.

Theorem 6.4. $f(v)$ *always exists and is unique for any node v in $K(P)$.*

Proof: When $|lp(v)| > 0$, $lp(v)$ is a prefix of a pattern, and every pattern is represented by a unique path in $K(P)$ on which every one of its prefixes is spelled out. Otherwise, if $|lp(v)| = 0$ then $f(v)$ is the root. □

$f(v)$ is called the *failure function* for reasons that will become clear. Here's an example of $K(P)$ with the failure function $f(v)$ drawn in dashed arrows in (6.3).

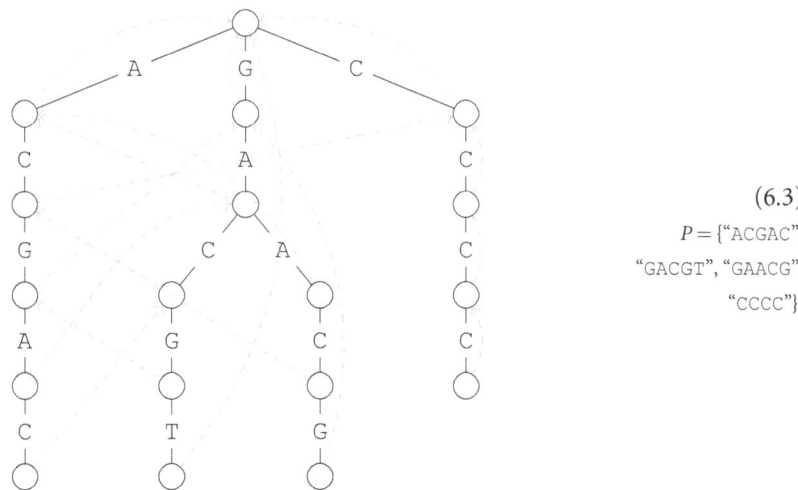

$$(6.3)$$
$P = \{\text{"ACGAC"},$
$\text{"GACGT"}, \text{"GAACG"},$
$\text{"CCCC"}\}.$

6.1.1   Search

Assume we have created a keyword tree from the patterns $\{P_i\}$ and have easy access to the function f. Also, for now, we make the assumption that no pattern is a substring of any other pattern. The first version of our search function is as follows:

1. Walk down string T and tree at same time, matching characters as in (6.4).

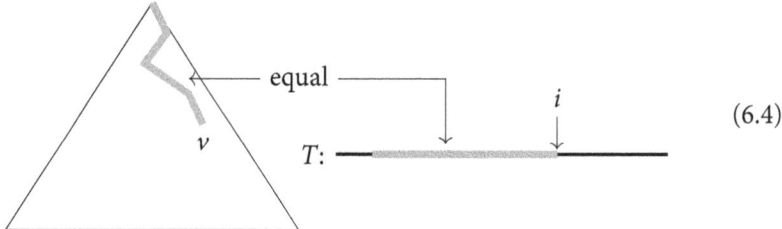

$$(6.4)$$

2. If you get to a node that represents a full pattern, report an occurrence. ($\leftarrow$ we will modify this step in Section 6.1.2.)
3. If you get stuck at node v, jump to node $f(v)$ and continue as in (6.5).

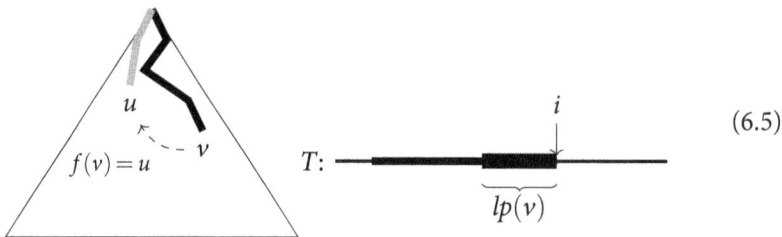

$$(6.5)$$

Getting stuck means that the next character in T doesn't equal a character represented by any edge leaving the current node. This approach correctly (up to one bug we will discuss in a moment) outputs the locations of the patterns since when we get stuck at a node v, we know we have just seen $L(v)$ in T. We can't continue from v, but we know the longest suffix of $L(v)$ that could still match a pattern is $lp(v)$. We jump directly to the node that represents $lp(v)$ as if we had just started searching at the start of $lp(v)$.

This idea is similar to the failure function from KMP, which we called *memo*. In KMP, we jump back to the latest position in the pattern that could still be extended from where we got stuck. Same idea here, but generalized to trees.[1]

6.1.2   Handling patterns contained in other patterns

We assumed that no pattern P_i is a substring of any other pattern P_j. The reason for this assumption was that we may spell out P_j (in step 1) without ever noticing that P_i is contained within it, because these two patterns could be in completely different parts of the tree (since

---

[1] The $f(v)$ links are also similar to the suffix links that we will see when we introduce suffix trees (Chapter 13), but instead of chopping off only 1 character like suffix links will, they chop off the characters needed so that we can still possibly complete a pattern.

they could have different prefixes). That is the situation represented by the top part of this diagram (6.6).

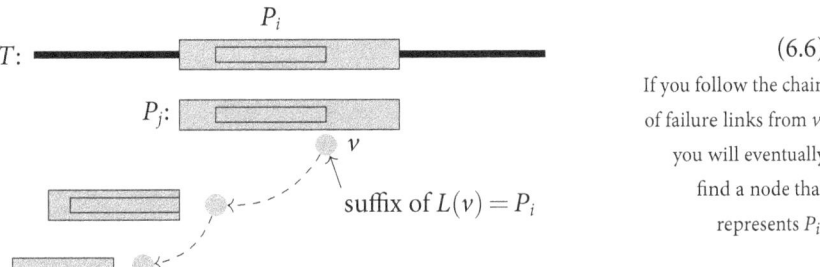

(6.6)

If you follow the chain of failure links from v, you will eventually find a node that represents P_i.

The fix for this is to realize that along the path for pattern P_j, we must have encountered a node that represents a string for which P_i is a suffix (since by assumption P_i is contained in P_j). Suppose that node was v. Then, if we follow the f links repeatedly from v, we will eventually reach a node that represents P_i. Why is this? Because at v, P_i is a proper suffix of $L(v)$ and also clearly the prefix of some pattern (P_i), and the failure function walks through, by definition, the proper suffixes of $L(v)$ that are also the prefix of some pattern. That is shown in the bottom parts of figure (6.6).

We could have several nested patterns contained in P_j that "end" at v, not just one. Following the failure function trail would walk through all of them (and possibly other nodes) from longest to shortest.

All of this means that when we reach a node v in the tree, we need to output all patterns P_a such that:

- P_a is represented by v (the normal case, which also handles the case when P_a is a prefix of P_j) or there is some node representing P_a that is reachable following failure links from v.

To do this efficiently, we augment our tree with additional links $o(v)$ that point to the next node that represents a full pattern along the $f^{(*)}(v)$ chain. (Here, $f^{(*)}$ denotes applying f multiple times.) These $o(v)$ links skip over nodes in the $f^{(*)}(v)$ chain that do not represent patterns. It's easy to add these links in when we compute the f links (see Exercise 6.2).

Now, the algorithm can be visualized as (6.7),

(6.7)

where we follow the solid edges down the tree as we match characters in T (these edges would all be labeled by a character from Σ, which we omit from the figure for clarity). At each node, we output an occurrence if the node represents a pattern (the solid nodes along the solid path). We *also* check at each node whether there is an o pointer leaving it (dotted arrows), and if so, we follow that chain of links and output occurrences for each node we reach. Following those links at each node (if they exist), does take time, but for each such link that we follow, we definitely output a pattern location, so following the links fits within the $O(z)$ part of our goal running time.

6.1.3   Running time

The runtime analysis of the Aho-Corasick algorithm is nearly identical to KMP. The index i into the string T is never decremented. It is incremented on a character match. Therefore, each character is matched at most once.

Every mismatch results in a "shift" of the leftmost possible starting point for a pattern of at least 1 so we can have at most $O(|T|)$ total mismatches.

Therefore the running time is:

$$O(\underbrace{\text{total length of patterns}}_{\text{build keyword tree}} + \underbrace{|T|}_{\text{search } T} + \underbrace{\text{\# positions output}}_{\text{output}}), \tag{6.8}$$

assuming we have f.

6.1.4   Computing f

The next piece of the puzzle is how to compute f. We compute f using a breadth-first search (BFS) of the keyword tree. That is, we compute $f(u)$ for each node that is $i - 1$ or fewer hops from the root before we compute $f(u)$ for nodes that are i hops from the root. Accordingly, assume we've computed $f(v)$ for all v at fewer than i hops from the root. We want to compute $f(u)$ for some u at i hops from the root. To do this:

- Let v be the parent of u and x be the character on the (v, u) edge. We already know $f(v)$ since it is fewer hops from the root than u.
- Traverse the chain of $f(v), f(f(v)), f(f(f(v)))$, etc. until you find a node w with a child edge (w, y) labeled x. Set $f(u)$ equal to y. If you reach the root, set $f(u)$ to the root.

Visually, this looks like (6.9).

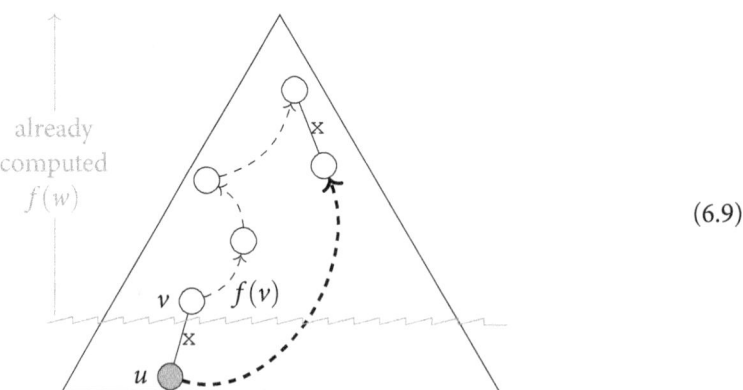

$$\tag{6.9}$$

The idea is that $f(v)$ is the longest proper suffix of $L(v)$ that matches a prefix of a pattern, $f(f(v))$ is the longest proper suffix of $L(f(v))$ that matches a pattern prefix, and so on. We want the longest (first encountered) one of those suffixes that can be extended with x since our $f(u)$ must point to a node that represents a string that ends with x.

Theorem 6.5 (Time to compute f). *Consider the path Q representing P_i from the root to a node v. The total time to compute $f(u)$ for nodes on Q is $O(|P_i|)$.*

Proof: We use $|lp(\cdot)|$ as a potential. Consider how $|lp|$ changes as we look at nodes along Q. (Note: the algorithm doesn't look at nodes of Q in isolation—it also looks at other nodes, interspersed with nodes on Q, and some of those may be shared with other patterns; we ignore this for this proof, and instead just myopically focus on this one path Q.)

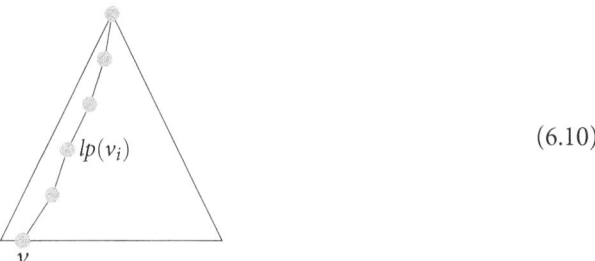

$$(6.10)$$

Let's track how $|lp|$ changes along this path. Whenever we follow an f link, we necessarily reduce $|lp|$ by at least 1. So each constant unit of work of following an f-link is paid for by "withdrawing" potential from $|lp|$. How much of this "potential" resource do we have available? $|lp(v)|$ increases by at most 1 compared to $|lp(parent(v))|$. So, $|lp|$ increases by at most $|Q|$ units total along this path. That is, the amount of \$ put into our "lp" bank account is at most $|Q|$. We debit this every time we have to follow an f link. Finally, $|lp|$ starts at 0 at the root, and it is never negative.

Said another way, each unit of work walking the f chain causes a reduction of at least 1 "lp units" and over the entire path there are at most $|Q|$ "lp units" available to spend, and we never go into debt. In a picture it looks like (6.11).

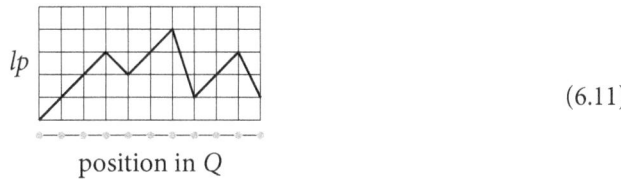

position in Q

$$(6.11)$$

The total "up" steps in (6.11) are at most $O(|Q|)$. Each "down" step pays for traversing an f link, and can never take us into negative values of lp. So the total "down" steps are also at most $O(|Q|)$. Therefore, the total work following the f links is $O(|Q|)$. The work just traversing the path from root to its end as part of the BFS is also $O(|Q|)$. □

As a corollary of theorem 6.5, the total running time to compute the fs over all nodes is $O(\sum_i |P_i|)$. Therefore, the running time is $O(\text{size of pattern set}) = O(\sum_i |P_i|)$, repeating this same argument for each P_i. Note that we did not prove that the time is $O(|K(P)|)$.

6.2   Wu-Manber—a suffix-based approach

If Aho-Corasick is KMP extended to multiple patterns, Wu-Manber [Manber and Wu, 1992] is sort of Rabin-Karp (Section 5.1) extended to multiple patterns, but with a dash of Boyer-Moore (Chapter 3) thrown in. Here's the idea: define a block size b and a hash function h that takes as input length-b strings. We preprocess our set of patterns into a hash table based on their *last* (suffix) blocks, putting pattern P_i in the bin $h(last(P_i))$, where $last(P_i)$ is the length-b suffix of P_i. This looks like (6.12).

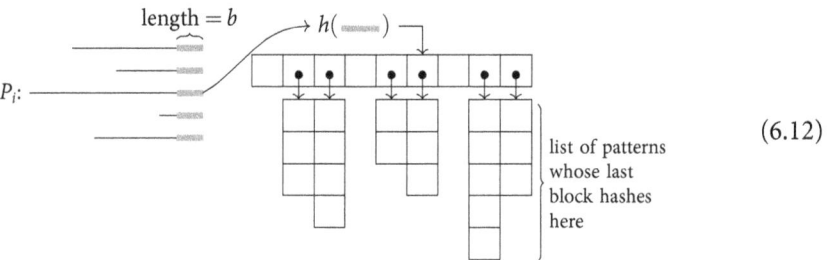

$$(6.12)$$

To search, we can walk down T left-to-right keeping a pointer i into a position at T. When we are at i, we check each pattern in $h(T[i-b+1,\ldots,i])$ to see if that pattern occurs ending at position i in T. If there is a match, the last block of the matching pattern P_i must hash to the same place as $T[i-b+1,\ldots,i]$ so we will never miss a pattern. We explicitly check whether there is a match at this position (so we will never report a false positive).

This would work if we just shifted by 1 after checking each pattern in the hash bin. But we can use the fact that, when we are at i, we know the block ending at i has a particular hash value. Therefore, we can shift each of the patterns so that the block ending at i shares a hash value with some block inside each of the patterns. That is, if the shaded bars represent the rightmost length-b substrings of the patterns that hash to the same value as the length-b string ending at i:

$$(6.13)$$

then we can add to i so that the rightmost of these blocks ends up over the block ending at i.

To facilitate this shift, we pre-compute a *GoodShift* array. Let B_{ij} be the block of length b ending at position j in pattern P_i. Then we compute the entries of that array using the definition in (6.14):

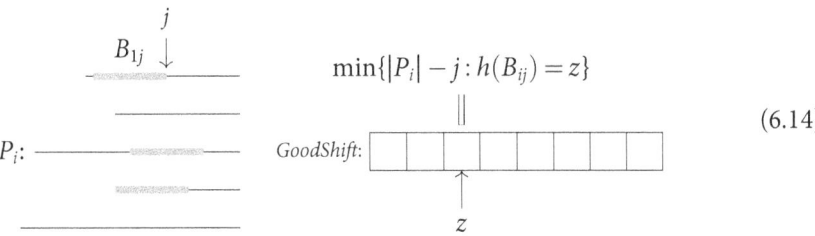

$$\tag{6.14}$$

so that *GoodShift*[z] contains the amount by which it is safe to shift given that we know T ending at i hashes to z with hash function h. This is like the Boyer-Moore good-shift rule (and similar to the KMP failure function). If the *GoodShift*[z] is 0, we fall back to the default of incrementing our pointer into T by 1.

This will never miss any occurrence of any of the patterns. The running time depends on the number of hash collisions. For non-adversarially designed patterns, this can often work well.

6.3   Summary and notes

This chapter has shown a couple of ways of searching for multiple patterns simultaneously. Our interest in these approaches is largely theoretical and historical: suffix trees (and suffix arrays and BWT) have largely subsumed Aho-Corasick in practice, though the idea of a "failure function" and the analysis we saw for that approach are still important for modern string algorithms. Ideas such as this can be used for more complex goals, as done in, for example, Ma et al. [2020]. Wu-Manber is more practically oriented; without strong guarantees about the runtime, its speed depends on the application.

> **Presentation Notes**
>
> Our presentation of Aho-Corasick follows that of Gusfield [1997].

6.4   Exercises

6.1   Write out the (straightforward) algorithm to build the keyword tree of a given set of patterns $\{P_1, \ldots, P_k\}$ in time proportional to the total size of all the patterns.

6.2   Explain how to compute $o(v)$ for every node while computing the $f(v)$ values for every node during the Aho-Corasick pre-processing.

6.3   *Two-Dimensional Exact Matching.* Suppose you are given a matrix $\mathcal{T} \in \Sigma^{m \times n}$ and a pattern matrix $P \in \Sigma^{p \times q}$, with $p < m$ and $q < n$. Design an algorithm that finds every submatrix in $\mathcal{T}$ that is an exact match of P in time $O(mn + pq)$. You may assume all rows of the matrices are distinct.

6.4   Give as efficient an algorithm as you can for computing the array *GoodShift* in the Wu-Manber algorithm.

Part II

Edit Distance

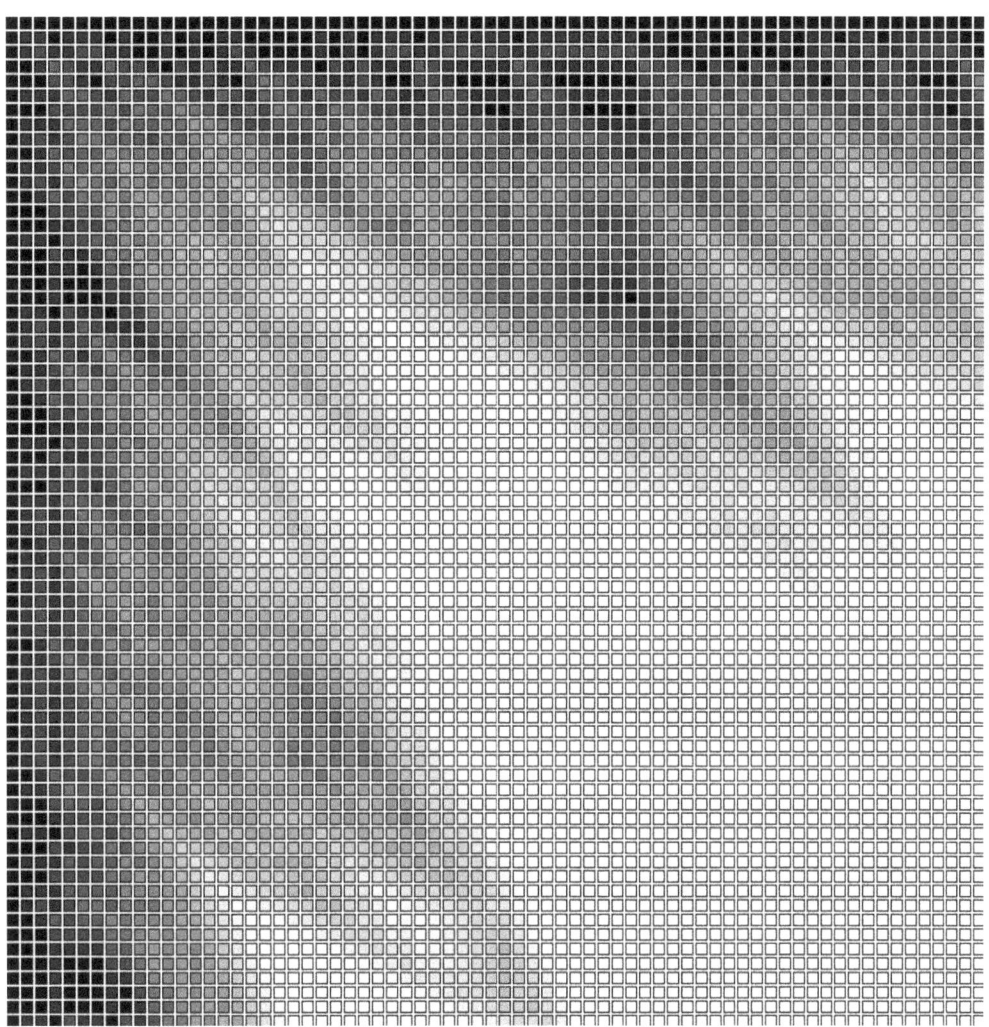

CHAPTER 7

Edit Distance for Inexact Matching

Requiring an exact match is often too restrictive. Instead, we need to allow some differences between the pattern and the text. For example, if we are comparing source code files, we don't expect to find the exact same sequence of characters. If we are comparing genomes or protein sequences from different individuals or different species, we expect some mutations to have occurred introducing some differences. If we are implementing a spell checker or an email clustering algorithm or sentiment analysis engine, we can't expect users to have spelled every word the same way. This motivates the inexact matching problem.

Inexact matching is one of the most important tools in computational biology. The program BLAST [Altschul et al., 1990, 1997] uses (sophisticated) variants of the algorithms we'll see in this chapter to compare DNA sequences with a vast database of known DNA sequences. BLAST is one of the most-cited papers in all of science.

As an example, here is part of a CTCF protein sequence from 8 different species.

```
H. sapiens      -EDSSDS-ENAEPDLDDNEDEEEPAVEIEPEPE---------PQPVTPA
P. troglodytes  -EDSSDS-ENAEPDLDDNEDEEEPAVEIEPEPE---------PQPVTPA
C. lupus        -EDSSDS-ENAEPDLDDNEDEEEPAVEIEPEPE---------PQPVTPA
B. taurus       -EDSSDS-ENAEPDLDDNEDEEEPAVEIEPEPE---------PQPVTPA
M. musculus     -EDSSDS-ENAEPDLDDNEDEEEPAVEIEPEPE--PQPQPPPPPQPVAPA
R. norvegicus   -EDSSDS-ENAEPDLDDNEEEEEPAVEIEPEPEPQPQPQPQPQPQPVAPA
G. gallus       -EDSSDSEENAEPDLDDNEDEEETAVEIEAEPE---------VSAEAPA
D. rerio        DDDDDDSDEHGEPDLDDIDEEDEDDL-LDEDQMGLLDQAPPSVPIP-APA
```
(7.1)

Each line is one string, and the strings are *aligned* in such a way to highlight their similarities and differences. The *gap* character "–" is a special character used to space things out—it's not part of the input sequences.

Using such an alignment based on inexact matching, one could:

- identify important sequences by finding conserved regions;
- find genes similar to known genes;
- estimate evolutionary distances (e.g., zebrafish is farther from humans than chicken);
- search databases of genetic sequences;
- obtain hints about protein structure and function;

along with many other tasks.

7.1   Edit distance and alignments

We look at the most widely used inexact string comparison problem: computing the edit distance.

Definition 7.1 (Edit distance). The *edit distance* between strings A and B is the smallest number of the following operations that are required to turn A into B:

- replace a character with another character in the alphabet (aka "substitution");
- delete a character;
- insert a character from the alphabet. ∎

For example, the edit distance between `riddle` and `triple` is 3 because `riddle` → `ridle` → `riple` → `triple`, and there is no shorter sequence of edit operations that makes that transformation.

The edit distance, denoted $ed(A, B)$, is symmetric: $ed(A, B) = ed(B, A)$. The reason for this is that each of the edit operations (replace, delete, insert) has an inverse that "undoes" that operation: replacing x with y can be undone by replacing y with x; inserting x can be undone by deleting x; and deleting x can be undone by inserting x. This means if we are given a chain of edit operations to transform A into B we can run the chain "backwards" to transform B into A.

In addition, the *order* of the operations in the chain doesn't matter: you can insert the characters you need to insert first, then delete the ones you need to delete, then do the replacements; or you can intermix those operations.

This symmetry and commutativity allows us to represent an edit chain using an *alignment* rather than a sequence of operations.

Definition 7.2 (alignment). An *alignment* between two strings A and B is a pair of strings A' and B' such that

1. $A' = A$ and $B' = B$ after removing all instances of the character "-",
2. $|A'| = |B'|$, and
3. there is no i such that $A'[i] = B'[i] =$ "-". ∎

Examples make this easier to see.

```
prin-ciple              prin-cip-le
|||| |||xx              |||| ||| |
princcipal              princcipal-
(1 gap, 2 mm)           (3 gaps, 0 mm)


misspell                prehistoric
||| ||||                   |||||||||
mis-pell                ---historic
(1 gap)                 (3 gaps)


aa-bb-ccaabb            al-go-rithm-
|x || | | |             || xx ||x |
ababbbc-a-b-            alKhwariz-mi
(5 gaps, 1 mm)          (4 gaps, 3 mm)
```

(7.2)

In (7.2), gap characters "–" have been inserted. When a gap has been inserted into B (but not A) this represents a deletion from A. When a gap has been inserted into A (but not B), that is an insertion into A to get to B. When the two characters that are stacked on top of one another are not the same (denoted by X in the examples), this is a substitution operation. Therefore, from the alignment, you can immediately derive a chain of edits. To solve the edit distance problem, we can instead try to find alignments that represent a small number of edit operations.

7.2 The string alignment problem

Problem 7.3 (String Alignment). *Let $a_1, \ldots, a_m$ and $b_1, \ldots, b_n$ be two strings. Given a real number g and a function $cost(x, y) : \Sigma \times \Sigma \to \mathbb{R}$, compute the alignment G that minimizes*

$$g \times (\# \text{ gap chars in } G) + \sum_{(a_i, b_j) \in G} cost(a_i, b_j), \tag{7.3}$$

where $(a_i, b_j) \in G$ indicates that character a_i is aligned to character b_j in the alignment. ◆

This problem slightly generalizes the edit distance since we are allowed to have different costs for aligning different characters. A typical setting might be $g = +1$, $cost(x, x) = 0$ and $cost(x, y) = +1$ when $x \neq y$. There are settings for the parameters that make the problem uninteresting or useless. It is also completely equivalent to fold the g parameter into the *cost* function, and instead have $cost(-, x) = cost(y, -) = g$ for any x and y. We could even generalize further by having different values of $cost(-, x)$ for different values of x—that won't make any difference to our algorithm.

An alternative way to view this problem is to think about putting edges between characters to minimize (7.3) as shown in (7.4).

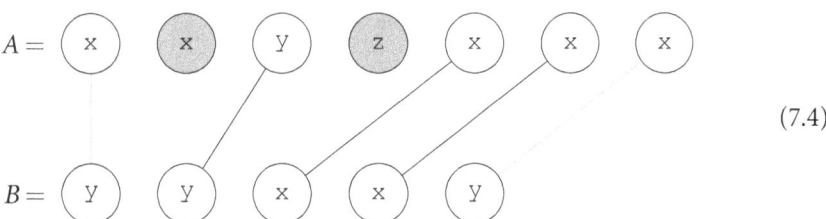

$$\tag{7.4}$$

We're allowed to match up letters between the two strings, so long as *the lines connecting the edges don't cross*. They can't cross because we're only allowed (by definition of the problem) to insert gap characters to space letters out.

7.2.1   Algorithm for minimum cost alignments

Consider our options for what we could do with the last characters a_m and b_n of two strings we are trying to align.

$$A = a_1 a_2 a_3 a_4 \ldots a_m$$
$$B = b_1 b_2 b_3 b_4 \ldots b_n$$

Our options are:

1. (a_m, b_n) are aligned to each other (either a match or mismatch);
2. a_m is aligned to nothing (a gap);
3. b_n is aligned to nothing (a gap);
4. a_m is aligned to some b_j ($j < n$) and b_n is matched to some a_k ($k < m$).

At least one of these four options must happen.

In fact, option #4 cannot happen. The reason is because the "no crossing rule" forbids this option. In other words, we can't "twist" the string to allow (7.5).

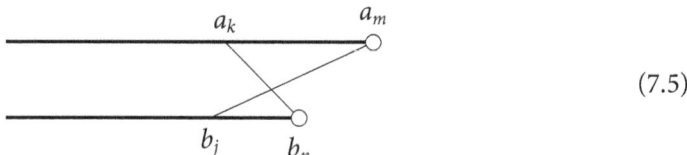

$$(7.5)$$

Only the first three options are possible, and at least one of them must happen. Option 1 is mutually exclusive with options 2 and 3, although both option 2 and option 3 could happen in the same alignment.

We turn these three cases into a recursive expression since once we decide how to handle the last character(s), we are left with a smaller problem of the same type. Define

• $OPT(i, j)$ to be the cost of the optimal alignment of the first i characters of A and the first j characters of B.

The cost of the optimal alignment of the strings A and B is therefore $OPT(m, n)$, where $|A| = m$ and $|B| = n$.

We can compute $OPT(i, j)$ using the following recurrence:

$$OPT(i,j) = \min \begin{cases} cost(a_i, b_j) + OPT(i-1, j-1) & \text{align } a_i, b_j \\ g + OPT(i-1, j) & a_i \text{ is not matched} \\ g + OPT(i, j-1) & b_j \text{ is not matched.} \end{cases} \quad (7.6)$$

The cost, $OPT(i, j)$, of aligning $a_1, \ldots, a_i$ to $b_1, \ldots, b_j$ is written in terms of problems that involve at least one shorter string. The key idea is that we don't know which of the three possibilities in the recurrence is the right one, so we will try them all.

We need some base cases to stop the recursion from going on forever:

$$OPT(i, 0) = i \times g,$$
$$OPT(0, j) = j \times g.$$

In other words, the only way to align the empty string (first 0 characters) to the prefix of A of length i is to align each character with a gap character. The cost of such an alignment is $i \times g$.

There are a few big ideas we've used so far. First, we abandoned trying to compute the actual alignment—we are focusing on computing the *cost* of the optimal alignment. We'll see later how to recover the alignment. Second, we added some parameters to make our problem more general—we're actually interested only in the alignment between the two complete strings, but we've generalized this to define the optimal between any two prefixes. Finally, we've written down a recurrence without worrying about how to actually compute it. But now we have to figure out a way to implement this recurrence.

7.2.2   Implementing the recursive algorithm

If we simply created a function `opt(int i, int j)` that recursively called itself using the recurrence directly, this would be extremely inefficient—exponential in fact. The reason is that many of the $OPT(i, j)$ subproblems would be solved more than once. Instead, we use the next big idea which is that there are only $O(nm)$ possible $OPT(i, j)$ subproblems, so we can instead solve each of those subproblems one by one.

We're ultimately interested in $OPT(m, n)$, but we will compute all other $OPT(i, j)$ for $i \leq m, j \leq n$ on the way to computing $OPT(m, n)$. Create an $(m + 1) \times (n + 1)$ matrix C, where entry $C[i, j]$ will store the value of $OPT(i, j)$.

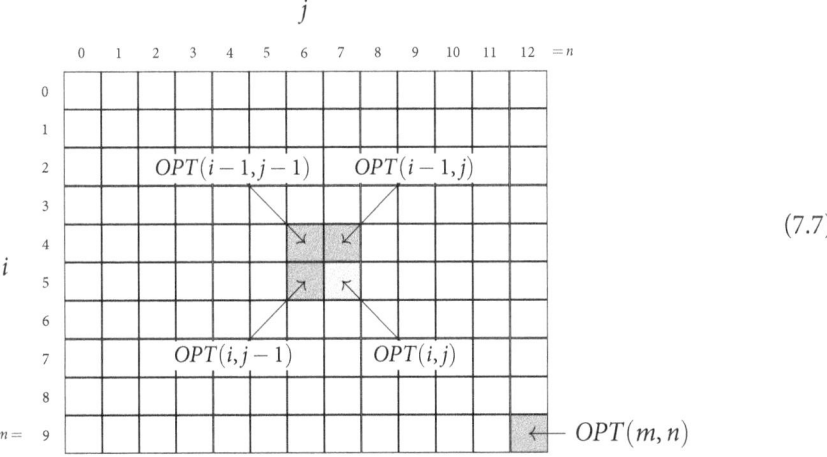

(7.7)

The first row and first column of this matrix are easy to fill in using the base cases. Consider an arbitrary cell (i, j). How can we fill it in? Looking at the recurrence, we see that the value of $OPT(i, j)$ depends on 3 other cells: $OPT(i-1, j)$, $OPT(i-1, j-1)$ and $OPT(i, j-1)$. If those cells were already filled in, it would be easy to fill in $OPT(i, j)$.

The algorithm. We can fill in the matrix in an order so that the subproblems we need to have computed have already been computed when we need them. We can do this by filling in rows from "left-to-right" and "top-to-bottom."

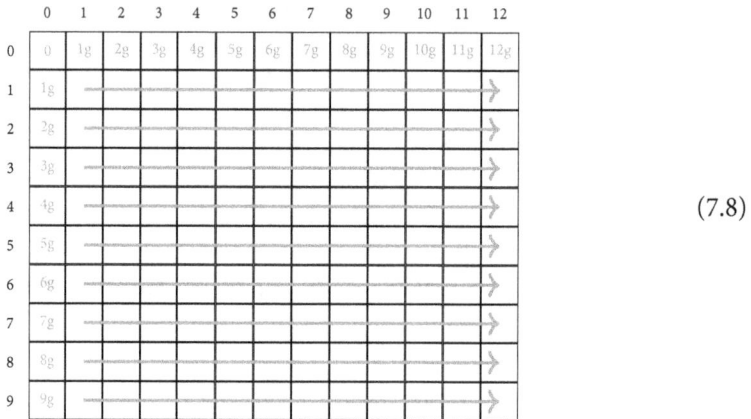

(7.8)

This leads to the following algorithm for computing the cost of the optimal alignment.

Algorithm 7.1. Compute score of optimal alignment.

function EDITDISTANCE(X, Y)
    ▷ *Initialize the base cases*
    for $i \leftarrow 0, \ldots, |X|$ **do** $C[i, 0] \leftarrow i \times g$
    for $j \leftarrow 0, \ldots, |Y|$ **do** $C[0, j] \leftarrow j \times g$
    ▷ *Fill in the rest of the matrix from "smaller" to "larger" subproblems*
    for $i \leftarrow 1, \ldots, |X|$ **do**
        for $j \leftarrow 1, \ldots, |Y|$ **do**
$$C[i, j] = \min \begin{cases} cost(a[i], b[j]) + C[i-1, j-1] \\ g + C[i-1, j] \\ g + C[i, j-1] \end{cases}$$
    return $C[|X|, |Y|]$

At the end of the algorithm, $C[|X|, |Y|] = C[n, m]$ contains the edit distance between the two strings ($= OPT(n, m)$). Why? By induction: *every* cell contains the optimal edit distance between some prefix of string A with some prefix of string B.

Running time. The running time of this style of algorithm is the number of subproblems times the time to compute each subproblem. Here, the number of entries in array $= O(m \times n)$, where m and n are the lengths of the two strings. Filling in each entry takes constant $O(1)$ time—we just need to take the minimum of 3 already-computed items. So the total running time is $O(mn)$.

7.3   Dynamic programming

The previous sequence alignment/edit distance algorithm is an example of *dynamic programming*. The main idea of dynamic programming is to break your problem into a reasonable number of subproblems and then solve the subproblems in an order so that when you need an answer to a subproblem, it's already been computed.

In order for dynamic programming to lead to an efficient algorithm, we need:

1. that the optimal value of the original problem can be computed from some similar subproblems;
2. there to be only a polynomial number of subproblems;
3. there to be a "natural" ordering of subproblems, so that you can solve a subproblem by only looking at "smaller" subproblems.

Dynamic programming is one of the most useful and general algorithm design techniques that applies in a very large number of circumstances.

7.4   Finding the actual alignment: traceback

We can compute the *value* of the edit distance using Algorithm 7.1. How do we find the actual sequence of operations? In other words how do we find the alignment? The technique we will use for finding the actual optimal solution is general and applies to any dynamic programming algorithm. This technique augments our algorithm to store *two* things in each cell: the value (as we have been doing) and an "arrow" that points to the subproblem that was used in the "winning" case (in this case, the one that is the minimum) in the recurrence.

For example, if cell (i, j) was filled in by matching the two characters, the cell (i, j) was computed using $(i - 1, j - 1)$.

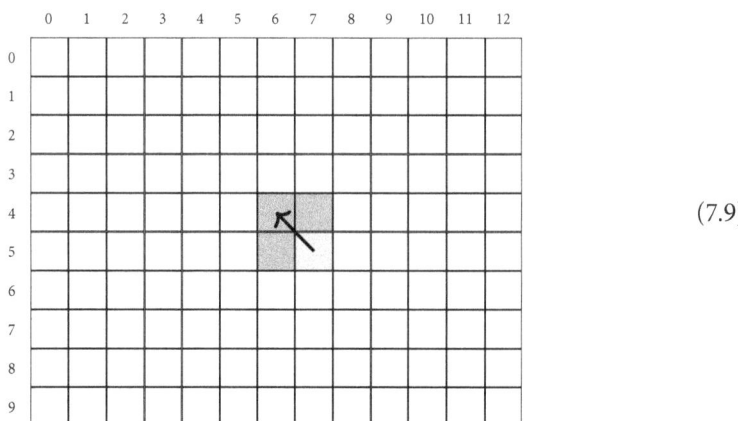

$$(7.9)$$

There might be ties in the min expression in the recurrence. This can happen when there are multiple optimal alignments. If that happens, we pick one of the minimizing cells arbitrarily to which to draw the arrow.

If we record the arrow for every cell, there will be a path that goes from the cell (m, n) to a cell in either the first row or first column of the matrix.

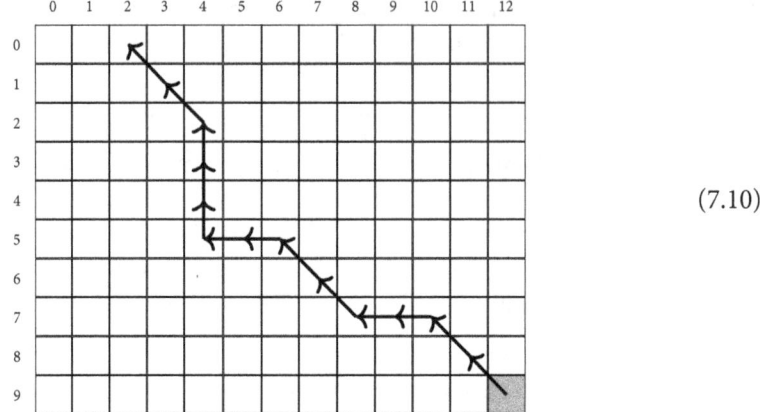

(7.10)

We can follow these pointers and build the alignment from its end to its start. Using the geometry from figure (7.10):

- If you follow a diagonal arrow, add both characters to the alignment.
- If you follow a left arrow, add a gap to the y-axis string and add the x-axis character. (We're "using up" an x-axis character without using up a y-axis character.)
- If you follow an up arrow, add the y-axis character and add a gap to the x-axis string.

Once you get to the edge of the matrix, add the appropriate number of gaps to reach cell $(0, 0)$.

Dynamic programming as shortest path in a graph: Another way of looking at the dynamic programming algorithm is to think of the problem as a graph.

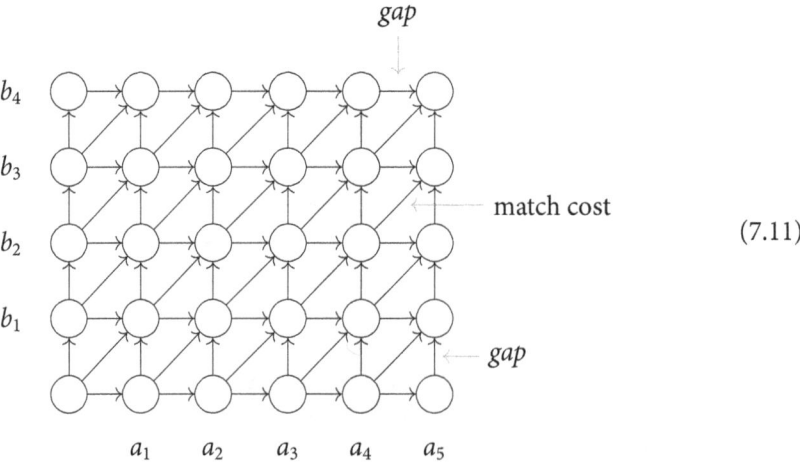

(7.11)

The weights (lengths) of the edges are the cost of "using" that traceback arrow. The optimal traceback path is the shortest path from $(0, 0)$ to (m, n). In fact, Dijkstra's algorithm for finding shortest paths is a generalization of the edit distance algorithm.

7.5   Local and semi-global alignment

We've seen up to this point how to compute the best alignment between two complete strings. This is called *global alignment*. What if we want to find the best-matching location of a string within a longer string? For example, we could have a gene sequence x that we want to find (approximately) within a much longer genome T. How could we do this? Of course, a simple way would be to compare x to every substring of T. But there are $O(|T|^2)$ substrings of T, so this approach is likely too slow. Instead, we modify our dynamic program.

7.5.1   Local alignment

We now consider a slightly different problem, called local alignment.

Problem 7.4 (Local Alignment). *Given two strings S and T, find substrings s in S and t in T with the maximum alignment score between them, where the score is given by the sum over a function* match(a, b) *that gives the score of putting character a atop character b (either of which may be a gap character "–"). ◆*

Visually, Problem 7.4 is looking for something like the central region in (7.12).

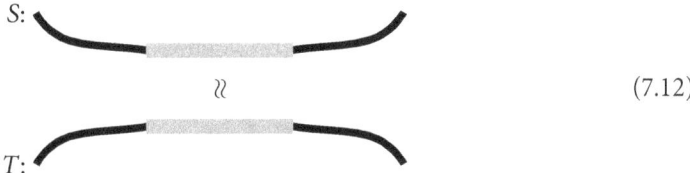

$$\tag{7.12}$$

One motivation for local alignment is the fact that proteins (and hence genes) are often made up of building blocks called "domains" that are reused in many genes. Hunting for genes that share building blocks would lead to the local alignment problem.

We have switched to a maximization view of the alignment problem. This is because if we phrase this as finding the substrings with the minimum edit distance then any two 1-character substrings of the same character will have 0 edit distance, which is an uninteresting solution. This is entirely about exposition and clarity, not a fundamental difference: we could have a match "score" be negative and achieve the same thing by minimizing. But it's more confusing to talk about negative scores.

To solve the local alignment problem, we redefine what each of our subproblems means. Instead of the best score between two prefixes, we let

-   $C[i, j]$ be best score between: some *suffix* of $S[1 \ldots i]$ and some *suffix* of $T[1 \ldots j]$.

That is, $C[i, j]$ is the best alignment score between some suffixes of the given prefixes.

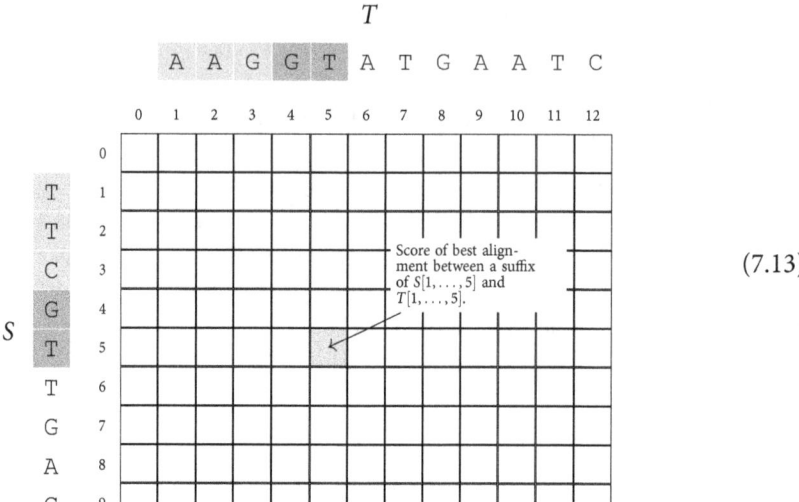

(7.13)

Given this definition, we compute $C[i, j]$ using the following recurrence.

$$C[i,j] = \max \begin{cases} C[i, j-1] + gap & (1) \\ C[i-1, j] + gap & (2) \\ C[i-1, j-1] + match(i,j) & (3) \\ 0 \end{cases}$$

(7.14)

Cases (1), (2), and (3) are the same as before. They represent a gap in S, a gap in T, and a match S and T. The new case, which is always 0, allows you to say the best alignment between a prefix of S and a prefix of T is the empty alignment. This lets us "start over" and is how the $C[i, j]$ entry represents alignments between suffixes: anything before the suffix that is a drag on the score can be discarded.

What are the base cases? The best alignment between the suffix of a string t and the suffix of an empty string is 0. So we initialize the first row and first column to be all 0s.

Where is the optimal solution? It's not necessarily at the bottom-right cell any more. Instead, it is the maximum value over every cell in the matrix. Why? If a cell (i, j) has a score M in it, there is some pair of suffixes of those prefixes that has that score. A "suffix of a prefix" is just another way to say "substring," so there are a pair of substrings with that score. The largest M represents the best scoring such pair.

To find the actual local alignment: start at an entry with the maximum score and traceback as usual. Stop when you reach an entry with a score of 0.

In (7.15) is an example local alignment between $S =$ "AGCGTAG" and $T =$ "CTCGTC", where the score of a match is 10, a mismatch -5 and a gap is -7.

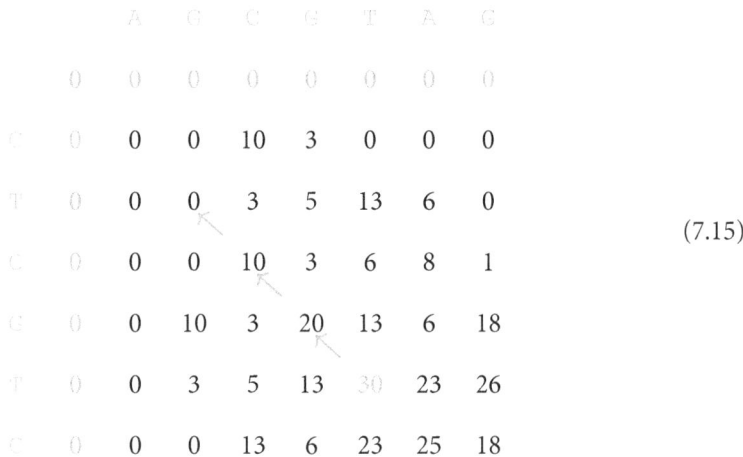

		A	G	C	G	T	A	C
	0	0	0	0	0	0	0	0
C	0	0	0	10	3	0	0	0
T	0	0	0	3	5	13	6	0
C	0	0	0	10	3	6	8	1
G	0	0	10	3	20	13	6	18
T	0	0	3	5	13	30	23	26
C	0	0	0	13	6	23	25	18

$$(7.15)$$

In (7.16) here's another example local alignment when S is "bestoftimes" and $T =$ "soften" with the same parameters as in (7.15).

		b	e	s	t	o	f	t	i	m	e	s
	0	0	0	0	0	0	0	0	0	0	0	0
s	0	0	0	10←3	0	0	0	0	0	0	10
o	0	0	0	3	5	13	6	0	0	0	0	3
f	0	0	0	0	0	6	23	16	9	2	0	0
t	0	0	0	0	10	3	16	33	26	19	12	5
e	0	0	10	3	3	5	9	26	28	21	29	22
n	0	0	3	5	0	0	2	19	21	23	22	24

$$(7.16)$$

7.5.2   Semi-global alignment

Let's generalize even a bit more:

Problem 7.5 (Semi-global alignment). *Given two strings S and T, and a scoring function* $score(a, b)$, *find the maximum scoring alignment between S and T where the gaps at the starts and ends of one sequence come for free.* ◆

Visually, this is (7.17).

$$(7.17)$$

Using the same kind of reasoning we used for local alignment, we can get no-cost spaces at the ends of any combination of the strings.

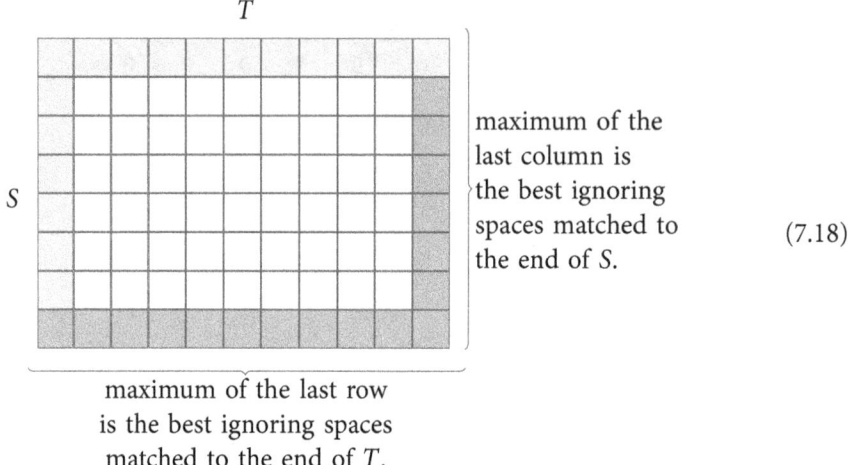

T

S

maximum of the last column is the best ignoring spaces matched to the end of S.

(7.18)

maximum of the last row is the best ignoring spaces matched to the end of T.

To get free gaps at the start of sequences, we set the initial row or column (the light shaded cells) to 0 as we did in local alignment.

To get free gaps matched to…	What to do
only the end of S	take max of rightmost column
only the end of T	take max of bottommost row
start of S	set leftmost column to 0
start of T	set topmost row to 0
ends of both T and S	take max of all entries in the matrix

You can combine these in other ways as well. For example, to allow free spaces matched to the start of T and both ends of S, set top row and leftmost column to 0 and take the max of the rightmost column.

7.6   Summary and notes

We've given an $O(nm)$ algorithm to compare two strings and determine their optimal alignment, which immediately gives both the value of the edit distance and the edit operations that can be used to transform one string to the other. $O(mn)$ is not the optimal running time for this problem. We'll see some techniques to speed it up later.

We've also seen how you can manipulate the meaning of the cells by changing how you initialize the first row or column, adding extra cases to the recurrence, or changing where you look for the optimal answer in the matrix.

The algorithm for edit distance [Wagner and Fischer, 1974] has been discovered several times in several contexts. The fundamental edit distance metric is sometimes called the

Levenshtein distance after the author of Levenshtein [1966]. The underlying DP approach was described in several papers for application to biological sequences, with Smith and Waterman [1981] focusing on local alignment, Needleman and Wunsch [1970] on global alignment, and Gotoh [1982] modifying an earlier algorithm to achieve the $O(mn)$ run-time. Nussinov and Jacobson [1980] applied and extended the DP approach to predict the structure of RNA molecules.

Presentation Notes

A similar discussion of edit distance is likely contained in any algorithms textbook [e.g, Cormen et al., 2009; Kleinberg and Tardos, 2006]. It is also covered in essentially every textbook focusing on string algorithms [e.g., Gusfield, 1997; Durbin et al., 1998; Crochemore and Rytter, 2003; Crochemore et al., 2007] or genomic analysis [Mäkinen et al., 2015; Jones and Pevzner, 2004; Pevzner, 2000].

7.7   Exercises

7.1   In the dynamic programming table for edit distance (aka global alignment) with the cost of insertion, deletion, or mutation each $= 1$, are the entries in each row monotonically increasing? If so, why? If not, why not?

7.2   True or false: an optimal path that stays within k of the diagonal can't have more than k gaps.

7.3   Consider the strings $a = $ "nonsense" and $b = $ "oneness". Suppose that a gap costs 5, a match costs -1, and the cost of aligning two mismatching characters x and y is given by the function $\text{cost}(\cdot, \cdot)$:

$$\text{cost(n,o)} = \text{cost(n,s)} = \text{cost(n,e)} = 2$$
$$\text{cost(o,e)} = \text{cost(o,s)} = \text{cost(s,e)} = 1.$$

Provide a well-labeled dynamic programming table for the global alignment between a and b. Show traceback arrows. Find the optimal, lowest-cost alignment and its cost.

7.4   If you only want the value of the edit distance, how could you modify the algorithm to run in $O(\min\{n, m\})$ *space*?

7.5   Formalize the edit distance problem when don't-care symbols are allowed in both strings and show how to handle them in the dynamic programming solution.

7.6   Describe how to modify and use the DP for local alignment so that it solves the longest common substring problem. What is the running time?

7.7   *Forbidden pairs.* Suppose you are also given two strings A and B (of length n and m, respectively) and also a set of pairs (i, j) that indicate that position i in A is not allowed to be aligned with position j in B. Give an $O(nm)$ time algorithm to compute optimal alignments in this case.

7.8   *Shortest common supersequence.* The SHORTEST COMMON SUPERSEQUENCE between strings x and y is the shortest string z such that both x and y are subsequences of z. (x is a "subsequence" of z if the letters of x appear in order in z, not necessarily consecutively: at is a subsequence of act.)

By modifying the basic sequence alignment algorithm, give an efficient algorithm to find the shortest common supersequence for two strings x and y.

7.9   *Counting optimal traceback paths.* The possible traceback paths leading from cell (m, n) each correspond to an optimal global alignment. Therefore, the number of distinct optimal alignments can be obtained by computing the number of these traceback paths. Give an algorithm to compute the number of optimal traceback paths in $O(nm)$ time. Hint: your algorithm should itself be a dynamic programming algorithm.

7.10   *Limited mismatches.* Suppose you are given strings x and y, match, mismatch, and gap costs m, s, and g, and a small integer $M \geq 0$ that is your mismatch budget. Give an "efficient" dynamic programming algorithm that finds the best global alignment between x and y that uses no more than M mismatches (it can use as many gaps or *matches* as it needs to). Explain the running time of your algorithm.

7.11   *Splitting URLs.* In some languages, words are sometimes not separated by spaces. For example, in German, numbers are sometimes written together "Dreihundertfünfzigfünftousand" and compound words are often created: "Glückszahl" means "lucky (Glück) number (zahl)." Such compound words also often occur inside web addresses. We would like to decompose such strings into the component words that were used to form them. Assume you have a function **word**(s) that takes a string s and returns a score indicating how likely it is that s is an indivisible word. For example, in German, "zahl" would receive a high score, but "kszahl" would not.

Give a dynamic programming algorithm to break a given string $a = a_1, a_2, \ldots, a_n$ into words $w_1, \ldots, w_k$ to maximize $\sum_i \mathbf{word}(w_i)$. (Note that you are **not** given k as input.)

7.12   You run an ice cream business, and you want to place some advertisements in your local newspaper. There are two kinds of ads you can run, and you've noticed that Type-C works best on cold days (by promoting the good taste of your ice cream) and Type-W works best on warm days (by mentioning how cold and refreshing your ice cream is). Depending on the weather and which ad you run, you see a certain amount of increased profit that day.

	Cold	Warm
Type-C ad	+$75	+$50
Type-W ad	+$50	+$100

You have committed to running an ad every day. The cost of placing either a Type-C or Type-W ad is $10 per day. But the newspaper charges you a fee of $25 every time you **change** which ad you are running.

You are given a (perfectly correct) weather prediction for the next n days. Design a dynamic programming algorithm to select which ad to run on each of the next n days to maximize your profit.

Examples:

```
Input  = WWWCCCWCWCWCW
Output = WWWCCCWWWWWW
Profit = $895

Input  = WCW
Output = WWW
Profit = $220

Input  = CWWWWWC
Output = WWWWWWW
Profit = $530
```

CHAPTER 8

Edit Distance in Linear Space

We saw in Chapter 7 an algorithm for finding the edit distance and optimal alignments that ran in $O(nm)$ time and used $O(nm)$ space to align strings of length n and m. $O(nm)$ might be an ok space usage from a theoretical perspective, but for a 10 Gb genome, you'd need a huge amount of memory even if the other string was only a few hundred letters long. In practice, it is often the space that is the limiting factor for computing edit distances. Can we use less? Surprisingly, we can use $O(\min\{n, m\})$ space (plus the space to store the input) to find optimal alignments. That is achieved by Hirschberg's algorithm [Hirschberg, 1975], which combines dynamic programming with divide-and-conquer.

8.1 Using linear space to compute edit distance values

We can compute a number of different values in linear space using small modifications of our algorithm for edit distance. Remember the matrix that we constructed during that algorithm and what each cell means.

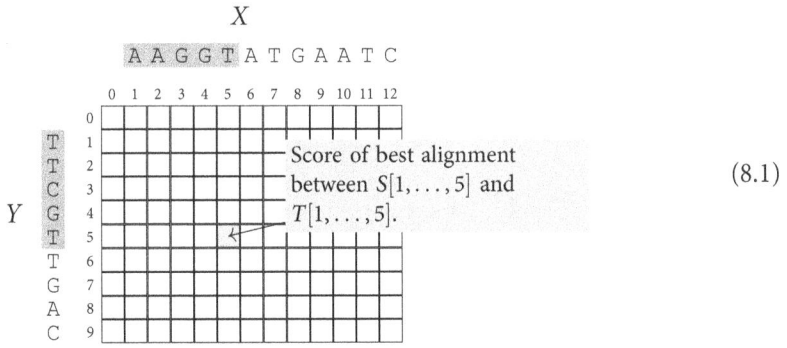

$$(8.1)$$

If you are only interested in the cost or score of an alignment, you need to use only $O(\min\{n, m\})$ space. When filling in an entry (light gray box in figure (8.2)) we only look at the current and previous rows.

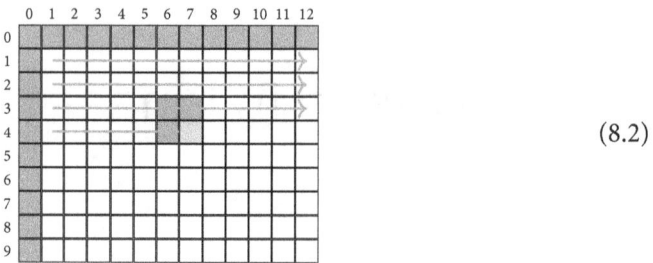

(8.2)

We only need to keep those two rows in memory. We reuse the memory for the current row and the previous row. Since which string we put on the rows vs. on the columns is our choice, we can put the smaller string on rows and use only $O(\min\{n, m\})$ space.

In fact, we can do much more in linear space. Given two strings X and Y, we can, in linear space and $O(nm)$ time, compute the *cost* of aligning

- every prefix of X with all of Y,
- all of X with every prefix of Y,
- a particular prefix of X with every prefix of Y,
- a particular suffix of X with every suffix of Y.

How can we do these things? We can use other entries in the matrix. For example, recall what each cell of the last row of the matrix represents: the score of an optimal alignment between all of Y and a prefix of X.

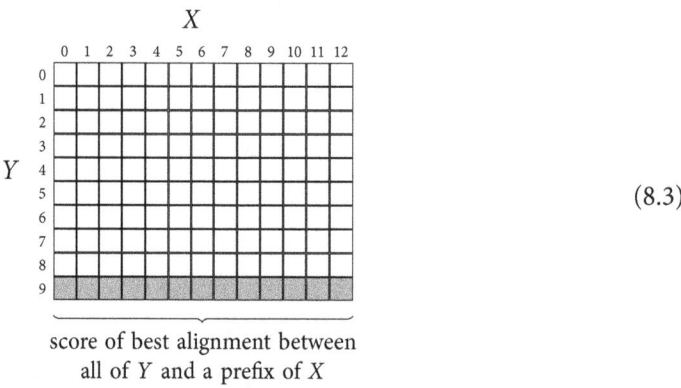

(8.3)

score of best alignment between
all of Y and a prefix of X

That immediately gives an $O(nm)$-time and $O(\min\{n, m\})$-space algorithm to compute the best score of every prefix of X with all of Y: we output the bottom row rather than just the bottom-right cell.

What if we fill in by columns instead of by rows? This order still computes the required subproblems before they are needed.

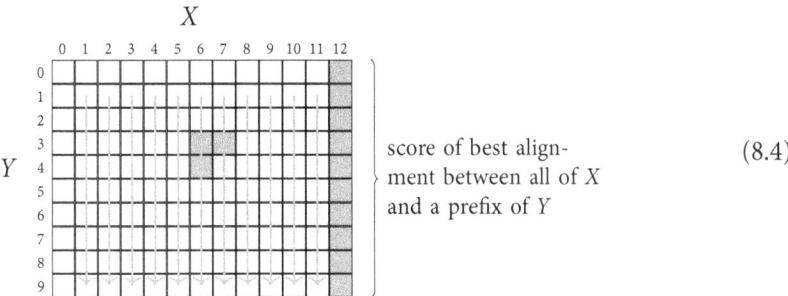

$$(8.4)$$

score of best alignment between all of X and a prefix of Y

The last column contains the scores of the best alignments between all of X and each prefix of Y.

What if we fill in the matrix by columns, but stop before we get to the last column? Then we are looking at some intermediate column.

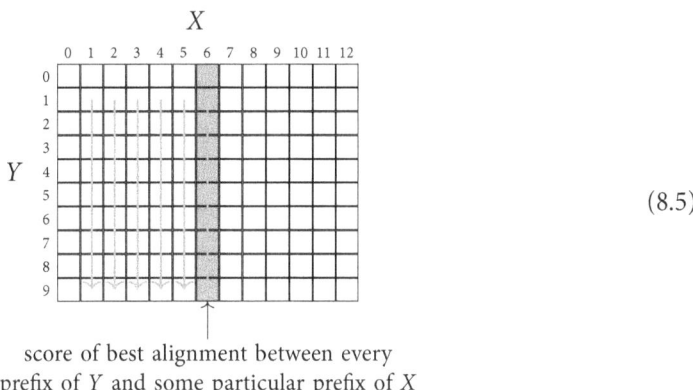

$$(8.5)$$

score of best alignment between every prefix of Y and some particular prefix of X

That column has the best scores between a particular prefix of X and all prefixes of Y.

What about computing the costs of the best alignments between X and all *suffixes* of Y? Our original dynamic programming algorithm broke our problem down by *prefixes*, but that was a somewhat arbitrary choice. We could have instead defined the subproblem to be based on smaller suffixes. To do this, define a new matrix B where:

- $B[i, j]$ is the best alignment between suffix $X[j \ldots]$ and suffix $Y[i \ldots]$.

This leads to the following matrix in (8.6).

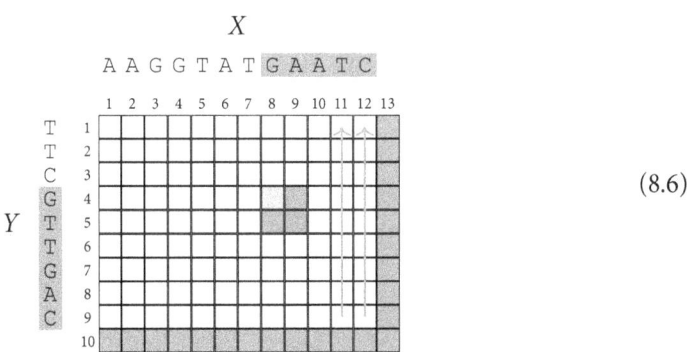

$$(8.6)$$

Each cell in this matrix represents the best alignment between a *suffix* of X and a suffix of Y. We can write the recurrence for this problem in the same way as we did for our first DP algorithm.

$$B[i,j] = \min \begin{cases} cost(X[i], Y[j]) + B[i+1, j+1] \\ gap + B[i, j+1] \\ gap + B[i+1, j] \end{cases} \qquad (8.7)$$

This uses the exactly the same reasoning as doing the "forward" dynamic programming (DP). Using this "backwards" DP, we can compute the scores of the optimal alignments between all of X and each suffix of Y (where are these values?). Similarly, by filling in by rows (starting from the first one), we can compute the scores of the best alignments between all of Y and each suffix of X (where are *these* values?).

The "backward" DP recurrence still writes the solution to one problem as dependent on the solution to some *smaller* problems. It just happens in this case, the smaller problems involve strings with larger indices.

8.2   Finding the actual alignment in linear space

The techniques in the previous section give us only the distances and not the actual alignment. We can't directly get the alignment since this requires the traceback arrows for every row (or column), and we didn't save those.

Can we find the actual alignment in linear space? Surprisingly, yes, we can output the optimal alignment in linear space. This will cost us some extra computation but only a constant factor. For such a dramatic reduction in space, it's often worth it. Usually, we can wait a bit longer but we can't obtain more memory for our machines as easily. In addition, simply having the system allocate and manage the memory of a huge matrix can take a lot of time, so in practice we may even save some time.

The idea of Hirschberg's algorithm is to use a divide-and-conquer algorithm to compute half alignments. Divide-and-conquer is another very general algorithmic design technique. Its idea is to split large problems into a few subproblems and then recursively solve each subproblem. We then merge the resulting answers. You've likely seen examples of these kinds of algorithms such as QuickSort or MergeSort.

How do we recover the traceback "arrows" in linear space? The first key idea is that we don't need all the arrows, we only need the arrows that are on the optimal path. Call the optimal traceback path the *ArrowPath*. But, of course, we don't know what the optimal path is. However, we do know that the optimal path crosses the $|X|/2$th column and uses some cell in that column. If we knew which cell was used in this column, we'd end up with two smaller problems.

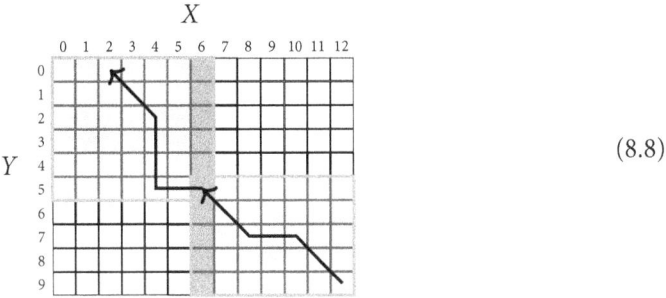

$$(8.8)$$

8.2.1   Hirschberg's algorithm

In the optimal alignment, the first $|X|/2$ characters of X are aligned with the first q characters of Y for some q as in (8.9).

$$
\begin{matrix}
 & \scriptstyle 1 & \scriptstyle 2 & \scriptstyle 3 & \scriptstyle 4 & \scriptstyle 5 & \scriptstyle 6 & \scriptstyle 7 & \scriptstyle 8 \\
X = & A & C & G & T & A & C & G & T \\
Y = & A & - & G & T & - & C & T & G
\end{matrix}
\underbrace{}_{q\,=\,3}
\tag{8.9}
$$

We don't know q, so we have to try all possible q. The key to running in $O(mn)$ time is finding a quick way to evaluate the alignment score that we would get for each q. To do that, we first define some helper functions.

Definition 8.1. Define the following two functions that run in linear space and return vectors of alignment scores:

- AllYPrefixCosts(X, i, Y) returns an array of the scores of optimal alignments between $X[1 \ldots i]$ and all prefixes of Y.
- AllYSuffixCosts(X, i, Y) returns an array of the scores of optimal alignments between $X[i \ldots n]$ and all suffixes of Y. ∎

These are implemented as described in Section 8.1 by returning a row or column of one of the DP matrixes. Since the algorithm is looking for a q in the middle column, we use the helper functions in Definition 8.1 as follows:

$$YPrefix \leftarrow \text{AllYPrefixCosts}(X, m/2, Y) \tag{8.10}$$

$$YSuffix \leftarrow \text{AllYSuffixCosts}(X, m/2 + 1, Y). \tag{8.11}$$

Now, the score for any q is equal to $YPrefix[q] + YSuffix[q+1]$. To see this, look at (8.9): $YPrefix[q]$ is the cost of the left side of the alignment and $YSuffix[q+1]$ is the cost of the right side. After the arrays in (8.10) and (8.11) are computed, evaluating the score of any choice of q can therefore be done in $O(1)$ time.

This leads to the following recursive algorithm, which we present in two parts.

Algorithm 8.1. Linear-space alignment.

Require: *ArrowPath* starts as an empty, global list of cells.
  function LINEARSPACEALIGN(*X*, *Y*)
       Use the code in Algorithm 8.2 to set *m*, *n*, *YPrefix* and *YSuffix*.

       ▷ *Find the q that minimizes the cost of the alignment:*
       $best \leftarrow \infty$
       $bestq \leftarrow$ **nil**
       for $q \leftarrow 0, \ldots, n$ **do**
            $cost \leftarrow YPrefix[q] + YSuffix[q+1]$
            if $cost < best$ **then**
                 $bestq \leftarrow q$
                 $best \leftarrow cost$
       Add $(m/2, bestq)$ to *ArrowPath*

       ▷ *Recursively find the cells in the subproblems*
       LINEARSPACEALIGN($X[1, \ldots, m/2], Y[1, \ldots, bestq]$)
       LINEARSPACEALIGN($X[m/2+1, \ldots, m], Y[bestq+1, \ldots, n]$)

The first step of this function is to initialize some of the variables using the following code.

Algorithm 8.2. Initialization for each LINEARSPACEALIGN call.

  $m \leftarrow |X|$
  $n \leftarrow |Y|$
  ▷ *If small, use $O(1)$ space with standard alignment:*
  if $n \leq 2$ **or** $m \leq 2$ **then**
       Use standard alignment
       Add traceback cells to *ArrowPath*
       return

  ▷ *Precompute the alignments*
  $YPrefix \leftarrow$ ALLYPREFIXCOSTS($X, m/2, Y$)
  $YSuffix \leftarrow$ ALLYSUFFIXCOSTS($X, m/2+1, Y$)

8.2.2   Proof of running time

Algorithm 8.1 in fact runs in the right time bounds. We can write down a recurrence $T(m, n)$ which is the time it takes to solve a problem using strings of length m and n. This recurrence is:

$$T(m, 2) \leq cm, \tag{8.12}$$

$$T(2, n) \leq cn, \tag{8.13}$$

$$T(m, n) \leq cmn + T(m/2, q) + T(m/2, n - q). \tag{8.14}$$

Here, c is some constant (that we don't know but that we can set to be big enough). The term $T(m/2, q)$ is the time to align the first half of X to the first q characters of Y, and the term $T(m/2, n - q)$ is the time it takes to align the second half of X to the remaining $n - q$ characters of Y. We're omitting $\lceil \rceil$ and $\lfloor \rfloor$ notations for clarity, since they don't affect the analysis. However, this still looks a little too complicated to analyze easily because we don't know what q is, and q is different each time through the recurrence.

Let's start by solving a simpler problem to get an idea about what the running time might be. Suppose for just a minute that both sequences have length n, and that we get a perfect split in half every time, $q = n/2$. Then (8.14) simplifies to:

$$T(n) \leq 2T(n/2) + cn^2. \tag{8.15}$$

This solves to $T(n) = O(n^2)$. (See Exercise 8.1.) This gives us some hope that the real recurrence solves to $O(mn)$. We now solve the original "too hard" recurrence by "guessing" that $T(n, m) \leq kmn$ for some constant k.

Theorem 8.2 (Hirschberg's running time). $T(m, n) \leq kmn$ for some k, where $T(m, n)$ is defined by (8.12)–(8.14).

Proof: By induction.

Base cases: If we choose $k \geq c$, then $T(m, 2) \leq c2m \leq k2m = kmn$. So when one of the sequences is small, we have $T(m, n) \leq kmn$.

Induction step: Assume $T(m', n') \leq km'n'$ for pairs (m', n') with *a product* smaller than mn. We then have:

$$T(m, n) \leq cmn + T(m/2, q) + T(m/2, n - q) \quad \text{by definition} \tag{8.16}$$

$$\leq cmn + kqm/2 + k(n - q)m/2 \quad \text{by induction hypothesis} \tag{8.17}$$

$$= cmn + kqm/2 + kmn/2 - kqm/2 \tag{8.18}$$

$$= (c + k/2)mn. \tag{8.19}$$

If we set $k = 2c$ (which is $\geq c$ so the base case still holds), we have $T(m, n) \leq 2cmn = kmn$, which is what we wanted to show. $\qquad \square$

In the proof manipulations, the key thing is that q cancels out, so it doesn't matter that q is different each time we execute the recurrence. We never explicitly figure out what c is. That is fine; we know some such c exists, and therefore our k exists.

8.3   Summary and notes

We have seen how we can compute the cost of an alignment easily in linear space and also how we can compute the cost of a string with all suffixes of a second string in linear space. We used that ability to create a divide-and-conquer algorithm for computing the actual alignment (traceback path in the DP matrix) in linear space. This still uses $O(nm)$ time. Hirschberg's algorithm was pointed out in Myers and Miller [1988] as being the right way to achieve space efficiency, and it was extended therein to handle affine gap penalties (see Chapter 10).

Presentation Notes

The linear-space algorithm is covered in both Gusfield [1997] and Jones and Pevzner [2004].

8.4   Exercises

8.1   Prove that $T(n) \leq 2T(n/2) + cn^2$ solves to $O(n^2)$.

8.2   It is possible for the traceback path to use more than 1 cell in the middle column. Why does Hirschberg's algorithm handle this correctly?

8.3   Describe how to compute B of (8.7) using the recurrence (7.6) for the regular edit distance.

CHAPTER 9

Faster Edit Distance via the "Four Russians" Trick

Recall the definition of edit distance.

Definition 9.1 (Edit Distance). The *edit distance* between strings x and y is the minimum number of insertion, deletion, and replacement operations required to convert x to y. Each insertion, deletion, and replacement operation affects one character. ∎

In this chapter, we'll look at the traditional edit distance where the cost of a mismatch, insertion, or deletion is 1 and the cost of a match is 0. In other words, we are computing exactly the number of operations needed to convert x to y. We also assume, as we usually do, that that alphabet Σ is a constant size.

Let $|x| = n$ and $|y| = m$. We saw a dynamic programming algorithm to compute the edit distance in $\Theta(nm)$ time. This is good, but it turns out for the traditional edit distance, this is not optimal.

The approach [Arlazarov et al., 1970] that we will use to get a faster running time—which also can be used in some other dynamic programming algorithms—is called the "Four Russians" speed up. It was first applied to the edit distance by Masek and Paterson [1980].

9.1   Dynamic programming blocks

The first key idea to this approach is to start with the standard dynamic programming (DP) method that we have already seen and decompose our DP matrix into "blocks."

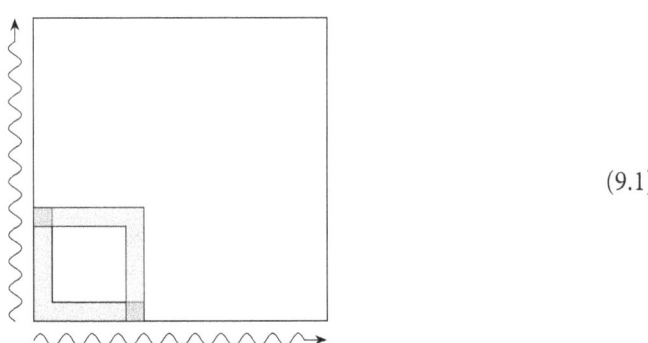

(9.1)

Let t be the side length of a block. We'll delay deciding how big these blocks are but we will assume for simplicity that $n = m$ (the strings are the same size) and the blocks exactly tile the matrix, with a single overlapping row and column between each adjacent pair.

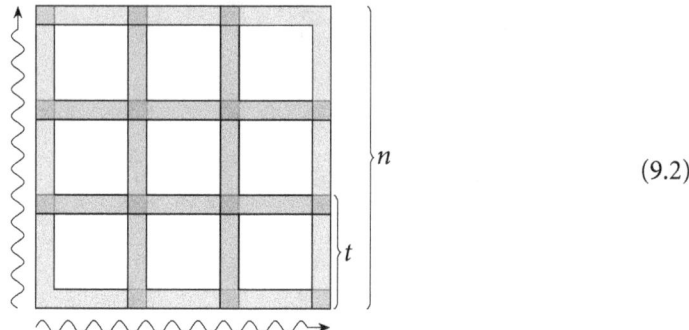

$$(9.2)$$

Next, imagine we have a function $f(bl, x', y')$ that takes as input the bottom row and left column (bl) of a block and the substrings of x and y that correspond to the block (x' and y'). f returns the values in the top row and right column of the block that would be computed according to the standard edit distance recurrence. Visually, the signature of f looks like (9.3).

$$f\left(\mathsf{L}, \wr, \sim\!\!\sim \right) \rightarrow \mathsf{\daleth} \qquad (9.3)$$

That is $f(bl, x', y')$ gives us the top row and right column of a block given its bottom row and left column and the relevant substrings of x and y.

If we can compute f faster than $\Theta(t^2)$, say in $O(t)$ time, we win. This is because we could replace the $\Omega(t^2)$ time it would take the standard recurrence to fill in the block with the faster $O(t)$ computation of f. The goal of the rest of this chapter is to see how to compute f in $O(t)$ time.

9.2   Precomputing f

The way we achieve this speed up is a combination of two ideas (1) we encode the inputs to f efficiently, and (2) we precompute the output of f for any possible input.

How many possible inputs are there to the function f? Naïvely, it seems like a lot. Something like:

$$(n+1)^{2t} |\Sigma|^{2t} \qquad (9.4)$$

since (a) every cell contains a number between 0 and n, and (b) there are $|\Sigma|^{2t}$ pairs of strings of length t. In more detail, there are $(n+1)^t$ possible values for an entire row or column. This is because each cell in those input rows or columns (and in fact in the entire matrix) is a number between 0 and n (inclusive), since the edit distance between any two strings of length n is at most n and at least 0. We also have $|\Sigma|^t$ possible substrings (since we have $|\Sigma|$ choices for each of t characters), leading to $O(|\Sigma|^{2t}(n+1)^{2t})$ possible inputs to f.

Computing the output for a possible input to f using the standard DP algorithm would take $\Theta(t^2)$, leading to a total running time of $\Omega((n+1)^{2t}|\Sigma|^{2t}t^2) = \Omega(n^2)$ to precompute f for every possible input, which is bad since it's not any faster than our original algorithm.

9.2.1   Offset encoding

The trick to making it work is realizing that in fact there are fewer possible functionally different inputs to f. The elements of the rows and columns in the input are not independent. Let D be the DP matrix. We use the fact that we are computing the standard edit distance to prove the following lemma.

Lemma 9.2. *In the DP matrix D for the standard edit distance, adjacent values of D in a row, column, or diagonal differ by at most 1.*

Proof: We prove this for a row i. The proof for columns and the diagonals is similar. Consider element q of row i. We have $D[i, q] \leq D[i, q-1] + 1$ since we can always insert a gap in the recurrence to achieve this score.

We need to argue that $D[i, q] \geq D[i, q-1] - 1$. If we do this, this will show $D[i, q-1] - 1 \leq D[i, q] \leq D[i, q-1] + 1$ and therefore show that entry $D[i, q]$ differs by at most ± 1 from the previous column. Repeating this argument for every q shows what we want for row i.

How do we see that $D[i, q] \geq D[i, q-1] - 1$? Consider $D[i, q-1]$ and $D[i, q]$. These are scores of optimal alignments between the same strings, except the second string has an extra character at position q. There are two cases for the optimal alignment for $D[i, q]$: either the character at q is aligned against a gap, or it is not aligned against a gap.

Case 1: If the character at q is aligned against a gap in the optimal alignment $D[i, q]$, then we have

$$D[i, q-1] < D[i, q], \tag{9.5}$$

since we eliminate a gap (of cost 1) by "throwing away" the character at q, and the optimal cost of $D[i, q-1]$ is definitely less that that remaining alignment, since the optimal alignment between prefix i and prefix $q-1$ is less than (or equal to) any such alignment.

Case 2: If the character at q is aligned against a character in the optimal alignment for $D[i, q]$, we can replace the character at q with a gap to get an alignment (that is probably not optimal) between prefix i and prefix $q-1$. This increases the cost by at most 1 since we may have turned a match (cost 0) into a gap (cost 1). If we turned a mismatch into a gap, the score is unchanged. So we have

$$D[i, q-1] \leq D[i, q] + 1, \tag{9.6}$$

since again, whatever the optimal alignment between prefix i and prefix $q-1$ is, it's definitely no worse than the alignment we construct by replacing the character at q with a gap.

Putting (9.5) and (9.6) together, we have

$$D[i, q-1] \leq D[i, q] + 1, \tag{9.7}$$

which is equivalent to $D[i, q] \geq D[i, q-1] - 1$, which is what we wanted to show. $\qquad\square$

Using the lemma, we can encode a row (or column) using the encoding that consists of an initial value followed by deltas that are in $\{-1, 0, 1\}$:

$$567767 \rightarrow \langle 5, 1, 1, 0, -1, 1 \rangle.$$

We use this to define the offset vector.

Definition 9.3 (Offset vector). An *offset vector* is the encoding of a row or column, except that the first entry is set to 0. For example:

$$567767 \rightarrow \langle 0, 1, 1, 0, 0, -1, 1 \rangle. \qquad\blacksquare$$

In other words, given the first value C and the offset vector, you can reconstruct the row or column. For example,

a	4	3	2	3
b	3	2	2	2
a	2	2	1	2
b	1	1	2	3
	d	b	a	d

1	$C+3$	$C+2$	$C+1$	$C+2$
1	$C+2$	$C+1$	$C+1$	$C+1$
1	$C+1$	$C+1$	C	$C+1$
0	C	C	$C+1$	$C+2$
	0	0	1	1

(9.8)

In (9.8), the corner value is a given C value and the first column and first row are encoded by the offset vectors $\langle 0, 1, 1, 1 \rangle$ and $\langle 0, 0, 1, 1 \rangle$.

Because the offset vectors specify the entire first row and column we have:

Theorem 9.4. *Given only the input offset vectors* $\llcorner$ *and the relevant substrings of x and y, we can compute the output offset vectors* $\urcorner$.

To do this, we can compute the offset vector of the last row and column by running the standard DP but keeping C as an unevaluated symbol. Continuing the example (9.8),

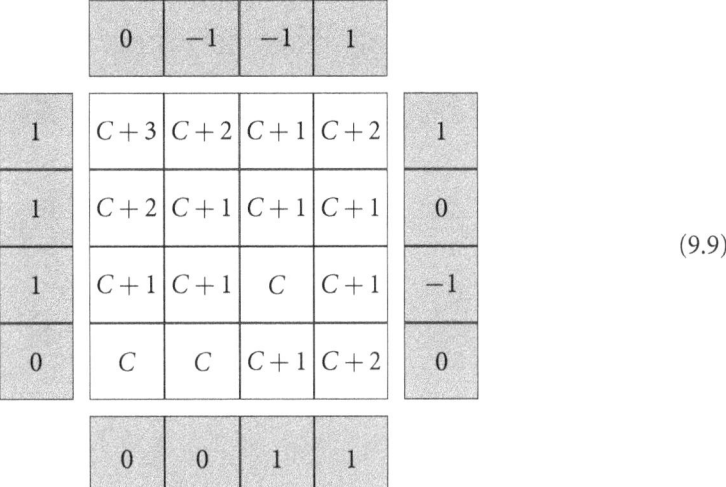

$$(9.9)$$

in (9.9) we can encode the "output" offset vectors as deltas from the topmost value in the first column and the bottommost value in the last column (which can themselves be computed from the input offset vectors).

To fill in the entire matrix, we can use this idea to compute the "output" offset vectors for each block, solving the bottom row of blocks left-to-right first, and proceeding up the matrix. The "output" offset vectors become the "input" offset vectors for later blocks. At the end, we have the offset vectors for the top row and right column, and we can propagate C along (say) the bottom row and right column to obtain the actual value in the $D[n, m]$ cell.

9.2.2   Number of possible inputs to *f*

Have we made any progress? We've seen that the offset vectors (along with the substrings for that block) encode the essential information about the input. How many such offset vectors are there?

There are $3^{2(t-1)}$ possible pairs of offset vectors. That means that there are $3^{2(t-1)}|\Sigma|^{2t}$ possible inputs to f (ignoring the C). Computing all values of $f(\cdot)$ now takes $O((3|\Sigma|)^{2t}t^2)$ time, where for each of the inputs we run the standard DP, but ignore C.

Setting $t = \log_{3|\Sigma|}(n)/2$, this becomes $O(n(\log n)^2)$ to compute all the possible values of f. This is some good progress. We've turned the $\Omega(n^2)$ precomputation into one that takes only $O(n(\log n)^2)$ time.

9.2.3   Storing *f* for quick access

We've got a pile of precomputed values for f. For a given input of (bl, x', y') we need to find $f(bl, x', y')$ from among this pile quickly. How can we do that?

We can think of the input to f as a vector of length $t + t + t + t$, where the first $2t$ entries encode the offset vectors in bl and the second $2t$ entries encode the substrings x' and y'. We can put these all into a *trie*. A trie is a tree where the data is stored on the edges of the tree and the path from the root to nodes is meaningful. The keyword tree we saw in Chapter 6 is an example of a trie. For storing f, each possible input is stored on a path in the trie.

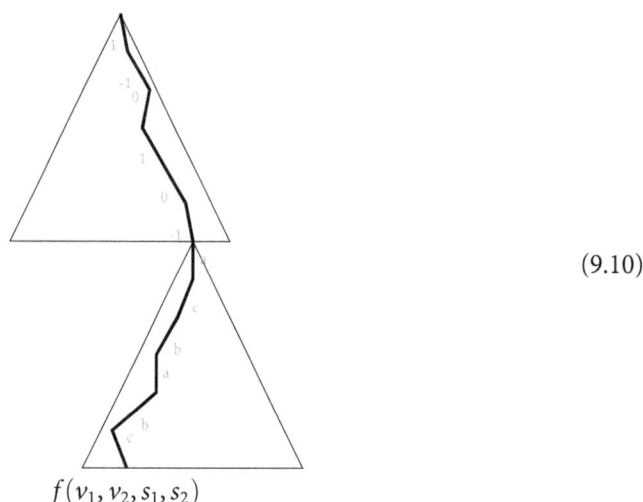

$$f(v_1, v_2, s_1, s_2)$$

(9.10)

The depth of this trie is at most $4t$ which is $O(t)$. At each leaf of this trie, we store the output offset vectors for the input that is spelled out by the path leading to that leaf. This means we can get the output offset vectors in time $O(t)$. Hence, we can compute the "output" of a block in time $O(t)$.

9.3   Total running time

Our total running time is the time to do our precomputation of f plus the time to solve each of the blocks. With $t = O(\log n)$, our precomputation time is $O(n(\log n)^2)$ as described in Section 9.2.2.

To solve each of the blocks, we have to compute the output vectors for each of the $O(n^2/t^2)$ blocks. Accessing $f(\cdot)$ to solve each block takes time $O(t)$, so our time to "fill in" the matrix is $O(tn^2/t^2) = O(n^2/t)$. Adding this to the precomputation, with $t = O(\log n)$, the total time is

$$O\left(n^2/\log n + n(\log n)^2\right) = O(n^2/\log n),$$

(9.11)

and we've saved a factor of $O(\log n)$.

In practice, it's often useful to take t equal to some constant instead of $\log n$. This doesn't give you an asymptotic speed up, but the algorithm now runs in time $O(n^2/t)$ so the constant factor can be better than our original $O(n^2)$ algorithm.

An aside about the RAM model: Our chosen t is $O(\log n)$. Normally, we assume we can operate on $O(\log n)$-sized objects in constant time. This is the RAM model, and you can justify it by thinking that $O(\log n)$ is the number of bits needed for a pointer if our computer has n bytes of RAM. Under this assumption, one does not need the trie of Section 9.2.3.

Instead, we just need an array, where entry i gives the value of f for input i. In the RAM model, we therefore save another factor of $t = O(\log n)$, taking constant time to solve each block, leading to $O(n^2/(\log n)^2)$ running time.

9.4   Finding the alignment

This gets us the score. How can we extract the actual alignment? Look at the arrows that would be stored in some block.

$$(9.12)$$

For clarity, not all the arrows are drawn in (9.12). But there is a path from each cell in the output row/column of the block back to some cell in the input row/column of the block. When we precompute this block, we can compute these arrows just as with the standard edit distance. Then, for each cell i in the output row/column we store a pointer $e(i)$ to the cell at the end of the arrow path—i.e., where in the input row/column would we end up if we followed the arrow path out of cell i? This adds $O(t)$ overhead in storage, and doesn't change our asymptotic space or runtime.

Using these $e(\cdot)$ pointers, we can find the *blocks* that are along the optimal traceback path: start at cell (n, n), jump to $e((n, n))$ and repeat until we fall off the matrix. We will cross at most $O(2n/t)$ blocks (Exercise 9.7). We then explicitly solve the full dynamic programming recurrence to recover the arrows, but just within these blocks. The total time to do this over all the encountered blocks is $O(t^2 \times 2n/t) = O(nt) = O(n \log n)$, for our choice of t. This is far smaller than the leading term in our runtime to compute the edit distance value, so it doesn't add to the asymptotic running time.

9.5   Summary and notes

It's still somewhat an open question about whether this is the fastest running time possible. Some evidence has been derived that obtaining a significant speed up would solve a major open problem in complexity theory and have deep and surprising implications.

Definition 9.5 (SETH (approximately)). The *Strong Exponential Time Hypothesis* is the assumption that there is some k for which the best algorithm for k-SAT is arbitrarily close to $O(2^n)$. ∎

The SETH essentially says that you can always construct a big enough SAT instance such that the best algorithm to solve the SAT instance takes time on the order of checking every possible setting of the instance's boolean variables. SETH is a hypothesis, so it's not yet proved—and may not be true. It's a stronger assumption than $P \neq NP$: $SETH \implies P \neq NP$ but the other way around does not hold.

Backurs and Indyk [2015] proved the following:

Theorem 9.6. *If the SETH is true, then there is no algorithm to compute the edit distance between length n strings in time $O(n^{2-\delta})$ for any constant δ.*

This says that, under a reasonable complexity assumption, it's unlikely we can do too much better than our $O(n^2)$ algorithm under a general worst-case analysis, although it does not rule out algorithms that perform better on average or on specific kinds of input.

The Four Russian's technique was generalized to many computations (in certain computational models) in Hopcroft et al. [1975].

> ### Presentation Notes
>
> The Four Russian's technique is covered in both Gusfield [1997] and Jones and Pevzner [2004].

9.6 Exercises

9.1 In the proof of Lemma 9.2, why can't we make the "easier" (and wrong) argument that each term in the recurrence adds at most 1 to the cost?

9.2 Does the "Four Russian's" idea still work when using a more general alignment cost function? Where does it work and where does it break down?

9.3 When precomputing the f function for a block, could we recursively use the Four Russian's trick? Pursue this idea and explain potential benefits and problems with it.

9.4 In our runtime analysis, we included the time to do the precomputation. This didn't affect our overall runtime because the precomputation was faster than the slowest step. But it would be nice if we could do the precomputation just once and use it for all future alignments of different strings. Is this possible? If so, why? If not, why not?

9.5 Describe the modifications you would have to make if (a) the blocks don't tile the DP matrix and (b) the strings are not the same length.

9.6 Explain how to compute $e(i)$ (Section 9.4) in $O(t^2)$ time.

9.7 Explain why the optimal traceback path will visit at most $O(2n/t)$ blocks.

CHAPTER 10

General and Affine Gap Penalties

Let's now look at how to handle more general gap penalty functions within the edit distance. This addresses the concern that often, inserting or deleting several consecutive characters should be considered "easier" than inserting the same number of characters in random, non-adjacent places. This requires changing our cost function for a pairwise alignment and creating a new algorithm to compute it.

10.1 General gap penalties

Consider the two alignments in (10.1).

$$\begin{matrix} \text{A A A G A A T T C A} \\ \text{A - A - A - T - C A} \end{matrix} \quad \text{vs.} \quad \begin{matrix} \text{A A A G A A T T C A} \\ \text{A A A - - - - T C A} \end{matrix} \qquad (10.1)$$

These have the same score under the standard edit distance, but the second one could be seen as more plausible intuitively since a single deletion of "GAAT" from the first string could change it into the second. These get the same score in the standard definition of edit distance because a run of k gaps costs $g \times k$, where g is the gap cost. It would be more realistic to support a general gap penalty, so that the score of a run of k gaps is $gap(k)$ for some function gap, where typically $gap(k) \leq g \times k$. Then, the optimal will prefer to group gaps together.

Our previous dynamic programming approach would no longer work with such general gap penalties because the score of the last character depends on details of the previous alignment.

$$\begin{matrix} \text{A A A G A A} | \text{T} \\ \text{A A A - - -} | \text{-} \end{matrix} \quad \text{vs.} \quad \begin{matrix} \text{A A A G A A T} | \text{C} \\ \text{A A A - - - -} | \text{-} \end{matrix} \qquad (10.2)$$

Instead, we need to know how long a final run of gaps is in order to give a score to the last subproblem.

A solution to this is to realize we need to have three different kinds of subproblems, stored in three different matrices. Assume we are aligning two strings x and y. Then we have the following subproblems:

- $M[i, j] =$ score of best alignment of $x[1 \dots i]$ and $y[1 \dots j]$ ending with a *character-character match or mismatch*.

- $X[i,j]$ = score of best alignment of $x[1 \ldots i]$ and $y[1 \ldots j]$ ending with *a gap in x*.
- $Y[i,j]$ = score of best alignment of $x[1 \ldots i]$ and $y[1 \ldots j]$ ending with *a gap in y*.

The reason we have these subproblems is that we have to know whether we are dealing with a gap run at the end of the sequence or not. We then follow the standard dynamic programming reasoning, writing the values of these subproblems in terms of other subproblems. This leads to the following set of recurrences. First, we handle the case where the subproblem ends in a match or mismatch.

$$M[i,j] = match(i,j) + \max \begin{cases} M[i-1,j-1] \\ X[i-1,j-1] \\ Y[i-1,j-1]. \end{cases} \tag{10.3}$$

The idea is that we use up a character in both x and y (with reward $match(i,j)$) and then the remaining alignment could end with an alignment between characters, a gap in X or a gap in Y. If we align the last characters, then any alignment is allowed prior to them.

Alternatively, we could align the end of x with a gap, leading to the following recurrence:

$$X[i,j] = \max_{1 \le k \le j} \begin{cases} M[i,j-k] - gap(k) \\ Y[i,j-k] - gap(k). \end{cases} \tag{10.4}$$

For the X and Y matrices, we have to try every possible length of gap run (that is the choice of k), and the gap can either start after a match (jumping to a M-type subproblem) or it can start after a gap run starting in the other sequence (jumping to Y).

$$X[i,j] = \max_{1 \le k \le j} \begin{cases} M[i,j-k] - gap(k) \\ Y[i,j-k] - gap(k). \end{cases} \tag{10.5}$$

The choice of k decides how long to make the gap run. We have to decide about the whole gap run at once in order to know how to score it since $gap(k)$ is a black box to us.

The case for Y is symmetric:

$$Y[i,j] = \max_{1 \le k \le j} \begin{cases} M[i,j-k] - gap(k) \\ X[i,j-k] - gap(k). \end{cases} \tag{10.6}$$

The final score of the optimal alignment is

$$\max\{M[n,m], X[n,m], Y[n,m]\}, \tag{10.7}$$

if our strings have length n and m. To find the actual alignment, we do the traceback as before, except that "arrows" can now point between matrices.

Running time. Assume $|x| = |y| = n$ for simplicity. Then we have $O(3n^2)$ subproblems. Now $O(2n^2)$ of those subproblems each take $O(n)$ time to solve since we have to compute the max over $\Omega(n)$ choices of k. This leads to a total run time of $O(n^3)$. So, for a black box gap function, we can compute the optimal alignment in cubic time.

10.2   Affine gap penalties

$O(n^3)$ for general gap penalties is usually too slow for all but pretty short sequences. We can still encourage spaces to group together using a special case of general penalties called *affine gap* penalties. In the affine gap setting, we assume that

$$gap(k) = gap\_start + k \times gap\_extend, \tag{10.8}$$

where *gap_start* is the cost of starting a gap, and *gap_extend* is the cost of extending a gap by one more space.

Recurrences. We can solve this with the same idea of using three matrices, but now we don't need to search over all gap lengths, we just have to know whether we are starting a new gap or not. That's where the speed improvement comes from. This leads to the following modified recurrences. For the M matrix, we have the same recurrence as the black box *gap* function:

$$M[i,j] = match(i,j) + \max \begin{cases} M[i-1,j-1] \\ X[i-1,j-1] \\ Y[i-1,j-1] \end{cases} \tag{10.9}$$

since this recurrence doesn't involve any gaps at all. For X, we have now:

$$X[i,j] = \max \begin{cases} gap\_start + gap\_extend + M[i,j-1] \\ gap\_extend + X[i,j-1] \\ gap\_start + gap\_extend + Y[i,j-1]. \end{cases} \tag{10.10}$$

The rationale here is that if the rest of the alignment ends in a match (first case) and we want to add a gap character in x, we should pay both the *gap_start* and *gap_extend* costs (the first to start the gap, and the second to add the gap character). Similarly, if the rest of the alignment ends in a gap in y (last case), then we need to start the gap here in x. On the other hand, if the rest of the alignment ends in a gap in x (second case), then we don't need to pay a gap starting cost—since the gap has already started—we just need to pay a gap extension cost.

The case for Y is symmetric to X:

$$Y[i,j] = \max \begin{cases} gap\_start + gap\_extend + M[i-1,j] \\ gap\_start + gap\_extend + X[i-1,j] \\ gap\_extend + Y[i-1,j]. \end{cases} \tag{10.11}$$

We can think of all of this logic as a finite state machine, where the states are M, Y or X.

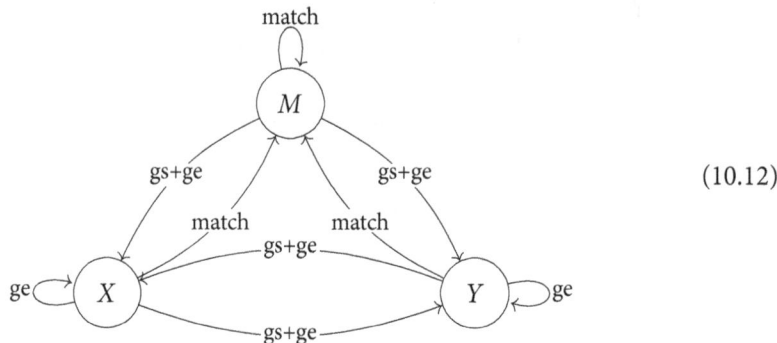

(10.12)

The "reward" of the transition is the amount we add in the recurrence (where in the figure *gap_start* and *gap_extend* are abbreviated *gs* and *ge*).

Base cases. What about the base cases? We use the definitions of the subproblems to reason about what the base cases should be:

- $M[0, i] =$ "score of best alignment between 0 characters of x and i characters of y that ends in a match" $= -\infty$ because no such alignment can exist. You can't have a match if you have no characters. $M[i, 0] = M[0, i]$ by the same reasoning.
- $X[0, i] =$ "score of best alignment between 0 characters of x and i characters of y that ends in a gap in x" $= gap\_start + i \times gap\_extend$ because this alignment must look like:

      ----
      yyyyyyyyyy
- $X[i, 0] =$ "score of best alignment between i characters of x and 0 characters of y that ends in a gap in x" $= -\infty$ because this situation is not allowed. It would look like this:

      xxxxxxxxxxxx-
      ----

  and you can't align a gap against a gap.

$Y[0, i]$ and $Y[i, 0]$ are computed using the same logic as $X[i, 0]$ and $X[0, i]$.

Running time. We have $O(3mn)$ subproblems. Each one takes constant time now since we don't have to maximize over a k variable. So the total runtime is $O(mn)$, which is back to the run time of the basic algorithm. But since we pay a little extra (*gap_start*) to start a gap, we're more likely to group together gap characters, so this is a more realistic score of the alignment between two sequences when "block" edits are more likely.

10.3   Why do we "need" 3 matrices?

A common question is whether we really need these three matrices, which seem to add a lot of complexity. Of course, the question about whether we *need* three matrices is hard to answer—perhaps someone will come up with some creative algorithm that does not require this complexity. But the naïve approach of just using one matrix does not work.

Suppose we try to get rid of two of the matrices with the following (wrong) recurrence:

$$M[i,j] = \max \begin{cases} M[i-1,j-1] + match(x[i],y[j]) \\ M[i-1,j] + gap + (start \text{ if } Arrow[i-1,j] \neq \leftarrow) \\ M[i,j-1] + gap + (start \text{ if } Arrow[i,j-1] \neq \uparrow). \end{cases} \qquad (10.13)$$

The (wrong) intuition here is "well, we only need to know whether we are starting a gap or extending a gap, and the arrows coming out of each subproblem tell us how the best alignment ends, so we can use them to decide if we are starting a new gap." Why doesn't this work?

The reason this does not work is that the best alignment for strings $x[1 \ldots i]$ and $y[1 \ldots j]$ (viewing them as complete strings) doesn't have to be used in the best alignment between $x[1 \ldots i+1]$ and $y[1 \ldots j+1]$ when we have affine (or general) gap penalties. Here's an example when that can happen, with $match = 10$, $mismatch = -2$, $gap = -7$, $gap\_start = -15$. If we look at the optimal alignment between "CART" and "CAT" it is:

$$\begin{matrix} \text{C} & \text{A} & \text{R} & \text{T} \\ \text{C} & \text{A} & \text{-} & \text{T} \end{matrix} \qquad OPT(4,3) = 30 - 15 - 7 = 8. \qquad (10.14)$$

If we use this optimal alignment as a subproblem when trying to construct the optimal alignment between "CARTS" and "CAT" we get the wrong solution:

$$\begin{matrix} \text{C} & \text{A} & \text{R} & \text{T} & \text{S} \\ \text{C} & \text{A} & \text{-} & \text{T} & \text{-} \end{matrix} \qquad WRONG(4,3) = 30 - 15 - 7 - 15 - 7 = -14, \qquad (10.15)$$

when the correct optimal is:

$$\begin{matrix} \text{C} & \text{A} & \text{R} & \text{T} & \text{S} \\ \text{C} & \text{A} & \text{T} & \text{-} & \text{-} \end{matrix} \qquad OPT(5,3) = 20 - 2 - 15 - 14 = -11. \qquad (10.16)$$

This is why we need to keep the X and Y matrices around. They tell us the score of "alternative" optima with constraints about which subsequences end with gaps. The intuition for this occurring is that later gaps might "pull" earlier gaps later in order to group them together.

10.4 Summary and notes

We've seen that we can solve alignment of two length n strings with arbitrary gap penalty function in $O(n^3)$. Good, but too slow for long sequences. To achieve the goal of grouping gaps, a particular form of the gap function called "affine gaps" can be solved in $O(n^2)$ time, and this still prefers grouping gaps together.

Presentation Notes

A similar discussion of affine gap penalties is likely present in any string algorithms text book; it certainly is discussed in Gusfield [1997]; Mäkinen et al. [2015]; Crochemore et al. [2007].

10.5 Exercises

10.1   What are the base cases of the M, X, and Y recurrences for the general black box *gap* penalty function?

10.2   We saw how to handle the affine gap function

$$gap(k) = gap\_start + k \times gap\_extend. \tag{10.17}$$

Suppose you are aligning DNA strings and instead your $gap(k)$ function is:

$$gap(k) := \begin{cases} gs + k \times ge_1 & \text{if the gap is surrounded by } \begin{matrix} A & \cdots & T \\ A & & T \end{matrix} \\ gs + k \times ge_2 & \text{otherwise} \end{cases} \tag{10.18}$$

where gs, ge_1 and ge_2 are user-given parameters. We assume $ge_1 < ge_2$. In other words, runs of gaps that are surrounded by specific matches cost less than other runs of gaps. Give an $O(nm)$-time dynamic programming algorithm for this gap function.

10.3   Suppose we want to find the best global alignment with affine gaps, but not allow a gap in one sequence to immediately follow a gap in the other. For example, we do not allow alignments of the form:

```
abab--ababab          ababab--abab
ababbb--abab          abab--bbabab
```

Describe how to modify the dynamic programming recurrences to enforce this requirement, no matter what scoring parameters the user chooses.

CHAPTER 11

Output-Sensitive Algorithms for Edit Distance

The $O(nm)$-time algorithm for edit distance is often far too slow to use in practice for large sequences. Comparing two 100-million-letter human chromosomes, even with tricks like the Four Russians (Chapter 9), isn't feasible using the standard DP algorithm directly.

In this chapter, we'll look at several approaches to make large-scale alignment more practical by having runtime that is dependent on the edit distance. With the DP algorithm of Chapter 7, even if two strings are equal, the runtime is the product of the length of the strings. The algorithms of this chapter instead have runtime $O(km)$, where k is the edit distance, and m is the length of one of the strings. For pairs with small edit distance, this is a significant win.

11.1   Banded alignment

A *diagonal* of the DP matrix D for edit distance is a sequence of cells starting at $(0, i)$ or $(i, 0)$ for some i and proceeding to successive cells by adding 1 to both the x and y index: $(1, i+1), (2, i+2), \ldots$ until reaching an edge of the matrix. Number the diagonals of the DP matrix so that vertex (x, y) is on diagonal $x - y$. In other words, we call a diagonal starting $(0, i)$ "diagonal $-i$" and a diagonal starting at $(i, 0)$ is called "diagonal i".

Theorem 11.1 (Diagonals). *Under the standard edit distance where matches cost 0, mismatches and insertions and deletions each cost 1, every cell on diagonal d of D has value $\geq |d|$.*

Proof: In the traceback to $(0, 0)$ from a cell on diagonal d there must be at least d steps toward diagonal 0, each corresponding to an insertion or deletion, which have cost 1. □

This observation leads to the *banded alignment* algorithm that is useful when we know that there will be at most k differences (mismatches, insertions, deletions) between P and the text T we want to match.

We can use the contrapositive of Theorem 11.1 to limit what part of the matrix we need to look at: if there is an alignment of edit distance $\leq k$, then the traceback path will only involve cells of value $\leq k$, and these must be on a diagonal d with $-k \leq d \leq k$.

This also means that, for cells on the boundary of this band, the edit distance recurrence need not look at the cells outside of the band (since if that cell contributed to the minimum, they would be on the traceback path). Hence, we can compute the values for the DP matrix as usual, except ignoring options in the recurrence that involve cells outside of the band. This takes $O(km)$ time where m is the length of the shorter sequence.

The cell (m, n) that represents a complete alignment between the two strings must be within the band—otherwise there would be no way to align the sequences with $\leq k$ differences. If the value obtained for that cell is $\leq k$, then the traceback path will give an optimal alignment of k or fewer differences. If the value v in that cell is $> k$ then the edit distance must be $> k$, but v is not necessarily the correct edit distance.

Since we can tell whether our guess k on an upper bound of the edit distance is correct, we can start with a small k (just big enough so that cell (m, n) is in the band). If we don't successfully align the sequences with a value $\leq k$, we can double k, and try again. We can repeat this until we find the optimal alignment. This runs in $O(km)$ time (Exercise 11.2).

11.2   Myers' Minimum Edit Script algorithm

Define the following variant of the edit distance problem:

Problem 11.2 (Minimum Edit Script). *Find the minimum number of insertions and deletions (indels) to make two strings A and B equal.* ◆

The minimum edit script problem is equivalent to our standard edit distance with indels costing 1, matches costing 0, and substitutions (mismatches) disallowed. Because a substitution can be simulated with a deletion and an insertion, the minimum edit script can give us an upper bound on the edit distance between two strings. We could then run the banded alignment algorithm with this upper bound and be assured we will find the optimal. We will see an algorithm from Myers [1986] for this problem.

As with other edit distance variants, we can turn this into a shortest-path problem on a graph. Specifically, create a vertex set that is the grid $\{0, \ldots, |A|\} \times \{0, \ldots, |B|\}$. Add horizontal and vertical directed edges between adjacent rows and columns. Finally, add a diagonal edge $(i, j) \to (i+1, j+1)$ if the corresponding positions match: $A[i+1] = B[j+1]$.

This looks like this (11.1).

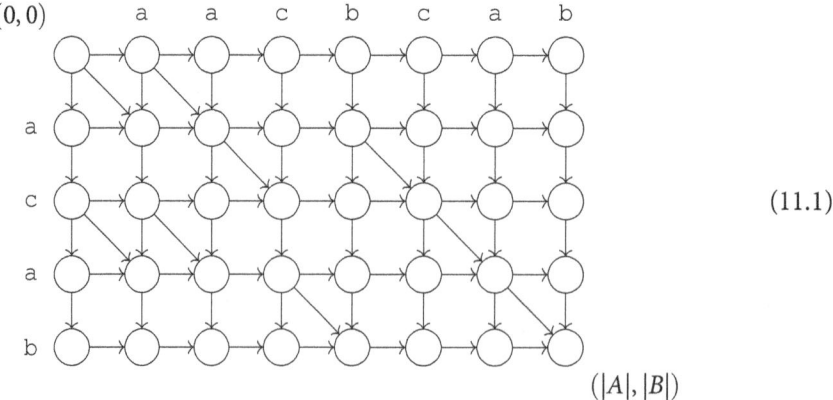

(11.1)

Problem 11.2 corresponds to finding a path from $(0,0)$ to $(|A|,|B|)$ that uses the minimum number of non-diagonal edges (since each non-diagonal edge corresponds to an insertion or deletion, just as with the standard edit distance formulation).

The algorithm is going to focus around a specific class of paths.

Definition 11.3. A $\bar{D}$-*path* is a path starting at $(0,0)$ that uses exactly D non-diagonal edges. ■

If there is a $\bar{D}$-path that ends at $(|A|,|B|)$, then the minimum edit script must have $\leq D$ indels. This motivates the idea of trying to find $\bar{D}$-paths starting with small D and increasing D until we find a path that reaches $(|A|,|B|)$. For example, if there is a $\bar{0}$-path to $(|A|,|B|)$, then the number of indels is 0, and we're finished. However, it could be that our longest $\bar{0}$-path along the main diagonal "gets stuck" (because there are missing diagonal edges) before it reaches $(|A|,|B|)$, and then we need to try to find a $\bar{1}$-path. As with many of the algorithms we've seen, the trick to doing this quickly is to use the computation we did for earlier values of D when computing the paths for $D+1$.

For any particular value D, for each diagonal $k \leq D$, there is some node on diagonal k that we can reach on a $\bar{D}$-path and that is farthest from $(0,0)$. Call this node $F[D,k]$.

$F[D,k]$ can be computed from $F[D-1,\cdot]$ by noticing that $F[D,k]$ must use exactly 1 more non-diagonal edge than some $\overline{D-1}$-path. That means $F[D,k]$ must jump to diagonal k from one of the adjacent diagonals.

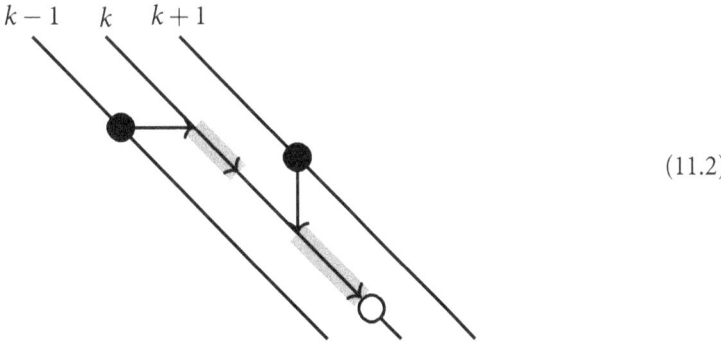

$$(11.2)$$

To compute $F[D,k]$ we can select whichever point $F[D-1,k-1]$ or $F[D-1,k+1]$ is farther from the origin, use a horizontal or vertical edge to hop over to diagonal k, and then run down diagonal edges as far as possible.

Algorithm 11.1. Minimum edit script.

1: **for** $d \leftarrow 0, \ldots, |A|+|B|$ **do**
2:     **for** $k \leftarrow -d, \ldots, d$ **do**
3:         Compute $F[d,k]$ by looking at $F[d-1,k-1]$ and $F[d-1,k+1]$.
4:         **if** $(|A|,|B|)$ is the endpoint of $F[d,k]$ **then**
5:             **return** d

When $k = -d$ or $k = d$ in Line 3 we only have one option for the previous diagonal to look at because the $-d - 1$ and $d + 1$ diagonals haven't been considered yet.

Running time. Let m and n be the lengths of A and B. If the minimum script has D operations, then Algorithm 11.1 will stop when $d = D$. Therefore, the outer loop will run $O(D)$ times. For each d, the inner loop runs $O(d)$ times, so the two loops together take $O(D^2) = O(D(m + n))$ time (since $D \leq m + n$), except for the statement in Line 3. That statement has an implicit loop in it because we have to run down the diagonal to compute the new $F[d, k]$.

But, since we stopped at $d = D$, we only looked at $O(D)$ diagonals, meaning we explored only the band $[-D, D]$ of diagonals. This region that we explored contained at most $O(D \min\{m, n\}) = O(D(m + n))$ diagonal edges (if every possible diagonal edge was present). Finally, along each diagonal k, $F[\cdot, k]$ only increases over time, and therefore, once the algorithm uses a diagonal edge, it will never use that diagonal edge again. The total time to execute Line 3 over all iterations is also at most $O(D(m + n))$. Therefore, the total runtime of the algorithm is $O(Dm')$, where m' is the length of the larger sequence.

Since $F[d, k]$ only depends on $F[d - 1, \cdot]$, the space usage for Algorithm 11.1 can be made to be only $O(m + n)$ by storing just a single position for each diagonal.

11.3   Landau-Vishkin*

We'll see now the Landau-Vishkin algorithm [Landau and Vishkin, 1986] for finding all places where a pattern matches a text with at most k differences including insertions, deletions, and substitutions that runs in time $O(k|T|)$. The algorithm is in many ways similar to the algorithm of the previous section—again, a refactored dynamic programming approach that works along the diagonals.

This section, toward the end, uses the ability to solve the Longest Common Extension problem (LCE) in constant time (after preprocessing) that we will introduce in Section 14.2. When the time comes, you may want to skim that section.

11.3.1   The k differences problem

Let P be a pattern of length m, and let T be a text of length n. Again consider the standard edit distance problem with matches costing 0, mismatches, insertions, and deletions costing 1. We want to solve the following problem:

Problem 11.4 (*k* Differences). *Find all locations of P in T where P occurs, allowing at most k mismatches, insertions, or deletions.* ◆

Let D be the DP matrix for the edit distance between two strings and look at the diagonals of D, where diagonal d is the set of cells (i, j) with $i - j = d$.

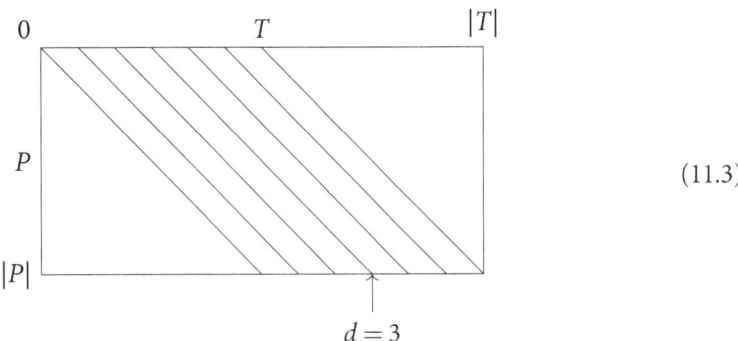

$$(11.3)$$

Theorem 11.5 (Diagonals are non-decreasing). *Along any diagonal d, $D[i,j] \geq D[i-1, j-1]$; in fact, $D[i,j] - D[i-1,j-1]$ is always 0 or 1.*

Proof: Suppose, for contradiction, that $D[i,j] < D[i-1,j-1]$. Let A be the alignment realizing $D[i,j]$. Let (a,b) be the pair of last letters of each of the two strings. If (a,b) were aligned to each other, their removal either reduces the alignment score or leaves it unchanged, and creates an alignment of prefix i and prefix j with score $< D[i-1,j-1]$, contradicting the optimality of D. If a were aligned to some character c and b were aligned to a gap, again their removal yields an alignment for the smaller prefixes of lower cost, contradicting optimality. The case where b was aligned to a character and a to a gap is symmetric.

For the second half of the statement, if $D[i,j]$ differed by more than 1 from $D[i-1,j-1]$, a smaller value of $D[i,j]$ could be obtained by using a mismatch from $D[i-1,j-1]$. □

In pictures, the proof tells us that along a diagonal, we have blocks of increasing values.

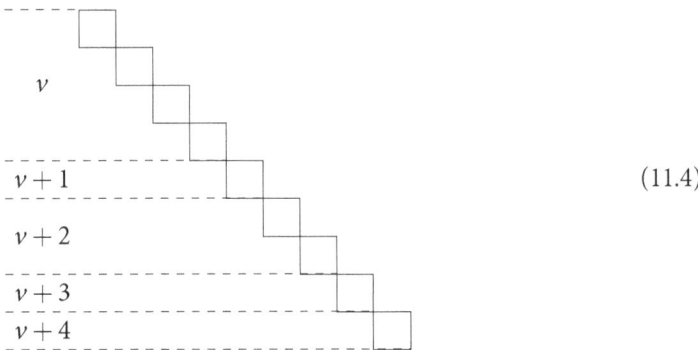

$$(11.4)$$

11.3.2 An alternative dynamic program

To solve Problem 11.4, we reparameterize the DP algorithm. We explain this first in relation to the standard DP matrix D, but then we'll see that we don't need D to compute the values needed in the reparameterization. This will lead to a new way to look at the DP and a fast solution introduced by Landau and Vishkin [1986] to Problem 11.4.

Definition 11.6 (Landau-Viskin L values). For a given edit distance DP matrix, define $L[d, e]$ to be the largest row index of D along diagonal d that that has value e. ∎

In other words, $L[d, e]$ is the largest row index where $D[i, j] = e$ and $i - j = d$. At first, this seems like an odd thing to do: we're defining what looks like a DP matrix L but the entries are not edit distances but instead are row indices of the original DP matrix. To see why this might be useful, note that if $L[d, k] = |P|$ for some d and k, then there is an alignment between P and a substring of T of edit distance at most k. In other words, since the row indices correspond to lengths of prefixes of P, we have a match if there is some diagonal where we can match all of P against a part of T with k edits.

We will see how to compute $L[d, e]$ by defining it in terms of other $L[d', e']$ values. Let's start with this partially defined recursive function.

Algorithm 11.2. Incomplete function to compute L values.

> **function** LDE(d,e)
>   ▷ *Base cases here*
> $row \leftarrow \max \begin{cases} \text{LDE}(d, e - 1) + 1 \\ \text{LDE}(d + 1, e - 1) + 1 \\ \text{LDE}(d - 1, e - 1) \end{cases}$ ▷ (*)
>   **while** $P[row + 1] = T[row + 1 + d]$ **do**
> $row \leftarrow row + 1$
>   **return** row

LDE computes $L[d, e]$ for a given d, and e by calling itself to compute other $L(d', e')$ values. At this point it should *not* be clear why this is correct, and we haven't defined any base cases, so the procedure is not completely defined yet. It is similar to Myers' recursive Algorithm 11.1: we jump from an adjacent diagonal and run down a diagonal as much as possible. But note that $L(d, e)$ only calls $L(d', e - 1)$ for some d's. Hence, we can turn this into a dynamic program by computing L for $e = 0, \ldots,$. (A pure recurrence would be a bit awkward to write, since it would involve both the max and the sum specified by the **while** loop.)

Ignoring the base cases for a moment, why is the procedure in Algorithm 11.2 correct? The key is the max on line (*): it finds a row of diagonal d for which $D = e$. In other words, after line (*), *row* is equal to some row that contains e along the diagonal d. It's not immediately clear why this is true—we will have to prove it in Theorem 11.7.

But assuming you believe it for now, the entry returned by the max is that starting point on diagonal d. The **while** loop then simply runs down the diagonal as long as the pattern and text match. By Theorem 11.5, all these e values must be consecutive along the diagonal, and so we will find the one with the largest row using the sequential search of the **while** loop.

Theorem 11.7. *After line* (*), row *is such that the entry of D on row* row *and diagonal d equals e.*

Proof: For now, let's reindex our DP matrix D so that $D[row, d]$ means the entry on row *row* along diagonal d of D. Why, after line $(*)$, do we know that $D[row, d] = e$? Consider 3 cases, depending on which term in line $(*)$ is the largest:

Case: $L(d, e-1) + 1$ is largest in the max. Then $D[row - 1, d]$ is by definition equal to $e - 1$, and $D[row, d]$ can't be equal to $e - 1$ because otherwise $L(d, e - 1)$ would have been larger. Hence, $D[row, d]$ must be e. (It can't be *larger* than e because we could achieve e by using the diagonal case of the edit distance DP recurrence.) See the following figure.

$$L[d, e-1] \;\text{-----}\; \boxed{e-1} \;\text{-----}\qquad\qquad (11.5)$$

Case: $L[d+1, e-1] + 1$ is the maximum. Then $D[row - 1, d + 1]$ is by definition $e - 1$. Since $L[d + 1, e - 1] + 1$ is the max, we must have $L[d, e - 1] + 1 \leq L[d + 1, e - 1] + 1$ (otherwise the first case would have been the max). Hence, $L[d, e - 1] \leq L[d + 1, e - 1]$, but if $L[d, e - 1] = L[d + 1, e - 1]$, then this is just the previous case. Hence, we can assume that $L[d, e - 1] < L[d + 1, e - 1]$, and therefore that diagonal d must equal or exceed e before row $L[d + 1, e - 1]$. Hence, $D[row, d]$ must equal e by Theorem 11.5 (since e is achievable by using the vertical case of the edit distance DP recurrence.) See figure (11.6).

$$L[d+1, e-1] \;\text{-----}\; \boxed{\geq e}\;\boxed{e-1}\qquad\qquad (11.6)$$

Case: $L[d - 1, e - 1]$ is the maximum. This case is symmetric to the previous one. $D[row, d - 1]$ is by definition $e - 1$. $D[row, d] = e$ is achievable using the horizontal case of the edit distance DP. $D[row - 1, d]$ must equal e since if it were $< e$, $L[d, e - 1] + 1$ would have been greater than or equal to $L[d - 1, e - 1]$ (and if it were equal, we could apply the first case). See figure (11.7).

$$L[d-1, e-1] \;\text{-----}\; \begin{matrix}\boxed{\geq e} \\ \boxed{e-1}\;\;e\end{matrix}\qquad\qquad (11.7)$$

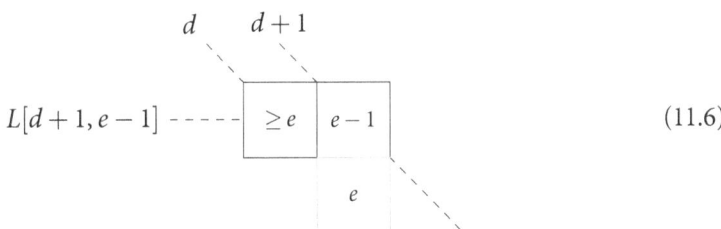

$\square$

Base cases. The base cases occur when $e = 0$. There are two types of base cases, depending on d. For $e = 0$ and $d \geq 0$, instead of line $(*)$, we start *row* at 0. The **while** loop will correctly increment *row* to the correct value of $\text{LDE}(d, 0)$.

For $e = 0$ and $d < 0$, diagonal d starts at row d in the middle of P (see (11.3)), so instead of line $(*)$, we start *row* at d. The **while** loop will once again increment as long as there is an exact match.

11.3.3   The Landau-Vishkin algorithm

The key idea of the algorithm is that the **while** loop in the computation of LDE can be replaced with something much faster. What is that loop doing? It's computing the longest exact match starting at position *row* of P and at position $row + d$ of T. That is exactly the longest common extension (LCE) problem from Section 14.2. There we see how we can pre-process two strings to solve instances of the LCE problem in $O(1)$ time. So that's what we do.

Algorithm 11.3. Faster L computation with LCE.

function $\textsc{FastLde}(LCE, d, e)$
$$row \leftarrow \max \begin{cases} L(d, e-1) + 1 \\ L(d+1, e-1) + 1 \\ L(d-1, e-1) \end{cases}$$
$row = row + LCE(row, row + d)$
return *row*

Now, $\textsc{FastLde}$ runs in constant time assuming we're using it to fill in DP matrix L since we've given it the ability to answer LCE queries in constant time. We can then solve Problem 11.4 by repeatedly calling the DP-ized version of $\textsc{FastLde}$.

Algorithm 11.4. Landau-Vishkin.

function $\textsc{Landau-Vishkin}(P, T)$
  Let S be the suffix tree of $T\$P$.
  Process S to create a function LCE that answers LCE queries between P and T in $O(1)$ time.
  for $e \leftarrow 0, \ldots, k$ **do**
    for $d \leftarrow -n, \ldots, n$ **do**
      if $\textsc{FastLde}(LCE, d, e) = |P|$ **then**
        return $d + |P|$

Constructing the LCE function takes $O(|T| + |P|)$ time. Each $\textsc{FastLde}$ runs in $O(1)$ time, and we make $O(kn)$ calls to it, giving a total running time of $O(kn)$.

11.4   Summary and notes

The three algorithms here provide successively more refinements of the main idea that for similar strings, only the region around the main diagonal of the DP matrix is important. Banded alignment just assumes we know or can successively guess the appropriate band. Myers' algorithm of Section 11.2 no longer requires knowing a prior bound on the number of differences, supporting indels, and giving an upper bound on the edit distance. Landau-Vishkin (Section 11.3.3) follows the same idea to handle indels and substitutions.

Myers [1999] presents an alternative way to speed things up: by exploiting the fact that computers can manipulate bit vectors very quickly, similar to the motivation of the Shift-And algorithm (Section 5.2) for exact matching. His *bit parallel* algorithm is used in many high-performance implementations of the edit distance.

> Presentation Notes
>
> Our presentation of Myers' and Landau-Viskin's algorithms follows their original exposition in the papers that introduced them.

11.5   Exercises

11.1   Give an example where the banded alignment algorithm puts a value $> k$ in the final cell and where this value is not the true edit distance.

11.2   Prove that the total runtime of the banded alignment is $O(km)$.

11.3   Prove that if D is even, then a $\bar{D}$-path can only end on an even diagonal and that if D is odd, a $\bar{D}$-path can only end on an odd diagonal.

11.4   Prove that a $\bar{D}$-path must stay in and end in a band of diagonals between $[-D, D]$.

11.5   Give pseudocode for line 3 of Algorithm 11.1.

11.6   Let a and b be two different cells in the DP matrix, and let $p(x)$ be the traceback path starting at cell x. Can $p(a)$ and $p(b)$ cross, meaning intersect and then stop intersecting?

11.7   Rewrite Algorithm 11.4 to include the base cases.

CHAPTER 12

Aligning Multiple Strings

In this chapter, we'll see how to align more than one string simultaneously to each other. This is the *multiple sequence alignment* problem. Aligning multiple sequences is especially useful in genomics, where the alignment can reveal those portions of the strings that have been conserved (unchanged) during evolution and which portions have changed. There are a few ways to define this problem, but all reasonable definitions are NP-complete[1], so we'll see an approximation algorithm for one formulation.

12.1   Multiple sequence alignments

Often we have a large collection of sequences we want to align. For example, the set of genomes of a new coronavirus or the set of sequences of the influenza virus over time.

(12.1)

We can extend our pairwise alignment problem to:

Problem 12.1 (Multiple Sequence Alignment). *Let* $cost(x_1, x_2, \ldots, x_p)$ *be a function giving the cost of aligning characters* $x_1, x_2, \ldots, x_p$, *where* $x_i \in \Sigma \cup \{-\}$ *in a column.*

Given strings $S_1, S_2, \ldots, S_p$, *find places to insert gap characters "$-$" into the strings to minimize the sum of the cost of the columns.* ◆

In a picture, we have (12.2).

$$cost(X_1, X_2, \ldots X_p) = cost(\;$$
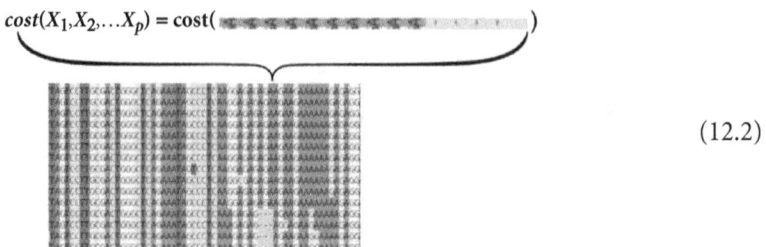

$$(12.2)$$

Suppose you had just three sequences; then we could apply the same dynamic programming idea as in the algorithm for sequence alignment for two sequences, but now with a 3-dimensional matrix.

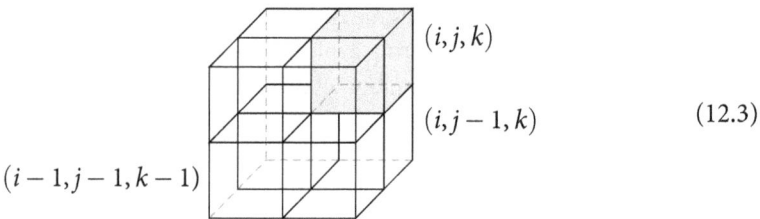

$$(12.3)$$

If our DP matrix is A, this leads to the following recurrence:

$$A[i,j,k] = \min \begin{cases} cost(x_i, y_j, z_k) + A[i-1, j-1, k-1] \\ cost(x_i, -, -) + A[i-1, j, k] \\ cost(x_i, y_j, -) + A[i-1, j-1, k] \\ cost(-, y_j, z_k) + A[i, j-1, k-1] \\ cost(-, y_j, -) + A[i, j-1, k] \\ cost(x_i, -, z_k) + A[i-1, j, k-1] \\ cost(-, -, z_k) + A[i, j, k-1], \end{cases} \qquad (12.4)$$

where we have a row for every possible pattern of gap characters being inserted.

For the case $p = 3$, this leads to $O(n^3)$ subproblems, each takes 2^3 time to solve (there are 2^3 possible patterns for inserting gap characters). This leads to an $O(n^3)$ time algorithm for 3 sequences. But for p sequences, we have $O(n^p)$ subproblems, each taking 2^p time for the max and c_p to compute *cost* (for some function c_p). This gives a $O(c_p n^p 2^p)$ algorithm, which is exponential in p, which is very slow. Even $O(n^3)$ is often too slow for the length of sequences encountered in practice.

12.2   Progressive alignment heuristic

A common heuristic used to overcome the exponential running time is to build a multiple sequence alignment up from pairwise alignments. We start with an alignment between one of the sequences S_c and some other sequence.

$$
\begin{array}{ll}
S_c & \texttt{YFPHFDLSHGSAQVKAHGKKVGDALTLAVGHLDDLPGAL} \\
S_1 & \texttt{YFPHFDLSHG-AQVKG--KKVADALTNAVAHVDDMPNAL}
\end{array} \tag{12.5}
$$

To add a third sequence S_2, we use the optimal pairwise alignment between S_c and S_2 as a guide. If that alignment looks like:

$$
\begin{array}{ll}
S_c & \texttt{YFPHF-DLS-----HGSAQVKAHGKKVGDALTLAVGHL----DDLPGAL} \\
S_2 & \texttt{FFPKFKGLTTADQLKKSADVRWHAERII----NAVNDAVASMDDTEKMS}
\end{array} \tag{12.6}
$$

we add S_2 to our growing MSA using that alignment and insert gaps into the previous sequences as needed to obtain a new S_c, S_1, S_2 alignment.

$$
\begin{array}{ll}
S_c & \texttt{YFPHF-DLS-----HGSAQVKAHGKKVGDALTLAVGHL----DDLPGAL} \\
S_1 & \texttt{YFPHF\underline{-}DLS\underline{-----}HG-AQVKG--KKVADALTNAVAHV\underline{----}DDMPNAL} \\
S_2 & \texttt{FFPKFKGLTTADQLKKSADVRWHAERII----NAVNDAVASMDDTEKMS}
\end{array} \tag{12.7}
$$

The underlined gaps are those that were inserted into S_1 to accommodate the S_2-S_c alignment. The (S_c, S_2) alignment provides the "glue" that connects the letters of S_c and S_2 (in the manner of (7.4)).

We then continue, using the S_c–S_i optimal pairwise alignment for each i to extend our MSA. This requires that our MSA *cost* function be decomposable into cost functions that evaluate only pairs of characters since we must be able to compute pairwise alignments between S_c and the other sequences.

There are many variants of progressive alignment, which we will not cover.

12.3   Approximation algorithm for one MSA formulation

Let's now develop an approximation algorithm that computes the MSA under one particular class of cost functions. Being an "approximation algorithm" means that we will be able to prove that—while the algorithm may not find the optimal alignment—there is a bound on how far off from the optimal the found alignment will be. This algorithm, which we call the Star algorithm, was introduced by Gusfield [1993].

A multiple sequence alignment (MSA) implies a pairwise alignment between every pair of sequences:

Definition 12.2 (d_M). $d_M(S_i, S_j) =$ the cost of the alignment between S_i and S_j as implied by MSA M. ∎

The alignment implied by M for S_i and S_j need not be optimal if we were just aligning those two strings in isolation. Here's an example of that happening with a set of small strings {"AT", "A", "T", "AT", "AT"} and a match cost of -1, a mismatch cost of 1, and gap cost of 2:

$$
\text{Optimal MSA:} \quad
\begin{array}{l}
\texttt{AT} \\
\texttt{A-} \\
\texttt{-T} \\
\texttt{AT} \\
\texttt{AT}
\end{array}
\tag{12.8}
$$

but the optimal alignment between "A" and "T" in isolation is $\begin{smallmatrix} \text{A} \\ \text{T} \end{smallmatrix}$ with a cost of $+1$. The optimal alignment between "A" and "T" is to align them with no gaps, but the other strings in the MSA "pull them apart."

We now assume that d_M is evaluated using some function $cost(u, v)$ that gives the cost of aligning character u against character v. In other words, $d_M(S_i, S_j) = \sum_{(u_i, v_i) \in M} cost(u_i, v_i)$, where the sum is taken over the columns of the alignment M.

We say a cost function obeys the triangle inequality if

$$cost(u, v) \leq cost(u, z) + cost(z, v). \qquad (12.9)$$

This is a very reasonable assumption since it probably doesn't make sense if two substitutions were cheaper than making one equivalent substitution (although there are settings where the triangle inequality doesn't make sense). We assume our *cost* function obeys the triangle inequality.

Definition 12.3 (*SP-Score*). *SP-Score*$(M) = \sum_{i<j} d_M(S_i, S_j)$. In other words, *SP-Score* is the sum of all the scores of the pairwise alignments implied by M. ∎

We can think of the *SP-Score* (which was discussed in Carrillo and Lipman [1988], among other places) as the sum of the weights of the edges of the complete graph, where the weights are given by d_M.

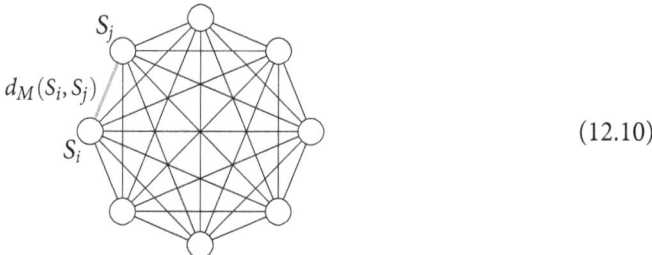

(12.10)

We can give now an algorithm that uses the *SP-Score* function for evaluating a MSA:

Algorithm 12.1. The Star MSA algorithm.

1. Build all $O(p^2)$ optimal pairwise alignments.
2. Let $S_c = $ the string in $S_1, S_2, \ldots, S_p$ that is closest to the others. That is, choose S_c to minimize $\sum_{i \neq c} a(S_c, S_i)$ where $a(x, y)$ is the cost of the optimal pairwise alignment between x and y.
3. Progressively align all other sequences to S_c.

Instead of focusing on the complete *SP-Score*, the algorithm focuses on the "star" centered around S_c.

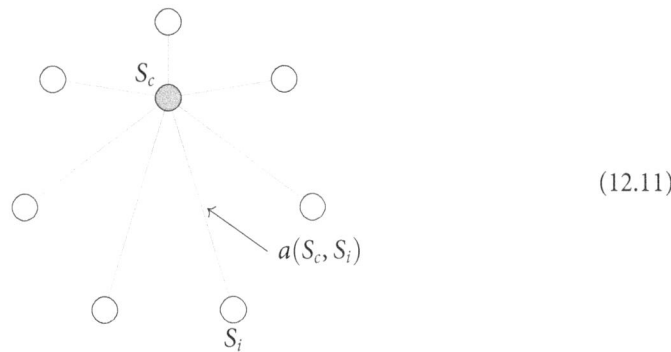

$$(12.11)$$

Since we assumed our *cost* function obeys the triangle inequality, one might expect that the edges that are *not* in this star can't be too much bigger than the edges that are in the star. This intuition is true, and leads to a proof that the Star algorithm can't be more than two times more than the optimal score. But note the difference between $a(S_i, S_j)$ and $d_M(S_i, S_j)$: a is the optimal *pairwise* alignment, while d_M is the pairwise alignment *induced* by the optimal MSA. The star centered at c is small according to a, but we have to do some math to show that this leads to an algorithm that isn't too far off from the optimal that d_M measures.

Theorem 12.4 (Star 2-approximation). *Let STAR be the cost of the result of the Star algorithm under the SP-score, and let OPT be the cost of optimal multiple sequence alignment (under SP-score). If the $cost(u, v)$ function underlying the SP-Score satisfies the triangle inequality, then $STAR \leq 2OPT$. In other words,*

$$\frac{STAR}{OPT} \leq 2. \tag{12.12}$$

Proof: We will prove two statements for some value B:

$$STAR \leq 2B, \tag{12.13a}$$
$$OPT \geq B. \tag{12.13b}$$

This implies that $\frac{STAR}{OPT} \leq 2$ as desired. Each of the proofs will be "one line" (for a generous definition of "line").

Let's prove the first statement (12.13a). Recall that $d_M(S_i, S_j)$ (from Definition 12.2) is the cost of the pairwise alignment between S_i and S_j implied by MSA M. Also recall that $a(S_i, S_j)$ is the score of the optimal pairwise alignment of S_i, S_j. Let $STAR$ be the MSA returned by the Star algorithm, and consider $d_{STAR}(S_i, S_j)$, which is the cost of the alignment between S_i and S_j induced by the MSA $STAR$. We then have:

$$2 \cdot STAR = \sum_{ij} d_{STAR}(S_i, S_j) \qquad \text{by defn} \tag{12.14a}$$

$$\leq \sum_{i \neq j} (d_{STAR}(S_i, S_c) + d_{STAR}(S_c, S_j)). \qquad \text{triangle inequality} \tag{12.14b}$$

Because the *STAR* alignment is optimal for pairs involving S_c:

$$= \sum_{i \neq j} \left(a(S_i, S_c) + a(S_c, S_j) \right) \tag{12.14c}$$

$$= \sum_{i \neq j} a(S_i, S_c) + \sum_{i \neq j} a(S_c, S_j). \tag{12.14d}$$

These sums above are the same, and each term appears $\leq p$ times:

$$\leq 2p \sum_i a(S_i, S_c). \tag{12.14e}$$

So $STAR \leq p \sum_i a(S_i, S_c)$. If we choose $B = \frac{p}{2} \sum_i a(S_i, S_c)$ then we've shown $STAR \leq 2B$.
   Let's prove the second statement (12.13b):

$$2 \cdot OPT = \sum_{ij} d_{OPT}(S_i, S_j). \qquad \text{by defn of } \textit{SP-Score} \tag{12.15a}$$

Because the optimal pairwise alignment must be less than or equal to any alignment, it definitely must be less than the pairwise alignment implied by the *STAR* MSA:

$$\geq \sum_{ij} a(S_i, S_j). \tag{12.15b}$$

The sum of the cost of all pairwise alignments is the sum of the cost of p different stars, each centered on a different sequence. The Star algorithm chose the smallest weight of those stars when picking S_c. Therefore:

$$\geq p \sum_i a(S_i, S_c). \tag{12.15c}$$

This last step can be visualized by noting we're summing up all the edges (to get $2OPT$), which is a union of stars, and the star centered on S_c has the smallest weight.

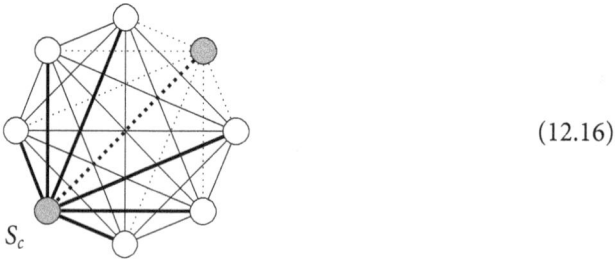

$$\tag{12.16}$$

The dotted star will have higher weight than the thick star that corresponds to S_c. So, we have $OPT \geq \frac{p}{2} \sum_i a(S_i, S_c) = B$.

Finally, putting everything together:

$$\frac{STAR}{OPT} \le \frac{p \sum_i a(S_i, S_c)}{\frac{p}{2} \sum_i a(S_i, S_c)} = \frac{2B}{B} = 2. \tag{12.17}$$

$\square$

12.4 Summary and notes

Multiple sequence alignments (MSAs) are a fundamental tool. They help reveal subtle patterns, compute consistent distances between sequences, etc. Applying the same DP idea as pairwise alignment leads to an exponentially slow algorithm for MSA for general numbers of sequences. The quality of MSAs is commonly measured using the SP-score: sum of the scores of the pairwise alignments implied by the MSA using an underlying pairwise $cost(u, v)$ function. Assuming this metric and assuming the $cost$ function obeys the triangle inequality, we can obtain a 2-approximation via the Star alignment algorithm.

There are many other possible objective functions for multiple sequence alignment (SP-score is discussed in Carrillo and Lipman [1988]). Some MSA approaches assume (or use) an evolutionary tree relating the sequences to define the objective.

> Presentation Notes
>
> The Star algorithm (and the bound) is due to Gusfield [1993].

12.5 Exercises

12.1   Give an example (strings and cost function) where progressive alignment of Section 12.2 gives a suboptimal alignment.

12.2   Give a class of examples (strings and cost function) where the Star algorithm returns an alignment that is as close to 2 times the optimal as you can.

Part III

Data Structures

CHAPTER 13

Suffix Trees

13.1 The fundamental string data structure

Consider a string T of length t. Our goal is to preprocess T, constructing a data structure that will allow various kinds of queries on T to be done efficiently. The most basic example of which is simply this: given a pattern P of length p, find all occurrences of P in the text T. What performance are we aiming for?

- The time to find all occurrences of pattern P in T should be $O(p + k)$ where k is the number of occurrences of P in T.
- Moreover, ideally we would require $O(t)$ time to do the preprocessing, and $O(t)$ space to store the data structure.

Suffix trees, invented by Peter Weiner [Weiner, 1973], are a solution to this problem, with all these ideal properties, and the ability to do much more. Many problems have simple algorithms once you have the suffix tree in hand.

13.2 Tries

The first piece of the puzzle toward understanding suffix trees is a *trie*, which is a tree that supports storing (*key, value*) pairs (like a binary search tree) where the *key* is broken up and stored along edges of the tree. In our case, the *key* is some string, so each edge of the tree is labeled with a character of the alphabet. Each node then implicitly represents a certain string of characters that is the *key* for the value stored at that node. Specifically, a node v in the tree represents the string of letters on the edges that we follow to get from the root to v. The root represents the empty string. Since our alphabet is assumed to be constant sized, we can use an array of pointers at each node to point to the node's children.

We can add a bit to each node that indicates whether the path from the root to this node is a full key—if the bit is set, we say the node is marked. We've seen an example of this concept under another name in Chapter 6: the *keyword tree*. In a keyword tree, the keys were the strings in our set.

To determine if a pattern (key) P occurs in our set we traverse down from the root of the tree following one character of P at a time until we either (1) walk off the bottom of the tree, in which case P does not occur, or (2) we stop at some node v. We now know that P is a prefix of some key in our set. If v is marked, then P is in our set, otherwise it is not.

This search takes $O(p)$ time because each step simply looks up the next character of P in an array of child pointers from the current node, which takes constant time assuming that $|\Sigma| = O(1)$.

13.3   Defining suffix trees

Our first attempt to build a data structure that solves this problem is to build a trie that stores all the strings that are suffixes of the given text T. It's going to be useful to avoid having one suffix match a prefix of another suffix, so we will append a special character denoted $, which occurs nowhere else in T, to the end of the text T. This character is by definition lexicographically less than any other character.

For example, if the text were $T = \texttt{banana\$}$, the suffixes of T are then as follows.

```
banana$
anana$
nana$
ana$
na$
a$
$
```

Since we appended the $ sign to the end, we're not including the empty suffix here. The trie of these suffixes would be the one in (13.1).

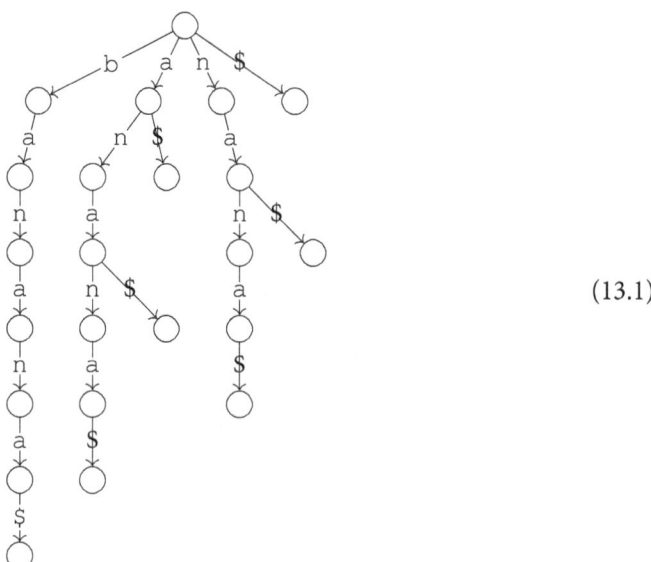

$$(13.1)$$

We can store at each node the count of the number of leaves in the subtree rooted at that node. Given a pattern P, we can count the number of occurrences of P in T in $O(|P|)$ time: we walk down the trie, and when we run out of P, we look at the count of the node v we're sitting on. It's our answer.

The space to store this tree could be as large as $\Omega(t^2)$. To see this, consider a pattern of the form $s = a^n b^n$, where exponentiation indicates repetition. Such a tree would look like (13.2) (for $s = $ "aaabbb").

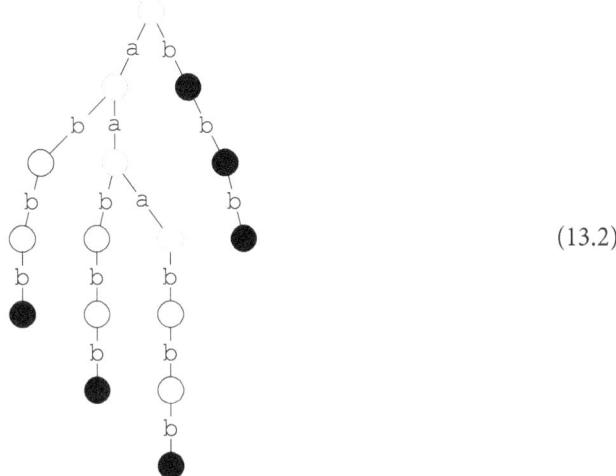

(13.2)

For clarity, instead of drawing the "$\$$" edges, we've filled nodes that would have a "$\$$" child. For general n, we have $n + 1$ nodes (including the root) in a path of "a"s, and $n + 1$ paths of n "b" nodes. This leads to $n(n + 1) + n + 1 = O(n^2) = O(t^2)$ nodes, which is not very efficient.

This approach is also unsatisfactory since it does not tell us where in s a pattern occurs, just the number of times patterns occur. Finally, it will also take too long to build the trie—at least time $O(t^2)$ since we might have to create $\Omega(t^2)$ nodes.

Shrinking the tree to $O(t)$ space. We can create a "compressed" tree of size only $O(t)$. Since no string occurs as a prefix of any other, we can divide the nodes of our trie into *internal* and *leaf* nodes where the leaf nodes represent complete suffixes of T.

The leaf nodes have no children. We can have each leaf node point to the place in T where the given suffix begins. We do this by recording the position that the suffix starts in the leaf node that spells out that suffix.

Moreover, if there is a path in the trie where each node has a single child, we compress that path into a single edge that represents the entire path. The string of characters corresponding to that path must occur in T, so we can represent it implicitly by a pair of pointers into the string T. An edge is now labeled with a pair of indices into T instead of just a single character (or a string). Here's an example for $s = $ abaaba$\$$. If we label edges with the substrings, we obtain the following tree in (13.3).

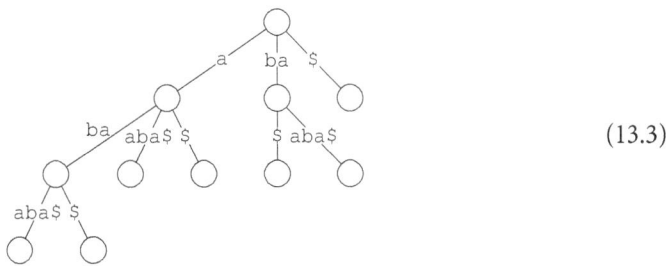

(13.3)

Converting this tree to use indexes into the string (using 1-based indexing) to label the edges, we obtain the suffix tree in (13.4).

(13.4)

This representation uses $O(t)$ space. (We count pointers as $O(1)$ space.) Why? Each internal node now has degree at least 2, hence the total number of nodes in the tree is at most twice the number of leaves (Exercise 13.1). But each leaf corresponds to some suffix of T, and there are $O(t)$ suffixes.

Building the tree—preview. What about the time to build the data structure? Let's first look at the naïve construction, by adding suffixes into it one at a time. To add a new suffix, we walk down the current tree until we come to a place where the path leads off of the current tree. (This must occur because the suffix is not already in the tree.) This could happen in the middle of an edge, or at an already existing node. In the former case, we split the edge in two and add a new node with a branching factor of 2 in the middle of it. In the latter case we simply add a new edge from an already existing node. In either case the process terminates with a tree containing $O(t)$ nodes, but the running time of this naïve construction algorithm is $O(t^2)$ since for each of the $O(t)$ suffixes we did $O(t)$ work. We will see how to create the tree faster than this in Chapter 15: in fact, it is possible to build a suffix tree on a string of length t in time $O(t)$.

13.4   Applications of suffix trees

Suffix trees can do exact search in time proportional to the length of the query string, once the tree is built. There are many other applications of suffix trees to problems on strings. Gusfield [1997] discusses many of these. We'll mention just a few here, and we'll see others in the next chapter.

13.4.1   Simple queries

Suffix trees make it easy to answer common (and less common) kinds of queries about strings. For example, it's easy to:

- Check whether P is a suffix of T: follow the path for P starting from the root and check whether you end at a leaf.
- Count the number of occurrences of P in T: follow the path for P; the number of leaves under the node you end up at is the number of occurrences of P. If you are

going to answer this kind of query often, you can store the number of leaves under each node in the nodes without increasing the asymptotic space usage.

- Find the lexicographically (alphabetically) first suffix: start at the root, repeatedly follow the edge labeled with the lexicographically (alphabetically) smallest letter.
- Find the longest repeat in T. That is, find the longest string r such that r occurs at least twice in T. To do this, we search over all nodes to find the deepest node (farthest from the root measured by number of characters along the path to that node) that has ≥ 2 leaves under it. We can do this in $O(t)$ time with a tree traversal, since our tree has only $O(t)$ nodes.
- Find the *maximal everywhere extension*[1] Here, we're given a text T and a string q, and we want the longest string that follows every instance of q. For example, if $T = $ "aretheythey?" and $q = $ "th" then the maximal everywhere extension is "ey" since every place "th" occurs, it is followed by "ey". To do this, walk down the suffix tree for T following the path spelled by q. If you stop in the middle of an edge, the string from where you stop to the next node is the maximal everywhere extension (otherwise the maximal everywhere extension is the empty string).

13.4.2 Longest common substring of two strings

Given two strings S and T, what is the longest substring that occurs in both of them? This is the *longest common substring* between those two strings. For example, if $S = $ boogie and $T = $ ogre then the longest common substring is og. How can one compute this efficiently? Before applying suffix trees, spend a moment to think how you might do this without such a data structure—maybe via local alignment or a naïve search approach (Exercise 7.6). Before the advent of suffix trees, an algorithm of Karp et al. [1972] gave an $O(n \log n)$ time solution. Suffix trees provide a faster algorithm.

Construct a new string $U = S\%T$. That is, concatenate S and T together with an intervening special character that occurs nowhere else (indicated here by "%"). Let n be the sum of the lengths of the two strings. Construct the suffix tree for U. Every leaf of the suffix tree represents a suffix that begins in S or in T. Mark every internal node with two bits: one that indicates if this node contains a prefix of a suffix (substring) of S, and another similarly for T (note that both bits could be set to 1 if there is a shared substring). These bits can be computed by depth-first search in time linear in the size of the tree (and hence linear in $|U|$). Take the deepest node in the suffix tree (in the sense of the longest string in the suffix tree) that has both bits set to 1. This tells you the longest common substring. The total time to solve this is $O(|S| + |T|)$ to build the tree (which we will see later) plus $O(|S| + |T|)$ to find the deepest double-marked node, so $O(|S| + |T|)$ total time. This is far faster than either local alignment or a naïve approach.

---

[1] Not to be confused with the longest common extension problem in Chapter 14.

13.4.3 Generalized suffix trees

Suppose you are given a database of strings $\mathcal{S} = \{s_1, \ldots, s_k\}$. You want to be able to quickly tell which strings in $\mathcal{S}$ contain a given query string q.

To do this, you can build a suffix tree $\mathcal{T}$ for the string

$$s_1 \#_1 s_2 \#_2 \ldots s_k \#_k, \tag{13.5}$$

where the $\#_i$ are a series of k unique special characters that do not appear anywhere else. We then prune the labels associated with the edges going into each leaf node, removing any text after the first # symbol.

For example, if $\mathcal{S} = \{aat, tag, gat\}$ then we build the suffix tree for "aat$\#_1$tag$\#_2$gat":

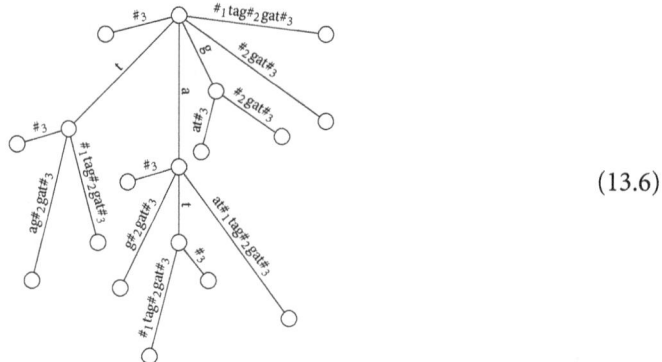

$$(13.6)$$

and then prune the edge labels to obtain (13.7).

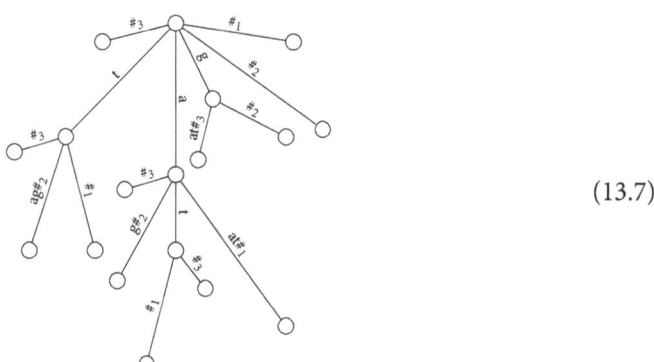

$$(13.7)$$

Of course, we would do this pruning by adjusting the start/end indices on the edges rather than editing any strings. In fact, often we can logically prune and not change the tree at all since the extra characters often won't affect how we use a generalized suffix tree.

Given a generalized suffix tree, we can find the strings that contain a query q by following the path represented by q in the suffix tree. If you spell out q ending at a node u, you report i if you find a leaf associated with string i (that is containing $\#_i$) under u.

13.4.4 All pairs suffix-prefix matches

Problem 13.1 (All Pairs Suffix-Prefix Matches). *Find all the longest suffix-prefix exact matches between each pair in a set of strings $S = \{s_1, \ldots, s_k\}$ of total length m.*

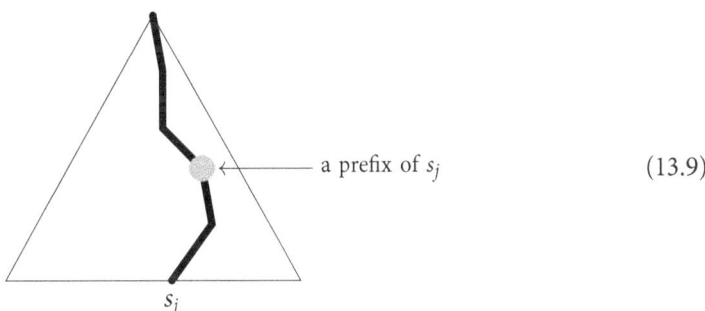

$$(13.8)$$

The answer to this problem can be represented by $\binom{k}{2}$ integers, so that is what we are seeking to output. This means we have to take at least $\Omega(k^2)$ time since that is how long it takes just to output the answer. We also must take $\Omega(m)$ since we have to read the input strings. We will see an algorithm that takes $O(m + k^2)$.

The algorithm builds a generalized suffix tree $\mathcal{T}$ for the set of strings S that are given as input. At each node u we additionally store a list $C(u)$ of indices of the strings in S for which the string represented by u is a suffix. For example, $C(u) = \{1, 5, 7\}$ if and only if the string represented by u is a suffix of $s_1, s_5,$ and s_7.

Consider the path in $\mathcal{T}$ that represents the entirety of string s_j. Each node on this path represents a prefix of s_j. Let v be some node on that path. If $C(v)$ contains i, then v must also represent a suffix of s_i.

a prefix of s_j

s_j

$$(13.9)$$

Hence, the deepest node v^* on the s_j path that contains i in $C(v^*)$ represents the longest suffix-prefix overlap between s_i and s_j.

The last piece of the puzzle is to develop an algorithm to find all of these nodes in a single traversal of the tree. Here is such an algorithm:

Algorithm 13.1. All-pairs suffix-prefix alignments.

1. Do a DFS on the generalized suffix tree $\mathcal{T}$.
2. Maintain k stacks during the DFS that you update as follows:
   - When entering node v, push v onto all stacks in $C(v)$.
   - When backtracking out of v, pop all stacks in $C(v)$.
3. When you reach a node corresponding to a full string (i.e., a leaf that corresponds to a complete string in S), output the depth of the nodes at the top of each stack.

At any point in the algorithm, each stack i will have at its top the deepest node v with $i \in C(v)$ on the current DFS path from the root.

Theorem 13.2 (Suffix-Prefix Running time). *The runtime of the preceding algorithm for all pairs suffix-prefix exact matching is $O(m + k^2)$.*

Proof: Building the generalized suffix tree takes $O(m)$ time. The $C(v)$ sets can be saved as you are building the suffix tree (when you add in a suffix for i, add it to $C(v)$).

There are $O(m)$ indices in the union of the $C(v)$ lists, so the total pop/push events take $O(m)$. There are k nodes at which you will output something (the nodes corresponding to full strings), and each output takes $O(k)$ time, leading to $O(k^2)$ time for output. $\square$

13.5   Summary and notes

We've seen what is arguably the most important data structure for strings: the suffix tree. It is *not* the most practical way to solve most problems (we will see suffix arrays and other data structures that are faster in practice). But all these other more practical approaches are built on the ideas of the suffix tree.

Why *suffix* trees? The key point is that any substring is the prefix of some suffix—so questions about substrings can be approached by talking about prefixes of suffixes. The suffix tree is natural in a language that reads left to right. Of course, you could instead build a *prefix* tree in the same way, with everything reversed, but this doesn't typically gain you anything over thinking of the suffix tree of the reversed string.

> **Presentation Notes**
>
> Owing to their fundamental place in string analysis, suffix trees are covered in many string and bioinformatics algorithms textbooks, including Gusfield [1997], Jones and Pevzner [2004], Mäkinen et al. [2015], Crochemore et al. [2007], Crochemore and Rytter [2003], and others.

13.6   Exercises

13.1   Complete the proof that a suffix tree for a string of length n has $O(n)$ nodes.

13.2   Give an algorithm to count the number of *distinct* substrings of a string x in $O(|x|)$ time.

13.3   Let $\{S_1, S_2, \ldots, S_k\}$ be a set of strings. Give and algorithm that outputs all pairs (i, j) such that S_i is a substring of S_j. Your algorithm should run in time $O(z + \sum_i |S_i|)$, where z is the number of output pairs.

13.4   Describe how to compute the $C(v)$ for every v that is needed in the "All pairs suffix-prefix matches" algorithm.

13.5   A *k-cover* $C_k(S, T)$ is a decomposition of the string T into non-overlapping substrings $T = s_1, s_2, \ldots$ such that each $|s_i| \geq k$ and each s_i is a substring of S. In other words, it expresses T as the concatenation of substrings of S of length $\geq k$. Give an $O(|S| + |T|)$-time algorithm to compute $C_k(S, T)$, if one exists. Your algorithm should output a sequence of pairs (i, j), which give the start and endpoints of substrings in S, in the order they must be concatenated to spell out T.

13.6   Let $A(p, q)$ be the number of characters that match when two strings p and q of length k are compared (without allowing insertions or deletions). Let x and y be strings of length n, and let k be an integer. Each of x and y contain $n - k + 1$ substrings of length k (called k-mers). We want to compute $A(p, q)$ for each pair p, q of k-mers where p comes from x and q comes from y. Give an algorithm to do this in $O(n^2)$ time.

13.7   Let S be a string of length n. Give an $O(n)$-time algorithm to find the longest repeated substring of S such that at least two copies of the substring do not overlap in S.

13.8   Let S be a sequence of n letters arranged along a circle. We want to cut the circle to turn the sequence into a string S' so that we obtain the alphabetically first such string. For example, look at (13.10).

$$(13.10)$$

Give an $O(n)$-time, $O(n)$-space algorithm to find such a cutting place. (This is the "minimum string rotation" problem, considered and extended in, e.g., Booth [1980]; Shiloach [1981]; Iliopoulos and Smyth [1992]; Babenko et al. [2016].)

13.9   Let T be a suffix tree for string S. Let $\mathbf{str}(u)$ be the string that is spelled out when walking from the root of T to a node u. A node in T is called *left diverse* if the occurrences of $\mathbf{str}(u)$ in S are not always preceded by the same character. For example, if $S = ababacb$ then the node representing ba is **not** left diverse since ba is always preceded by a. But the node with $\mathbf{str}(u) = b$ *is* left diverse because sometimes b is preceded by a and sometimes by c. Give an $O(|S|)$ algorithm to identify all the left-diverse nodes in T.

13.10   Suppose A and B are two strings. A *maximum exact match* (MEM) between A and B is given by positions i in A, j in B and a length ℓ where the length-ℓ substrings starting at i and j are equal and can't be extended left or right with a longer match. Give an $O(|A| + |B|)$-time algorithm to find all the MEMs between A and B. (Hint: eventually consider Exercise 13.9.)

13.11   Suppose A and B are two strings. A *maximum unique match* (MUM) between them is a substring that occurs exactly once in A and B that cannot be extended to the left or right to form a longer common substring. Give an $O(|A| + |B|)$ algorithm for finding all MUMs between A and B.

13.12   Suppose you want to print something out for every leaf of a suffix tree under a given node u (to which you have a pointer). If there are k leaves, show how to do this in $O(k)$ time.

CHAPTER 14

More Applications of Suffix Trees

We'll now see even more applications of suffix trees. Some of these applications build directly on the application ideas in the previous chapter. Another set of applications use a general tool that we will introduce called the longest common extension.

14.1   Lempel-Ziv compression

Suppose we are given a string S of length n that we would like to store in a small space. Lempel-Ziv [Ziv and Lempel, 1977] is one approach to this general compression problem.

Definition 14.1 (Lempel-Ziv Notation). Define the following notation:

$$b(i) := \text{the longest string in } S[\ldots i-1] \text{ matching a prefix of } S[i\ldots]. \tag{14.1}$$

That is, $b(i)$ is the longest extension starting at position i that matches a substring completely contained before i. Also define.

$$e(i) := |b(i)| \tag{14.2}$$

$$p(i) := \text{the position of } b(i) \text{ in } S[\ldots i-1]. \tag{14.3}$$

■

The following figure illustrates the terms defined in (14.1)–(14.3).

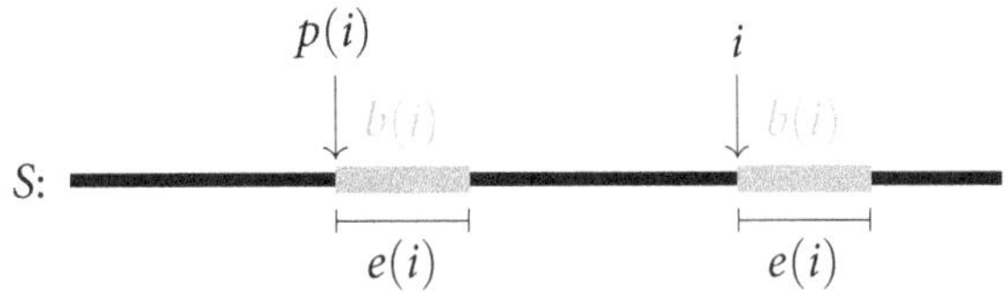

$$\tag{14.4}$$

Given these values, we can state the Lempel-Ziv algorithm:

Algorithm 14.1. Lempel-Ziv compression.

1. Walk down the string from left to right.
2. When at position i, output $(p(i), e(i))$ instead of $S[i \ldots i + e(i) - 1]$. If $e(i) = 0$, output $(S[i], 0)$ instead.
3. Skip ahead to $i \leftarrow i + e(i)$.

We need an efficient way to compute $p(i)$ and $e(i)$. To do this, we compute the suffix tree $\mathcal{T}$ of S and store a value $c(v)$ at each node in the suffix tree, where $c(v)$ is the smallest (earliest) suffix index that is present in the subtree rooted at v.

How does this $c(v)$ value help us? Consider the path S_i in $\mathcal{T}$ that represents the suffix $S[i \ldots]$ starting at position i.

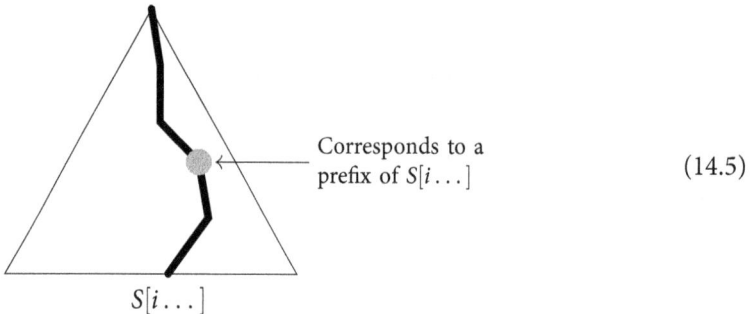

$$\text{(14.5)}$$

Every node along this path corresponds to a prefix of the suffix starting at i.
Define the set

$$V_i = \{v \in S_i : c(v) + depth(v) < i\}, \tag{14.6}$$

where $depth(v)$ is the length of the string represented by the path to v from the root. Every node $v \in V_i$ matches a prefix of $S[i \ldots]$ because it is on the S_i path and some instance of that prefix string is completely contained in $S[\ldots i - 1]$ since $c(v) + depth(v) < i$.

$$\text{(14.7)}$$

The deepest node in V_i represents $b(i)$. This gives the following algorithm to compute $p(i), e(i)$ at each stage of Lempel-Ziv:

Algorithm 14.2. Computing $p(i)$ and $e(i)$.

1. Walk down path spelling out $S[i \ldots]$.
2. Stop at node v when one of these conditions holds:
   - the next node u you would visit has $c(u) + depth(u) \geq i$, or
   - there is no path out of v that continues the match with S, or
   - you reach the end of S.
3. Return $p(i) = c(v)$, and $e(i) = depth(v)$.

Theorem 14.2 (Lempel-Ziv runtime). *Lempel-Ziv takes time $O(|S|)$ to compress a string S.*

Proof: It takes time $O(|S|)$ to build the suffix tree (Chapter 15). Via a bottom-up traversal of the suffix tree, we can compute $c(v)$ for every v.

It takes $O(e(i))$ time to search at position i, but each time you spend $O(e(i))$ time, you skip $e(i)$ letters of S, leading to a total of $O(|S|)$ time.   □

14.2 Longest common extension

Here's another problem where we can process T and then must answer queries about T on the fly. This problem seems a bit contrived at first—but in fact it can be used as a building block for several other types of questions.

Problem 14.3 (Longest common extension). *We are given strings S and T. In the future, many pairs of indices (i, j) will be provided as queries, where i is a position in S and j is a position in T. We want to quickly find: the longest substring of S starting at i that matches a substring of T starting at j.* ◆

(14.8) gives us a picture.

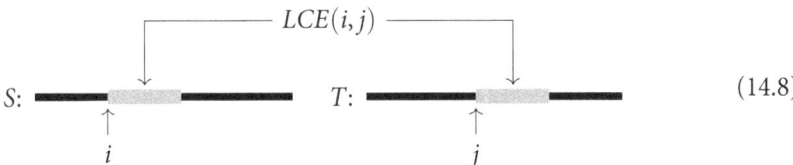

$$(14.8)$$

To answer these queries quickly, we need to introduce the concept of a *lowest common ancestor* (LCA) of two nodes in a tree.

Definition 14.4 (Lowest common ancestor). The *lowest common ancestor* (LCA) of two tree nodes u and v is the deepest node (measured as edge hops) that is on the path from the root to u and the path from the root to v. ∎

It's easy to preprocess the tree so that $LCA(u, v)$ can be found in $O(1)$ time if we are allowed to use $O((\text{size of the tree})^2)$ space: just create a 2D array A so that entry $A[u, v]$

gives the LCA of u and v. It turns out that you can get $O(1)$ answers to LCA queries in linear space with *linear* preprocessing time for any tree, which is a nice result by Gabow and Tarjan [1983]. We won't have time to go into how to do that, however. But now we can see how to use LCA to answer LCE queries:

Algorithm 14.3. Preprocessing for LCE.

1. Build a suffix tree R for $S\%T$. (Takes time $O(|S| + |T|)$.)
2. Preprocess R so that lowest common ancestors (LCA) can be found in constant time (as just discussed).
3. Build an array mapping indices i and j to the leaf nodes representing those suffixes.

When we get a LCE query consisting of two positions (i, j), we can answer it using the following algorithm:

Algorithm 14.4. Answering LCE query (i, j).

1. Find the leaf nodes for i and j. (Takes $O(1)$ time using the array.)
2. Return the string represented by the $LCA(i, j)$. (Takes $O(1)$ time using the LCA preprocessing.)

This will be the node in the suffix tree that corresponds to the place where the two suffixes of $S\%T$ starting at i (in S) and j (in T) diverge, which is exactly the longest common extension.

14.3   Finding palindromes

Suppose you want to find all the maximal even palindromes in T. An *even palindrome* is a string $\alpha\alpha^R$, where α^R means reversing the string. A maximal even palindrome is a palindrome that can't be extended to be longer in T. In pictures:

$$
\text{(14.9)}
$$

where ⬅ is the reverse string of ➡ and $x \neq y$. The trick here is to notice that when we reverse T, the character just to the right of the center of a palindrome at position i moves to position $|T| - i$ (starting indexing at 1) and the palindrome also flips around.

Algorithm 14.5. Find all maximal even palindromes in T.

1. Process T and T^R so that LCE queries can be solved in constant time (see previous section).
2. For every position i in T: Compute $LCE_{T,T^R}(i, |T| - i)$.

This takes $O(|T|)$ time to preprocess and execute the LCEs, giving a *linear* time algorithm to find all the maximal palindromes.

14.4 Finding all *k*-mismatch occurrences of a pattern

Say two strings u and v of the same length have *k mismatches* if they are equal in all but k positions (without insertions or deletions).

Problem 14.5. *Given a long string T and a shorter string P and an integer k, find all the positions in T that are the start of a string that has k mismatches with P.* ◆

This can be done in $O(k|T|)$ time using LCE. Let's see how we check whether index i in T is the start of a k mismatch of P.

Algorithm 14.6. Check if there is a k mismatch to P at position i in T.

Require: P and T preprocessed to support *LCE* queries in $O(1)$ time.
    function CHECKMISMATCH(i, P, T, k)
        $j \leftarrow 0$ ▷ *current position in P*
        $c \leftarrow 0$ ▷ *number of mismatches found so far*
        while $c \leq k$ **do**
            $lceij \leftarrow LCE_{P,T}(i,j)$
            $j \leftarrow j + lceij + 1$ ▷ *jump past the next mismatch in P and T*
            $i \leftarrow i + lceij + 1$
            if $j \geq |P| + 1$ **then**
                return true   ▷ *if we ran out of pattern*
            $c \leftarrow c + 1$
        return false   ▷ *we ran out of mismatches*

The operation of the algorithm looks like (14.10).

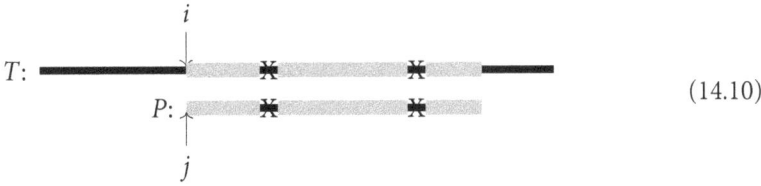

$$\text{(14.10)}$$

Algorithm 14.6 takes time $O(k)$ since each LCE query takes $O(1)$ time. Therefore, to check every position of T takes $O(k|T|)$. Building the suffix tree and the data structures to support $O(1)$ LCE queries takes $O(|T| + |P|) = O(|T|)$ since $|P| < |T|$. So the total running time is $O(|T| + k|T|) = O(k|T|)$.

14.5   Longest common substring with online queries

Suppose T is some large database of text, which we can preprocess, but then queries come later on, and we need to answer them efficiently by finding the longest common substring (LCS) between T and each query q. This problem is different than the one in Section 13.4.2 because we aren't allowed to preprocess q. To solve this problem, we need to introduce the concept of a *suffix link*.

Definition 14.6 (Suffix links). A *suffix link* is an extra pointer leaving each node in a suffix tree. If a node u represents string $x\alpha$ (where x is a character and α is a string) then u's suffix link connects from u to the node representing the string α. ∎

Every node represents the prefix of some suffix. Suppose node u represents the prefix of suffix i of length m. Then u's suffix link should point to the node on the path representing the prefix of length $m-1$ of suffix $i+1$. A tree with all the suffix links shown as dashed arrows is given in (14.11).

$$\text{(14.11)}$$

Every node has a suffix link. Why? If node representing string $x\alpha$ exists in our tree, then a node representing α must exist. The reason for this is since the node $x\alpha$ exists, it has two children y and z. That means both the strings αy and αz must exist in the string. The node represented by α must have 2 children too.

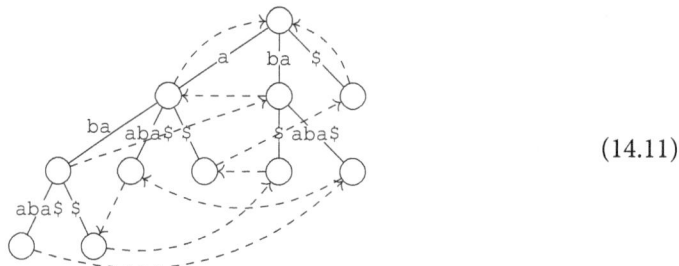

$$\text{(14.12)}$$

Using suffix links we can get an $O(|q|)$-time algorithm to find the longest common substring between T and q. The longest common substring starts at some suffix—we just don't know which suffix, so we try them all starting with suffix 1. Following a suffix link chops off the first character of q taking you to the place in the tree that you would have been at if the query had started with the second character of q, and so on through all the suffixes of q.

Algorithm 14.7. Longest common substring for online queries q.

1. Walk down the tree, following q.
2. If you hit a dead-end:
    (a) save the depth of the place where you stopped matching if it is deeper than the deepest dead end you found so far,
        i.   if you are at a node u, follow the suffix link out of it;
        ii. else, if you are in the middle of an edge, let α be the string from the previous node to where you are in that edge. Jump back to the parent of that edge, and follow the suffix link out of it to a node v, and then hop down, node-by-node until you reach where you would have been if you had spelled out α starting from v. (See (14.13).)
    (b) go to step 1 (starting from your current place in q and the tree).
3. When you exhaust q, return the string represented by the deepest place in the tree that you visited.

Step 2 deserves some extra explanation. In the case we hit a dead end in the middle of an edge, the situation looks like (14.3).

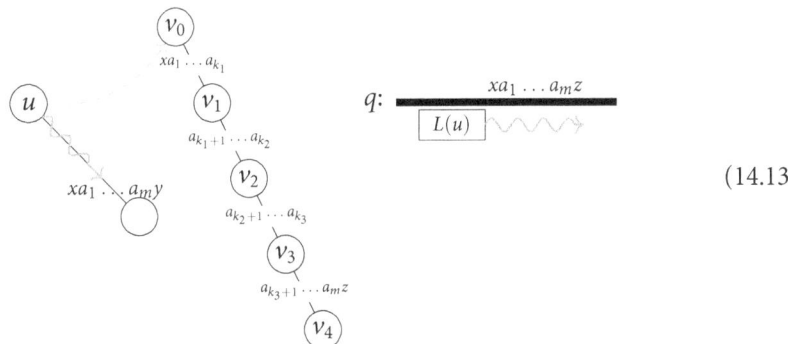

$$(14.13)$$

Here, the current node is u, and we match part of an edge leaving u by matching the characters $xa_1 \ldots a_m$ along that edge (for some m, which may be large), but then we mismatch q in the middle of an edge (in the figure, the edge has a y where q has a z.) In this case, the algorithm still follows the suffix link (dashed pointer) leaving u to end up at some node v_0. But now, if we naïvely start matching characters of q starting at v_0, we may compare lots of characters of q again.

Instead, we use the fact that edge lengths are easy to compute in a suffix tree (subtract the two numbers on an edge), and we know how far down we have to hop: it's the number of characters we matched along the edge leaving u (in this case $m + 1$ characters). We hop down from v_0 not by comparing characters, but by jumping down node-by-node until we've gone down the right number of characters. We find the right edge out of each v_i along this "hop down" path by looking at the character at the corresponding position in the string we matched along the edge leaving u.

We can show that all of these hops down take linear time, and that therefore the entire algorithm runs in $O(|q|)$. To do that, we're going to need the following lemma to limit how far back the suffix links can take us.

Lemma 14.7. *Let $n(u)$ be the* node depth *of node u—that is the number of nodes on the path from the root to u. Following a suffix link reduces the node depth by at most 1.*

Proof: Consider a node u and the node v pointed to by its suffix link. Let P_u and P_v be the paths from the root to u and v respectively. Each of the $n(u)$ nodes of P_u has a suffix link that *also* points to a node in P_v.

$$(14.14)$$

Since the lengths of the strings represented by each node in P_u are different, their suffix links must point to distinct nodes and these suffix links cannot "cross" since the strings represented by the nodes along each path are monotonically increasing in length.

At most one internal node on P_u could point to the root and therefore not have a corresponding internal node on P_v. □

Now, we can prove the running time:

Theorem 14.8. *After $O(|T|)$ preprocessing to build the suffix tree and suffix links, finding the longest common substring between T and q for a new string q takes $O(|q|)$ time.*

Proof: We take $O(|T|)$ time to build the suffix tree of T with the suffix links. (See Corollary 15.3.)

Algorithm 14.7 can be divided into phases, corresponding to each time step 1 is executed. Phase i is associated with computing the longest common substring starting at position i in q. Hence, there are at most $O(|q|)$ phases. When we end the phase, we follow a suffix link (possibly after backing up to the previous node). Over the course of the algorithm, this takes $O(|q|)$ time, since following the suffix link takes $O(1)$ time.

At the start of step 2, the algorithm is at a particular node, with a given node depth. By Lemma 14.7, each time we follow a suffix link, we reduce the node depth by at most 1.

Since every time we decrement the node depth by following a suffix link, we end the phase and, therefore, increment our "starting place" in q, the algorithm can decrement the node depth at most $|q|$ times. The node depth must always be between 0 and $|q|$, so if we never decremented our node depth, we could do at most $|q|$ downward hops. Since we can, in fact, decrement at most $|q|$ times, our node depth could be reduced by in total $|q|$. This means that the maximum total "room" we have for downward hops is $\leq 2|q|$, and therefore the total time for all the hops we make across all the times step 2 is executed in the algorithm is $\leq 2|q|$.

Finally, outside of the hops, each phase examines only one character that was examined in the previous phase, so the total time for those reexaminations is $O(|q|)$. The total time to examine the other characters of q is $O(|q|)$.

Therefore, the total work we have to do across the entire algorithm is $O(|q|)$ to follow the suffix links, $O(|q|)$ to compare the characters, and $O(|q|)$ to do all the hops down. Therefore, the total time is $O(|q|)$.   □

14.6   Summary and notes

These last two chapters just touch on a few of the applications of suffix trees. We've seen a variety of applications, including longest common extension (LCE), Lempel-Ziv compression, longest common substring, online longest common substring, exact search, palindrome finding, k-mismatch search. We saw more of k difference finding in Chapter 11, where we also used LCE. Lempel-Ziv will also play a role in Section 29.2, where we use it to construct genome graphs.

The suffix links introduced in Section 14.5 are important for many suffix tree algorithms, including the algorithm to efficiently construct suffix trees that we will see in the next chapter.

The algorithm for online longest common substring is actually a special case of the more general algorithm introduced by Chang and Lawler [1994] for computing the length of the longest substring starting at each position in T that occurs someplace in another string q. These lengths are called *match statistics*.

A line of work enables efficient search of text compressed using Lempel-Ziv-like algorithms [Bille et al., 2010; Farach and Thorup, 1995; Gawrychowski, 2013, 2011a,b; Farach and Thorup, 1998; Navarro and Tarhio, 2000; Kärkkäinen et al., 2003]. Bille et al. [2015] extend LCE queries to trees, and Birenzwige et al. [2020] improve the space usage of LCE queries (and sparse suffix trees).

> Presentation Notes
>
> The Lempel-Ziv variant we present is the simplified variant from Gusfield [1997], which also covers many other applications (including the ones described here).

14.7   Exercises

14.1   Prove that if a node u of a suffix tree has outgoing edges with strings starting with $C \subset \Sigma^*$ then the node *suffixlink*(u) has children labeled with strings starting with C (and maybe others).

14.2   You are given K strings $S_1, \ldots, S_K$ of total length n. Define $\ell(k)$ for $k = 2, \ldots, K$ to be the length of the longest substring that is common to at least k of the strings $S_1, \ldots, S_K$. For example, if the strings are `ration`, `natural`, `nation`, then $\ell(3) = 2$ (`at`), $\ell(2) = 5$ (`ation`). Give an $O(Kn)$-time algorithm to compute $\ell(k)$ for all $k = 2, \ldots, K$.

14.3   Prove that every node in a suffix tree except one has exactly one suffix link pointing to it.

14.4   A *k-mismatch palindrome* is a string xy where, $|x| = |y|$ and reverse(y) and x are the same in all but at most k positions. Give an $O(kn)$-time algorithm to find all the k-mismatch palindromes in a string S of length n.

CHAPTER 15

Suffix Tree Construction

In this chapter, we will see Ukkonen's algorithm [Ukkonen, 1995] for directly creating a suffix tree in linear-time. We first describe an algorithm to construct suffix *tries*. This algorithm will essentially be the same at its core as Ukkonen's algorithm to compute a suffix *tree*, but we will have to make some changes to handle the fact that not every trie node is explicitly represented in the tree.

15.1   Constructing suffix tries

Suppose we want to a build suffix trie for:

$$s = \text{``abbacabaa''}. \tag{15.1}$$

We will walk down the string from left to right:

$$\text{abbacabaa} \tag{15.2}$$

We will build successive suffix tries for

$$\underbrace{s[1\ldots 1], s[1\ldots 2], \ldots, s[1\ldots n]}_{\text{longer and longer prefixes of } s}. \tag{15.3}$$

To build a suffix trie for $s[1\ldots i]$, we will use the suffix trie for $s[1\ldots i-1]$ built in the previous step. This will "destroy" the previous suffix trie and produce the next one. Let $\mathcal{T}(s[1\ldots i])$ be the suffix trie for the prefix up to the ith position.

To convert $\mathcal{T}(s[1\ldots i-1]) \rightarrow \mathcal{T}([s[1\ldots i])$ we need to add character $s[i]$ to all of the suffixes. For example, when $i = 5$ and we're at the "c" character in abbacabaa, we need to add:

$$
\begin{array}{c}
\text{abba}\underline{\text{c}} \\
\text{bba}\underline{\text{c}} \\
\text{ba}\underline{\text{c}} \\
\text{a}\underline{\text{c}} \\
\text{c}
\end{array} \quad . \tag{15.4}
$$

The underlined suffixes (and the empty suffix) will already be represented in $\mathcal{T}(s[1 \ldots i-1])$, so all we have to do is find these suffixes and tack on our new letter (in this case a c).

Look at the suffixes that we want to extend:

$$
\begin{array}{r}
abba \\
bba \\
ba \\
a \\
\epsilon
\end{array}
\qquad (15.5)
$$

They are related by their suffix links (Definition 14.6). Following the suffix links starting from the longest suffix (the one represented by the deepest node—in terms of characters) will visit each of the suffixes that we need to extend.

We follow these links and "hang" the new character off of the nodes that we visit. If such an edge already exists, we stop. In our running example, this looks like (15.6).

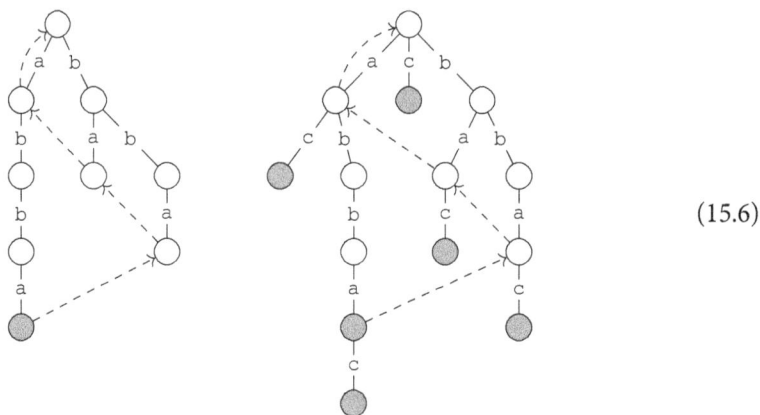

$$(15.6)$$

We also chain together the newly added nodes with suffix links.

The last piece is how to track the deepest node—since we need to start our suffix link traversal from there for each iteration. This is easy: the new deepest node is the node we just added below the previous deepest node, so we simply maintain a pointer to the deepest node and update it to the newly added child when we add it.

To summarize one iteration:

Algorithm 15.1. An iteration of suffix trie building.

- CurrentSuffix ← DeepestNode
- Repeat until you reach the root or the current node already has an edge labeled $s[i]$ leaving it:
  1. Add child labeled $s[i]$ to CurrentSuffix.
  2. Connect the previously added node to this new child via a suffix link.
  3. If new child is now the deepest node: DeepestNode ← newly added child.
  4. Follow CurrentSuffix's suffix link to set CurrentSuffix to next shortest suffix.

One tricky thing about this algorithm is understanding why you can stop the current iteration if you encounter a node that already has a child edge labeled $s[i]$. This is because if you already have a node for suffix $\alpha s[i]$, then you have a node for every smaller suffix. If the suffix $\alpha s[i]$ existed in the previous iteration, then any suffix $\beta s[i]$ must exist where β is a proper suffix of α.

Here's an example showing a step in the sequence of node additions.

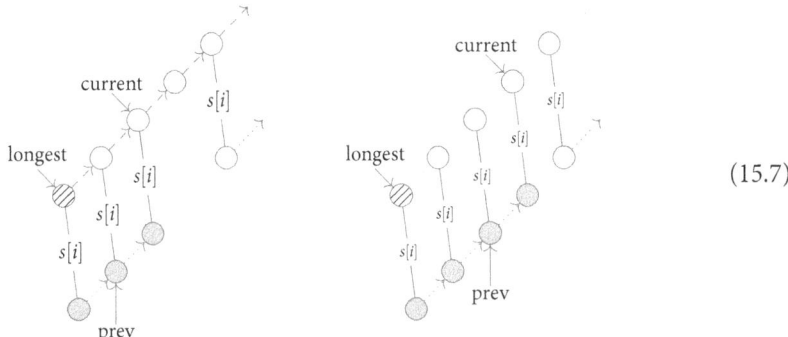

(15.7)

The dashed arrows show the path of suffix links that you are following. The shaded nodes are the nodes that are being added (with edges labeled $s[i]$ leading to them). The dotted arrows between the shaded nodes are the suffix links that are being added.

Here's a complete example for $s =$ "abaaba\$" in (15.8)–(15.11). The first three steps are shown in (15.8).

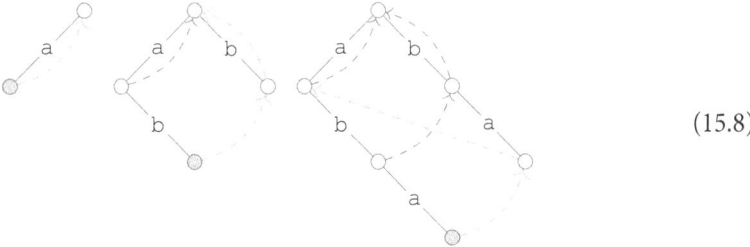

(15.8)

Note that in the 3rd step, there already is a path for suffix "a" so we stop at that point. We continue to extend, following the same procedure.

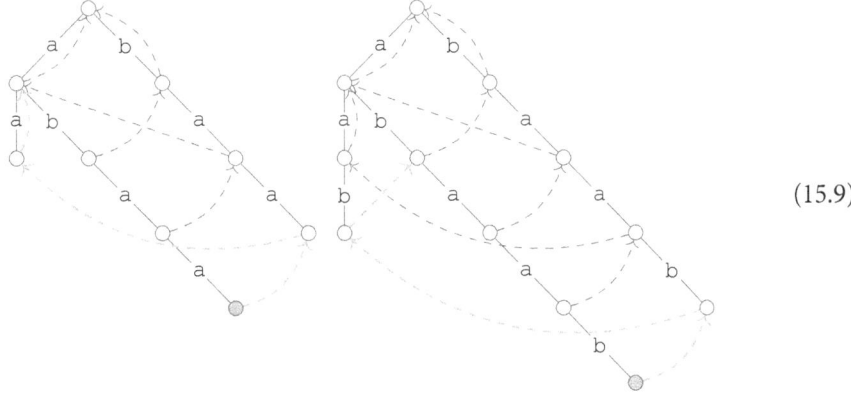

(15.9)

In the final two steps, we add the last character

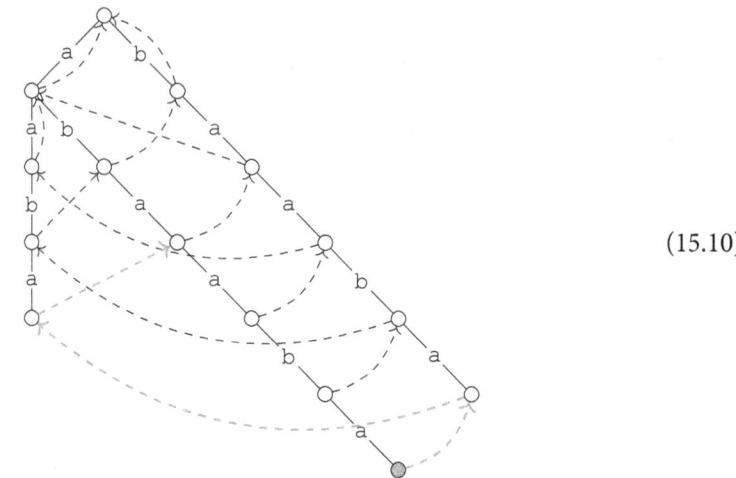

(15.10)

and the $ marker.

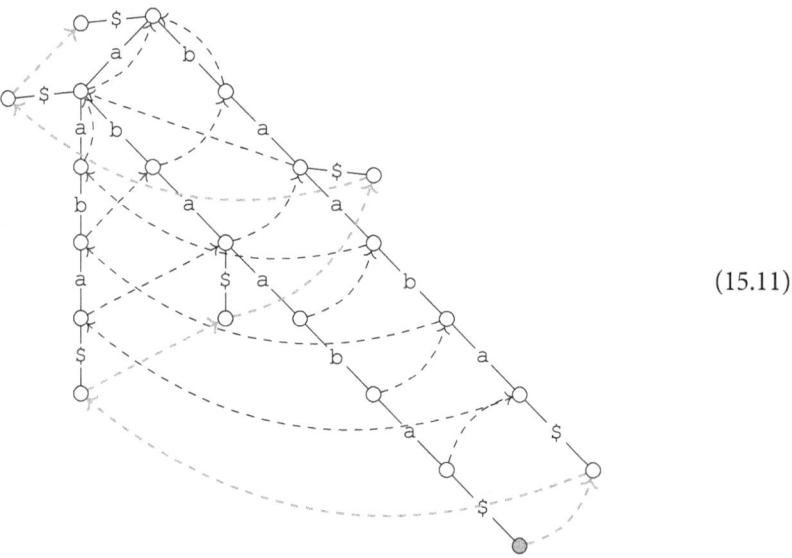

(15.11)

This completes the example.

15.2 Ukkonen's algorithm

Ukkonen's algorithm constructs a suffix tree for T (with the suffix links) in $O(|T|)$ time. It is essentially the same algorithm we saw in the previous section to construct suffix tries, but adapted to create the tree without creating all the trie nodes. We walk down the input string left-to-right. After we read each input character, we have a *phase* where we add all the new suffixes (that end with that character) by hopping from node to node following suffix links. We add suffix links as we create nodes, as in the trie algorithm.

As we build the tree, we will label each edge with substrings rather than creating nodes explicitly. These labels will be encoded using indices into our string, as they appear in the definition of suffix trees. For edges that enter leaves, we allow the special end index symbol "ρ" that always stands for the index of the end of the string we have the processed so far.

For every phase, at each location that we jump to in the tree, we have to determine how to update the tree, which will depend on what the tree looks like at that point. We then have to jump to the next suffix to extend, which is a little more complex because we may need to jump to the middle of an edge. Finally, at the end of a phase, we have to figure out where in the tree to start the next phase. We now see how to do each of these things.

15.2.1   Updating the tree

Let's say we're in the phase that is adding the suffixes for character $s[i]$. At this point, the tree contains the strings for all the suffixes of $s[1, \ldots, i-1]$. Imagine putting a finger on the end of one of those suffixes (which we want to extend with $s[i]$). The suffix might end on a node or in the middle of an edge. We have the following cases based on where that finger is. It could be on:

A   a leaf node,
B   an internal node without a child starting with $s[i]$,
C   an internal node with a child whose label starts with $s[i]$,
D   an edge, and the next letter matches $s[i]$, or
E   an edge, and the next letter does not match $s[i]$.

For each of these cases, we describe how to update the tree and how to move to the next location in the tree.

Case A. This occurs when we would have (in the trie algorithm) added a new leaf child of a current leaf. Instead, we extend the label of the edge into the existing leaf.

$$(15.12)$$

Once a node is created as a leaf, it will always be a leaf: there is no way for the algorithm to add a child to any existing leaf. Let's consider how the labels of the edges leading to leaf nodes change over time.

First, what should the labels of the leaf edges be after the algorithm is finished? Since each leaf represents a suffix, the edge for each leaf u must be labeled by $(x_u, |T|)$ (where x_u is some position). But this is true in each intermediate phase as well: after phase i, the leaf nodes should all be labeled (x_i, i). Ukkonen's algorithm exploits this by changing how the leaf edges are updated.

On creation of a leaf node, instead of labeling the edge by (x_u, y_u), Ukkonen's algorithm labels it with (x_u, ρ), where ρ is a pointer to a global integer that tracks the length of the string we've processed so far. To update all the leaf labels, the algorithm needs to only increment the global value pointed to by ρ.

Cases B and C. These are the same as with the trie algorithm: if we want to add a new leaf hanging off a node (case B) with $s[i]$ we do so. If some edge out of the node already continues on with $s[i]$, then the suffix we want is already represented, and we're done.

(15.13)

In case B, the suffix link we follow is unchanged from what would be followed in the trie algorithm: if our finger pointed at node u (solid in (15.13)), we follow u's suffix link to find the next point in the tree (and, as with the trie algorithm, we keep a pointer to the most recently added node (with diagonal lines in (15.13)), to link up to the next created node with a suffix link).

In case C, our suffix was already in the tree. Hence, all smaller new suffixes are already in the tree, so we terminate the phase.

Case D. In this case, our finger points to a character in the middle of an edge, and the next character happens to be $s[i]$, the one we want to add. We don't need to do anything, and we can terminate the phase, because the current suffix (and all smaller ones) already exist in the tree (same as case C).

(15.14)

We don't need to follow or create any suffix links because no new nodes were created and the phase terminates.

Case E. If our finger points to the middle of an edge, and the edge string does not continue with $s[i]$, we split the edge, adding a new node in the middle, and a new leaf with an edge labeled $s[i]$ leading to it.

(15.15)

We have to do something a little more complicated to find the next location in the tree after a case E update. This is because our new internal node (diagonal lines) doesn't have a suffix link because we just created it. Instead, we follow the suffix link of the parent of our new internal node. This will take us to a new place in the tree.

$$(15.16)$$

We walk down from this node not by comparing characters but by hopping from node to node since we know how far down we have to go ($|\gamma|$ characters). By storing edge lengths, we can do this in time proportional to the number of nodes instead of the number of characters. This takes us to either a node or the middle of an edge, and we can apply either case A, case B, or case E (the other cases can't occur). We can also add the suffix links for the two new nodes (diagonal lines and shaded) that we added in the previous location: they will point to the nodes involved in the case we apply at our new location (e.g., if we split the edge, the suffix link from the shaded node will go to the new leaf node and the link from the node with diagonal lines will go to the new edge-splitting node).

15.2.2 Starting each phase

In each phase of the algorithm, we start by extending the deepest leaf. Hence, the first extension is always case A. Therefore, within a phase there is an initial run of case A, followed by cases where one or two nodes are created (cases B and E; collectively label these N) or a leaf is extended, terminated by a case where the algorithm can stop (cases C and D; collectively labeled X):

AAAAAANNNANANNNX.

In the next phase, the location of all the suffixes we need to extend are at the location of the leaves we visited or created in those previous steps. Each of the N cases creates a new leaf, and in the next phase, these will become leaf extension (A) cases. But we can handle all of those updates at once by incrementing our global ρ variable. Hence, we can start our walking of the tree for the new phase at the location we terminated the previous phase (step X).

We maintain a pointer b that is updated to point to the place where we stopped during the last phase. We start each phase at node b rather than the deepest leaf. We handle all the extensions between the deepest leaf and b by incrementing ρ to increment all their edge labels at once.

> **Algorithm 15.2.** Ukkonen's algorithm (high level).
>
> $\rho \leftarrow 1$
> Create the two-node tree that represents $s[1]$
> $b \leftarrow$ the root
> **for** $i = 2, \ldots, |T|$ **do**
>   Increment ρ ▷ *handles all leaf cases*
>
>   Determine the case to extend suffixes by $s[i]$ at location b
>
>   **if not** case C **and not** case D **then**
>   Update tree as required by the case
>
>   Set b to the new location by following a suffix link and hopping down if needed (case E)
>
>   Create the suffix links from any nodes you created to the corresponding nodes near b

15.2.3 Example run of Ukkonen's algorithm

Let's work through an example with $s = \dfrac{\texttt{abaabba\$}}{\texttt{12345678}}$. The first two steps use case B: our pointer b is at the root, and no outgoing edge is labeled with either $s[1]$ or $s[2]$.

$$ (15.17) $$

In phase 3, where $s[3] = a$, we *do* have an edge leaving our current node that starts with a. So this is case C, and we terminate without changing anything except our pointer to the current suffix.

$$ (15.18) $$

When $i = 4$, $s[i] = \texttt{a}$, but the edge that our pointer is on continues $\texttt{ba...}$ after our current location. So this edge does not continue with the character we want, and therefore we need to add a branching point by creating a new node (case E).

$$i = 4$$

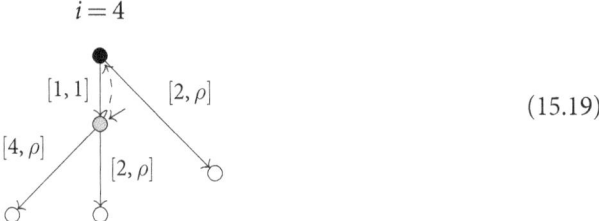

(15.19)

We add the suffix link (dashed arrow) from the new node to the root. Why the root? The parent of the new node is the root. Following the root's (imaginary) suffix link, we end up back at the root, and that's the target for the suffix link.

For $i = 5$, we're adding b, and we have an edge leaving the current node that starts with that letter, so we're in case C again.

$$i = 5$$

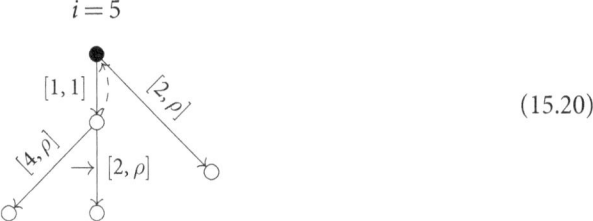

(15.20)

When $i = 6$, we have $s[i] = $ b. Our current pointer points to suffix aba since it points to one character past the start of $[2, \rho]$. The continuation of that suffix is a…, so we can't just advance the pointer along the edge. We're in case E and need to add a branch point.

$$i = 6$$

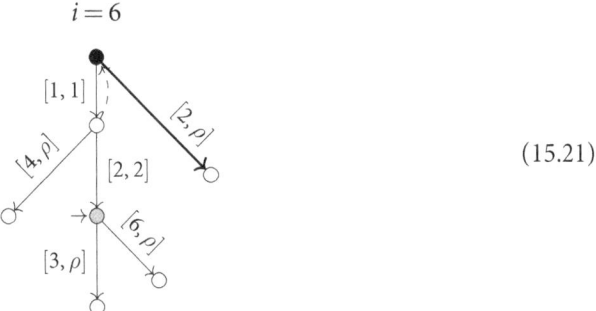

(15.21)

We're not done with $i = 6$ yet: we follow the suffix link for the parent of our new node (in this case ending up at the root) and then hop down the right number of characters (1 in this case). This takes us to the middle of the thick edge of (15.21). We can't extend, so we have to add a node, again using case E.

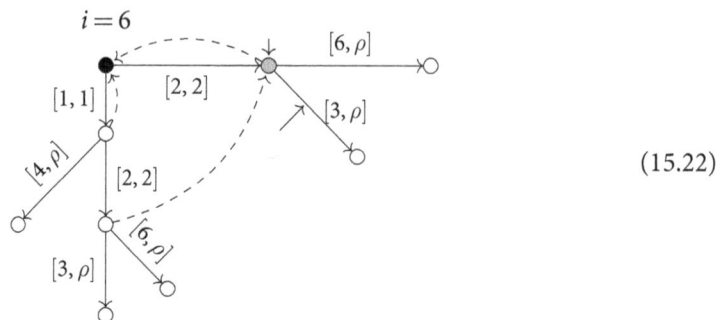

$$(15.22)$$

For $i = 7$, we're adding an a. Since $s[3] = $ a, we have an edge coming out of our current node that matches our current character. We advance our pointer to the middle of the edge and end using case C again. Our pointer ends up at the location of the arrow on the new $[3, \rho]$ edge.

Finally, at $i = 8$, we're adding a \$, which requires us to split the edge at this arrow using case E. We then apply case B twice following suffix links:

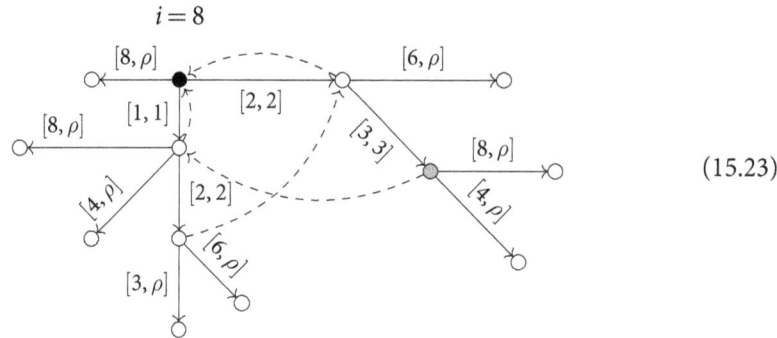

$$(15.23)$$

and we're done. See Exercise 15.2.

15.2.4 Running time of Ukkonen's algorithm

Theorem 15.1. *Ukkonen's algorithm takes time $O(|T| + hops)$, where hops is the number of downward hops for case E (as in (15.16)), taken over all iterations.*

Proof: There are $|T|$ total phases, each does a constant amount of work to increment the leaf-end value ρ to handle all the A-type extensions.

Each phase *starts* where the last phase stops (the X case), so there are at most $O(|T|)$ total explicit extensions:

$$
\begin{array}{c}
\text{AAAAAANNNANANNNX} \\
\text{NNNNANNAX} \\
\text{NNNNNNNNX}\,.
\end{array}
$$

$$(15.24)$$

In the diagram (15.24), each of the rows represents a phase of the algorithm (each of which starts where the previous phase ends), and each letter represents some operation on the tree (A is a leaf extension, N is a node creation case, and X is a stopping case). Each letter represents the extension of a suffix starting at a unique point in T, and each suffix starting point is associated with at most 2 letters. Hence, there are at most $2|T|$ letters in the complete trace of the running of the algorithm, and therefore at most $2|T|$ total explicit extensions.

Each of the N-type extensions takes $O(1 + hops')$ work to create one or two nodes and follow a suffix link, followed by some number $hops'$ of hops. The X-type case takes $O(1)$ time to recognize, so the total work for all these extensions is $O(|T| + hops)$. $\qquad\square$

Theorem 15.2. *The total number of downward hops (Theorem 15.1) during Ukkonen's algorithm is $O(|T|)$.*

Proof: Let ϕ be the current node depth at any point during the algorithm. Since phase $i + 1$ starts where phase i left off by using the b pointer, the node depth doesn't change at phase boundaries. It may decrease during a step when we follow a suffix link, and it increases when we hop down edges in case E.

At any type-N step, the node depth is first reduced by at most 2: 1 to walk back up to the node with the suffix link to follow, and 1 after following the suffix link, which reduces the node depth by at most 1 by Lemma 14.7, and then ϕ is increased by the number of hops. ϕ starts at 0 and is changed by at least $hops_i - 2$ in iteration i. Summing these changes over all extension phases:

$$\phi_{final} \geq 0 + \sum_{\text{extension } i} (hops_i - 2) \geq hops - 4|T|, \qquad (15.25)$$

$hops_i$ is the number of downward hops we do in iteration i and the $4|T|$ comes from the fact that there are at most $2|T|$ extensions (i.e., there are at most $2|T|$ letters in the complete diagram (15.24)).

ϕ is always $\leq |T|$ since the node depth can't be greater than the length of T. Therefore,

$$|T| \geq \phi_{final} \geq hops - 4|T| \qquad (15.26)$$

and therefore $5|T| \geq hops$ and $hops$ is $O(|T|)$. $\qquad\square$

Corollary 15.3 (Ukkonen running time). Ukkonen's algorithm constructs a suffix tree for T in $O(|T|)$ time.

Proof: Follows directly from combining Theorems 15.1 and 15.2. $\qquad\square$

15.3   Summary and notes

McCreight [1976] gave the first linear-time algorithm for constructing suffix trees, though this algorithm worked right-to-left. An optimal algorithm for construction of suffix trees with large alphabets was given in Farach [1997]. Kurtz [1999] explored practical optimizations of suffix trees to reduce the constant hidden behind their linear size.

> ### Presentation Notes
>
> Mingfu Shao gave extensive comments on an early draft of this chapter; he also provided the worked example of Section 15.2.3. Our presentation was informed by Gusfield [1997], who gives details of several other suffix tree construction algorithms.

15.4   Exercises

15.1   Assume you are going to be provided a string S one character at a time over a network. Interspersed with these characters are query strings q_i that indicate that you should print "YES" if q_i occurs as a substring in the part of S received so far, and "NO" if not. In other words, you are receiving a (possibly infinite) stream of the form:

$$s_1 \; s_2 \; q_3 \; s_4 \; s_5 \; s_6 \; q_7 \; q_8 \; q_9 \; s_{10} \; \cdots$$

For example, we know $S = s_1 s_2 s_4 s_5 s_6 s_{10} \ldots$, and we've been asked 4 queries about various prefixes of S. Let n_j be the number of characters of S received up to and including item j in the stream.

Describe how to modify Ukkonen's algorithm so that all of the following simultaneously hold: (1) when a query q_i is received, its answer can be printed in $O(|q_i|)$ time; (2) the total space used up through item i is $O(n_i)$; and (3) the total time to process the stream up to item i is $O(n_i + \sum_{q_j : j \leq i} |q_j|)$; that is the total time is linear in the total number of characters (in both S and the queries) received up to point i.

Your answer should be no more than one paragraph.

15.2   Redraw the tree of (15.23) to show the string labels on each edge and verify that the tree is correct.

15.3   True or false: all edges labeled $[x, \rho]$ when Ukkonen's algorithm is finished were created in the phase for $s[x]$. If true, say why. If false, give a counterexample.

CHAPTER 16

Suffix Arrays

Suffix trees are flexible and useful data structures for many string problems. They have the nice property that, asymptotically, they take linear space in the string they contain. The suffix tree of T takes space $O(|T|)$, but the big-O notation hides a lot of overhead: we have to maintain node structures, with pointers to encode the edges, plus some numbers on each edge. When dealing with really large strings, even a factor of 10 bigger than T could be a significant problem. That raises the question whether we can get a lot of the benefits of suffix trees without all that overhead. That is the goal of suffix arrays [Manber and Myers, 1990, 1993].

16.1 Suffix tree operations in less space

Imagine that you write down all the suffixes of a string T of length t. The ith suffix is the one that begins at position i. Now imagine that you sort all these suffixes and then write down the starting position of each suffix in an array in their sorted order. This is the suffix array. For example, suppose $T = $ "banana$", and we've sorted the suffixes:

$$
\begin{array}{ll}
6 & \texttt{\$} \\
5 & \texttt{a\$} \\
3 & \texttt{ana\$} \\
1 & \texttt{anana\$} \\
0 & \texttt{banana\$} \\
4 & \texttt{na\$} \\
2 & \texttt{nana\$} \quad .
\end{array}
\tag{16.1}
$$

The numbers to the left are the indices of these suffixes. The *suffix array* is the array of these numbers.

$$
\boxed{6 \mid 5 \mid 3 \mid 1 \mid 0 \mid 4 \mid 2}
\tag{16.2}
$$

This array can be computed in a straightforward way by sorting the suffixes directly. This takes $O(t^2 \log t)$ time because each comparison of two suffixes takes $O(t)$ time. In fact, the suffix array can be constructed in $O(t)$ time, which we will see in Chapter 17.

Definition 16.1 (Suffix Array). The suffix array $SA(S)$ of a string S is a permutation of the integers $\{0, \ldots, |S| - 1\}$, where $SA(S)[i]$ is the starting position of the alphabetically ith suffix. That is, if we sort all the suffixes of S, $SA(S)[i]$ is the starting location of the suffix that occurs at position i in that sorted list. ∎

It's often handy to have an auxiliary array called the "LCP" array around. Each successive suffix in this order matches the previous one in some number of letters. (Maybe zero letters.) This is recorded in the longest common prefix array, or the LCP array. In this case we have:

$$
\begin{array}{l}
\text{suffix array:} \quad \boxed{6 \;|\; 5 \;|\; 3 \;|\; 1 \;|\; 0 \;|\; 4 \;|\; 2} \\[6pt]
\text{LCP:} \qquad\qquad\;\; \boxed{0 \;|\; 1 \;|\; 3 \;|\; 0 \;|\; 0 \;|\; 2}
\end{array}
\tag{16.3}
$$

where $LCP[1]$ gives the length of the longest common prefix between the first and second alphabetically ordered suffixes. In general, $LCP[i]$ gives the length of the longest common prefix between the $SA(S)[i-1]$ and $SA(S)[i]$ suffixes.

16.2 Suffix array ↔ suffix tree

Suffix arrays are very closely related to suffix trees. So closely in fact we can convert between the two in $O(t)$ time.

Suffix Tree → Suffix Array. The suffix array can be computed from the suffix tree by doing an in-order traversal of the tree, where in-order means that we visit children in lexicographic (alphabetical) order. Here is the suffix tree for "`cattcat$`", where the edges leaving each node are sorted in alphabetical order based on the substring they represent.

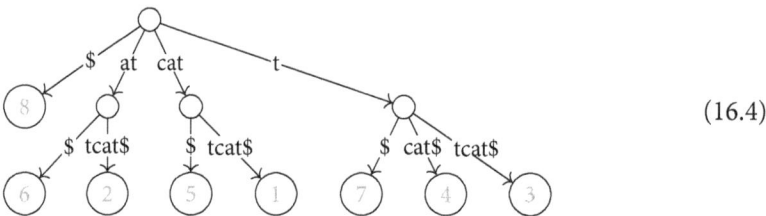

$$(16.4)$$

The numbers at each leaf give the starting position of the suffix represented by that leaf. Reading these left-to-right we obtain the following suffix array.

$$
\begin{array}{ll}
8 & \text{\$} \\
6 & \text{at\$} \\
2 & \text{attcat\$} \\
5 & \text{cat\$} \\
1 & \text{cattcat\$} \\
7 & \text{t\$} \\
4 & \text{tcat\$} \\
3 & \text{ttcat\$}
\end{array}
\tag{16.5}
$$

As we traverse the suffix tree in order, we output the indices of the suffixes as we encounter them. The traversal will visit each suffix in alphabetical order—which is exactly the order we need for the suffix array. This takes $O(|T|)$ time.

Suffix array → suffix tree. We add the suffixes one at a time into a partially built suffix tree in the order that they appear in the suffix array. At any point in time, we keep track of the sequence of nodes on the path from the most recently added leaf to the root. To add the next suffix, we find where this suffix's path deviates from the current path. To do this, we use the common prefix length value. We walk up the path until we pass this prefix length. This tells us where to add the new node.

A potential argument can be used to see that this process runs in linear time. Imagine a token on each of the edges on the path from the current leaf to the root. We use these tokens to pay for walking up the tree until we find the branch point where a new child is added. The tokens on the path pay for the steps we take up the tree. We'll need a new token for the edge that connects to the new leaf. We may also need another token in case we have to split an edge. So in all, at most two new tokens are needed to pay for the work. This proves that the running time is linear to construct a suffix tree from the suffix array (with the LCP array).

16.3 Searching for substrings using the suffix array

Consider the standard string search problem: we have the suffix array A for a string T, and we want to find where pattern P occurs in A.

We can do this in $O(|P| \log |T|)$ time using binary search with the suffix array: Maintain a range $[U, D]$ of candidate positions in the array; initially the range is the entire suffix array. Let $M(U, D)$ be the midpoint of that range. Repeatedly check whether P comes before or after the suffix at position $M(U, D)$ of the array, and update U and D accordingly. This takes $O(|P| \log |T|)$ since we're doing a binary search over $|T|$ elements, and each comparison of P against the string at suffix $M(U, D)$ takes $O(|P|)$ time.

How do we find *all* the occurrences of P in T? The thing to note is that these will all be adjacent in the suffix array since all the suffixes that start with P obviously start with the same sequence. We can find the range that starts with P in two ways: The first way: we could do two binary searches: one that takes the "left" range when there is a tie—which will find the start of the range—and one that takes the "right" range when there is a tie—which will find the end of the range. These searches will give you the range which contains suffixes starting with P. The second way: using the LCP array, we can walk left and right from a suffix that starts with P, continuing as long as the LCP is $\geq |P|$.

$O(|P| \log |T|)$ time to search is slower than using suffix trees, although not too much slower: we're often happy to pay a log-factor slowdown to have a much smaller data structure. In fact, however, we can do better. Next, we will see an $O(|P| + \log |T|)$ time algorithm that is almost as good as searching in a suffix tree.

16.4 Faster search

The simple binary search scheme described in Section 16.2 to find a pattern P in a text T did not use the fact that the things we were comparing were related strings rather than arbitrary entries in an array. Using this fact, one can get an $O(|P| + \log|T|)$-time search algorithm, which was presented in the original suffix array paper [Manber and Myers, 1990].

Throughout the following, we will conflate the string starting at position X with the suffix number X and also the position of X in the suffix array.

To explain the algorithm, we need to define $lcp(X, Y)$ as the length of the longest common prefix between suffix X and suffix Y. Our LCP array gives this value directly when $Y = X + 1$. Note that $lcp(X, Y)$ is exactly the depth of the LCA of suffix X and Y in the suffix tree. For now, assume we have easy access to these values.

Using the $lcp(X, Y)$ values, we can avoid repeatedly comparing characters of P during our binary search. Recall that our binary search is maintaining a range $[U, D]$. Now we also maintain u and d:

$$u = \text{length of the longest prefix of } U \text{ that matches a prefix of } P$$
$$d = \text{length of the longest prefix of } D \text{ that matches a prefix of } P$$

In a picture, if $P = $ "aaattwd", we might be in a situation like (16.6).

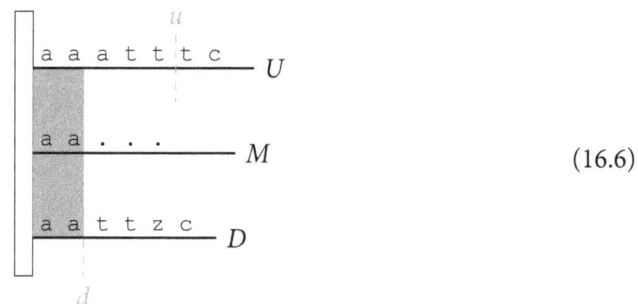

(16.6)

At the start, these can be computed by directly comparing P to the first and last suffix in the suffix array (which takes at most $O(|P|)$ time). Our binary search will now update u and d as well as U and D.

Suppose $u > d$ (the case where $u < d$ will be symmetric). Let M be the midpoint of the range $[U, D]$. We have three cases that we can use to update the binary search range quickly:

Case #1: $lcp(U, M) > u = lcp(U, P)$. Then the situation looks like (16.7),

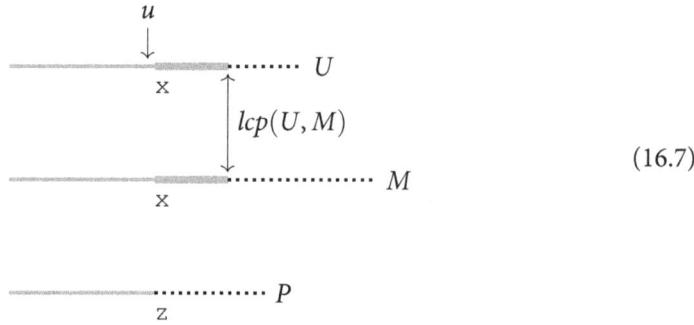

(16.7)

and we know that P comes after M since everything between M and U has something different than what is in P after position u. We need 0 new character comparisons to discover this. We set $U = M$, and d and u are unchanged.

Case #2: $lcp(U, M) < u = lcp(U, P)$. Then the situation looks like this:

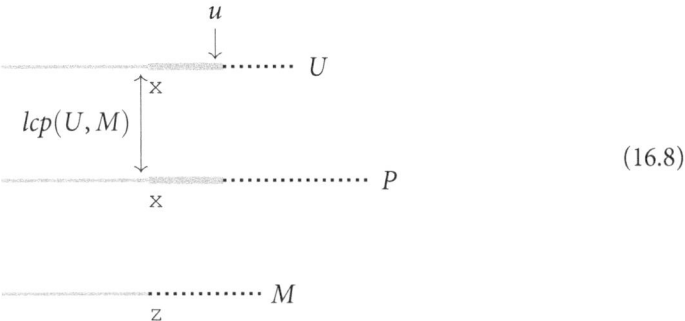

$$(16.8)$$

and we know that P must be before M since it agrees with U more than M. We need 0 new character comparisons to discover this. We set D to M. u is unchanged, and $d \leftarrow lcp(U, M)$.

Case #3: $lcp(U, M) = u = lcp(U, P)$. This looks like (16.9).

$$u = lcp(U, M)$$
$$\downarrow \cdots\cdots U$$
$$a$$
$$\cdots\cdots P$$
$$b$$
$$\cdots\cdots M$$
$$c$$

$$(16.9)$$

Here, we have no information about where P should go, but we know we can start comparing at position $u + 1$ in P. We do that until we see whether P is before or after M.

When $u < d$, we have cases 1, 2, and 3, but with the roles of u and d and U and D swapped. When $u = d$, we don't have any information about whether P is more like U or D, so we have one more case:

Case #0: $u = d$. Compare P to M starting from position $u + 1 = d + 1$ until you know whether P is before or after M.

To restate and summarize, we have the following algorithm:

Algorithm 16.1. Faster suffix array search.

1. Set the range $[U, D]$ to be the entire suffix array.
2. Compute u and d directly by looking at the first and last suffixes of the array.
3. Repeatedly do the following:
   - If $u = d$, apply case 0
   - If $u > d$, apply case 1, 2, 3 as appropriate
   - If $u < d$, apply case 1, 2, 3 as appropriate, but using D and d in place of U and u.

Theorem 16.2. *Assuming the required* $lcp(X, Y)$ *values can each be obtained in* $O(1)$*, the faster suffix array search algorithm 16.1 finds the location of an occurrence of P in* $O(|P| + \log|T|)$ *time.*

Proof: Only cases #0 and #3 actually compare any characters, and they always start comparing at $\max\{u, d\}$. If they match k characters of P, then at least one of u or d will be incremented by k, and we will never look at those characters again. Thus, *matching* characters takes at most $O(|P|)$ time.

The *mismatching* characters might be compared more than once, but there is only one mismatch per iteration, since we stop comparing as soon as there is a mismatch. Since there are still $O(\log|T|)$ iterations in the binary search, we spend at most $O(\log|T|)$ time comparing mismatching characters.

This leads to an overall running time of $O(|P| + \log|T|)$, nearly matching the running time of suffix trees. (The additive $O(\log|T|)$ factor can be removed with a more sophisticated algorithm and some additional data structures [Abouelhoda et al., 2004].)   □

16.4.1   Obtaining the $lcp(i, j)$ values

We can relate the $lcp(i, j)$ for arbitrary i and j to the LCP between suffixes that are adjacent in the suffix array:

Theorem 16.3. $lcp(i, j) = \min_{k=i,...,j-1} lcp(k, k+1).$

Proof: $lcp(k, k+1) \geq lcp(i, j)$ for all k in the range of the min because everything in this range shares the same $lcp(i, j)$ prefix at least.

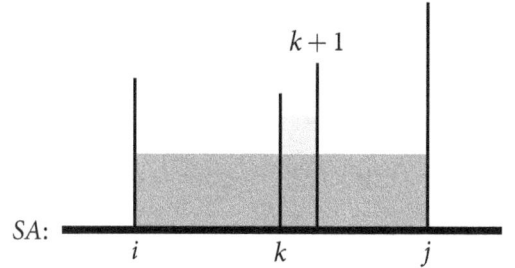

(16.10)

Every string in this range starts with the same $lcp(i, j)$ characters. Adjacent strings might share a longer prefix.

Let k^* be a k that achieves the minimum in the theorem. We therefore have $lcp(k^*, k^* + 1) \geq lcp(i, j)$ since the previous discussion holds for any k. Other consecutive pairs can have larger lcp, but not smaller by the minimality of $lcp(k^*, k^* + 1)$, so the situation looks like (16.11).

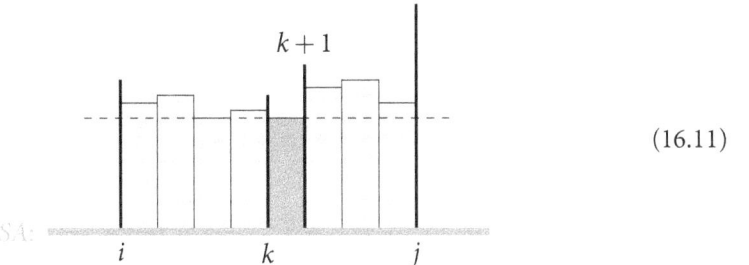

(16.11)

Therefore, by transitivity, $lcp(i,j) = lcp(k^*, k^* + 1)$, proving the theorem. □

The values $lcp(k, k + 1)$ are the values in our LCP array stored with the suffix array. Based on Theorem 16.3, we need a way to find the minimum over ranges in a list of integers. That is the RANGE MIN QUERY problem: given a list of integers, quickly find the minimum value in a query range $[i, j]$. This is a general problem with several solutions. But we can take a simpler approach here.

The first idea is that even though it looks like there are $O(|T|^2)$ relevant $lcp(i, j)$ values, in fact this is not the case. The only values that are relevant are the ones where i and j are the possible end points of a range encountered during a binary search. There are $O(n)$ of these for a binary search on a length-n array. (See Exercise 16.7.) You can see this visually if we plot the various ranges that $[U, D]$ could take on during a binary search:

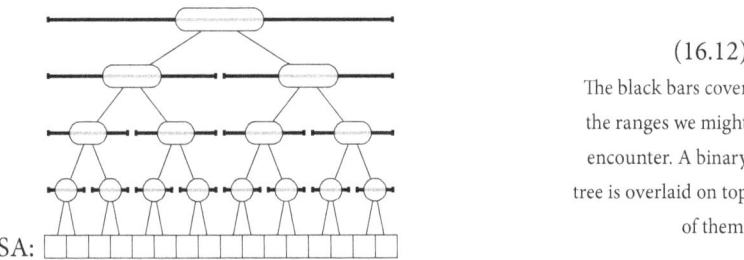

(16.12)

The black bars cover the ranges we might encounter. A binary tree is overlaid on top of them.

A binary tree is overlaid on top of these ranges in Figure (16.12); it has $O(|T|)$ nodes. The binary search will explore some root-to-leaf path of this tree. If we filled in each node with the min of the corresponding ranges, we would have the necessary lcp values at the ready. We can do this by filling in the minimums from the bottom to the top. The minimum for the ranges at the lowest level are the minimum of two suffix array values, and then—this is key—the minimum of a range at higher levels is the minimum of the values in its children. Therefore, with an $O(|T|)$ tree traversal, doing constant work (taking the minimum of two numbers) at each node, we can precompute all the lcp values we will ever need for any query.

We don't need to store (or construct) the tree, since it's always a complete binary tree. We can "traverse" it implicitly, and we can store values in an array. This adds $O(|T|)$ space to our suffix array data structure, but it's still only three lists of $O(|T|)$ integers.

16.5 Summary and notes

Suffix arrays let you search just as fast as suffix trees, but with much less overhead. We didn't see the fastest search algorithm, but the one we presented was pretty fast (and is the algorithm given when suffix arrays were introduced [Manber and Myers, 1990]). The proof of Theorem 16.3 and the discussion that followed it highlighted even more the correspondence between suffix arrays and trees: any range $[i, j]$ corresponds to a node of string depth $lcp(i, j)$ in the tree. We can think of the suffix array as storing the leaf order of the suffix tree and implicitly representing those internal nodes using ranges.

We also saw an $O(|T|)$ algorithm to construct the suffix array: use an $O(|T|)$ algorithm to construct the suffix tree, and then covert that into a suffix array. You can also obtain the LCP vector during this traversal (see Exercise 16.2). But this isn't too satisfying, since the reason we're exploring suffix arrays is to avoid creating the suffix tree. In the next chapter, we'll see a fast algorithm for constructing the suffix array directly.

> **Presentation Notes**
>
> Our presentation of the faster search algorithm was also informed by Gusfield [1997] in addition to its original description.

16.6 Exercises

16.1   Show the suffix array for `fiddledeedee$`.

16.2   Explain how to modify the suffix tree → suffix array algorithm to also output the LCP array without affecting its asymptotic runtime.

16.3   Explain how to initialize u and d in the search algorithm of Section 16.4.

16.4   Let S be a string of length n over an alphabet Σ. Suppose you are given the suffix array A for S and an array $C[a]$ that says how many occurrences of each $a \in \Sigma$ there are in S. Sketch an algorithm to reconstruct S from A and C, assuming you are *not* given S. Your answer should be fewer than 3 sentences.

16.5   Let $\Sigma = \{0, 1\}$ and consider binary strings of length n over Σ. Since there are only 2^n such strings but $n!$ permutations of $\{0, \ldots, n - 1\}$, there must be many such permutations that are not valid suffix arrays.

Of course, any suffix array must start with n (the "$" suffix; assuming 0-based indexing). Give a general construction (one that can work for any large enough n) to create a permutation π of $\{0, \ldots, n\}$ that starts with n but that cannot be a suffix array of any length-n binary string.

(Hint: Let $a_1, \ldots, a_k$ be the contiguous entries of a suffix array corresponding to suffixes starting with 0. What can you say about the location of the numbers $a_1 + 1, \ldots, a_k + 1$ in the same suffix array?)

16.6   Let A be a suffix array and let iSA be the inverse suffix array for A (meaning $iSA[j]$ is the index into A where suffix j is). Recall that $LCP[i]$ is the length of longest shared prefix of ith and $i + 1$st entries in A.

(a) Prove the following:

$$LCP[iSA[i + 1] - 1] \geq LCP[iSA[i] - 1] - 1 \qquad (16.13)$$

(b) Use the relation in (a) to give an $O(|A|)$-time algorithm to compute the LCP array from A.

16.7   Prove that there are at most $O(n)$ ranges that are possible to be considered during a binary search on an n-element array.

16.8   Let S be a string, let s_i denote the suffix of S starting at position i, and let A be the suffix array of S without adding any $\$$ to the end of the string. So, A is the sorted list of suffixes of S.

Define $pred_A(s_i)$ to be the suffix that occurs right before suffix s_i in the suffix array A. In other words, if $A[j] = i$, then $pred_A(s_i) = s_{A[j-1]}$.

Suppose s_t is the alphabetically first suffix (so that $A[0] = t$). Argue that for all k with $0 < k \leq t$,

$$LCP\left(pred_A(s_{t-k}), s_{t-k}\right) \leq k, \qquad (16.14)$$

where $LCP(x, y)$ is the longest common prefix of the strings x and y.

16.9   Let A be the suffix array for a string S. Suppose you are able to efficiently compute longest common extensions (Section 14.2) $LCE(i, j)$ for any pair of positions i, j in S. Give an expression for $LCP[k]$ (for any k) in terms of LCE.

16.10   Let A be the suffix array for a string S. Give an expression using the longest common extension (LCE) that computes in $O(1)$ time the minimum value of $A[i + 1, \ldots, j]$ for any indices i and j, assuming you can answer LCE queries in constant time.

CHAPTER 17

Suffix Array Construction

We'd like to be able to construct a suffix array quickly without the overhead of creating a suffix tree as an intermediate step. The direct "sort all the suffixes" approach takes $O(|T|^2 \log |T|)$ time, which is too slow. The key idea to improve on this direct algorithm is to use the fact that we are not sorting a collection of arbitrary strings, but rather strings that are related by the fact that they are all suffixes of the same string. There are a number of ways to take advantage of this so that you can do the string comparisons required in the sorting in less than $O(|T|)$ time. In fact, it's possible to construct the array directly in $O(|T|)$ time. We will first see a simpler $O(|T| \log |T|)$ algorithm.

17.1 An $O(|T| \log |T|)$ algorithm for suffix array construction

The first idea behind this algorithm is that if we could sort the suffixes by their prefixes of length 2, then we would be making progress. The second idea reduces the problem of sorting by longer and longer prefixes to the problem of sorting by things of length 2. This was the first algorithm proposed for creating suffix arrays [Manber and Myers, 1990]. It uses ideas from the Karp-Miller-Rosenberg algorithm (which we won't describe [Karp et al., 1972]).

Let's start with the first step of the algorithm: given a string $s = a_1 a_2 a_3 \ldots$ we sort its suffixes using only their first 2 characters as a key. This can be done by creating and sorting an array of triples:

$$\{(a_i, a_{i+1}, i)\}.$$

We pretend the string is padded at the end with an infinite string of $ characters.

Let's keep a running example of $s = $ "`cattcat$`". Sorting the suffixes just by their first 2 characters gives us the following (don't worry about the numbers to the left yet).

$$
\begin{array}{lll}
0 & \$\$ & \\
1 & \text{at} & \text{tcat\$} \\
1 & \text{at} & \$ \\
2 & \text{ca} & \text{ttcat\$} \\
2 & \text{ca} & \text{t\$} \\
3 & \text{t\$} & \\
4 & \text{tc} & \text{at\$} \\
5 & \text{tt} & \text{cat\$}
\end{array}
\qquad (17.1)
$$

The next insights are that (1) there are at most $|T|$ different 2-character prefixes in the string, and (2) we can use their sorted order as a *code* for 2-character groups in the string. The numbers to the left in (17.1) give this code (which can be computed by walking down the sorted list comparing 2-character prefixes). Using this code, we can re-write the string.

$$a = \text{``cattcat\$''} \tag{17.2}$$

$$a' = \text{``21542130''} \tag{17.3}$$

Comparing tuples (a'_i, a'_{i+2}) in this coded string is the same as comparing prefixes of length 4 in the original string since the coding obeys the same order as the 2-character groups. For example, the pair $(5, 2)$ represents the string "ttca".

So we repeat the process, now with a new "alphabet" of $\{0, \ldots, |T|\}$. On the next round, the "super-characters" will represent 4 original characters, and then 8 and so on, doubling each time until we are sorting by all $|T|$ characters on the $O(\log|T|)$th round.

There are $O(\log|T|)$ rounds, by the end of which we are sorting by the entire string. At each round we do linear work to create the code plus $O(|T|\log|T|)$ work to sort (each comparison in the sort must only compare tuples of 2 super-characters, which takes constant time). That gives us an $O(|T|\log^2|T|)$ algorithm.

17.1.1   Radix sort

We can reduce this to $O(|T|\log|T|)$ by using a radix sort. Once we are sorting "coded" strings, we know the alphabet is the set $\{0, \ldots, |T|\}$ and each item that we are sorting is a 3-tuple of (a_i, a_{i+r}, i) where r doubles each round. Sorting items with 3 "digits" in "base" $|T|$ can be done in $O(|T|)$ time using radix sort, removing one of the log factors.

Radix sort puts the items into buckets based on least-significant digit, flattens the list, keeping the ordering in each bucket, and repeats with the next-most significant digit, etc. Let's look at a general example of sorting a list of 3-digit items:

$$100, 123, 042, 333, 847, 892, 236.$$

In round 1, we group by the least-significant digit:

    [100] [042, 892] [123, 333] [236] [847].

In round 2, we scan through this list, putting items into new buckets based on their middle digit, keeping the order from the first bucketing:

    [100] [123] [333, 236] [042, 847] [892].

We repeat for the remaining digit:

    [042] [100, 123] [236] [333] [847, 892]

obtaining the completely sorted list.

Each pass through the data looks at each item, taking $O(n)$ time for a list of n items. Processing each item takes $O(1)$ time, since we can place it directly into the correct bucket. There are a total of $O(d)$ passes when the items have d digits.

For this suffix array algorithm, we have at most $O(|T|)$ different buckets (since each a_j value is in $\{0, \ldots, |T| - 1\}$), and we have to do only 2 rounds since each item has only 2 relevant "digits": a_i, a_{i+r} for some r. Therefore, we can sort in $O(|T|)$ time.

This leads to a total running time of $O(|T| \times \log |T|)$ for this suffix array construction algorithm, since we have $O(\log |T|)$ rounds of considering longer and longer suffixes, but now constructing the code and sorting the 2-digit coded prefixes takes $O(|T|)$ time each round.

17.2   A linear-time construction algorithm

We can take this "coding" idea one level further to obtain an $O(|T|)$-time algorithm to create the suffix array. There are a few linear-time suffix array construction algorithms. The one we will see is due to Kärkkäinen and Sanders [2003].

First, we ensure the text length is equal to 1 (mod 3) after padding with at least 1 special termination character $\$$. Assume string indices start at 0. We divide the suffixes conceptually into 3 groups:

- Group 0: Suffixes starting at positions $i = 0, 3, 6, 9, \cdots = (i \bmod 3 = 0)$
- Group 1: Suffixes starting at positions $i = 1, 4, 7, 10, \cdots = (i \bmod 3 = 1)$
- Group 2: Suffixes starting at positions $i = 2, 5, 8, 11, \cdots = (i \bmod 3 = 2)$.

This gives us the following groupings for "mississippi$", for example.

$$
\begin{array}{l}
\texttt{m i s s i s s i p p i \$ \$}
\end{array}
\tag{17.4}
$$

The basic outline of the algorithm is to recursively handle suffixes from the $i \bmod 3 = 1$ ($\_\_$) and $i \bmod 3 = 2$ ($\ldots$) groups and then merge the $i \bmod 3 = 0$ ($\_\_$) group after each recursion. We now describe the steps taken by the algorithm, which is called the "Skew Algorithm."

Step 1: create T'. We first create a new string T' that is the concatenation of $T[1 \ldots]$ (that is T with its first character removed) and $T[2 \ldots]$ (that is T with its first and second character removed). Suppose $T = \texttt{mississippi}$, then we have (17.5).

$$
T' = \boxed{\texttt{iss}} \boxed{\texttt{iss}} \boxed{\texttt{ipp}} \boxed{\texttt{i\$\$}} \boxed{\texttt{ssi}} \boxed{\texttt{ssi}} \boxed{\texttt{ppi}}
\tag{17.5}
$$

This puts the group-1 suffixes starting in the first part of T' and the group-2 suffixes starting in the second part of T. This at most doubles the size of T and takes $O(|T|)$ time. We conceptually divide T' into blocks of length 3, as shown in (17.5). Because of how we padded with $\$$, the final block of group-1 suffixes will contain a $\$$.

Step 2: encode T'. We then encode each group-1 or group-2 block of 3 characters using a new alphabet where if C_i and C_j are the codes for 3-blocks i and j then $C_i < C_j$ if and only if block i is lexicographically before block j (and $C_i = C_j$ if blocks i and j are the same 3 letters).

We can do this by sorting each of the 3-blocks using a radix sort (takes $O(|T|)$ time) and assigning the new code corresponding to the sorted order. This gives us a new coded string t.

$$T' = \boxed{\texttt{iss}\;\texttt{iss}\;\texttt{ipp}\;\texttt{i\$\$}\;\texttt{ssi}\;\texttt{ssi}\;\texttt{ppi}}$$

$$t = \quad \texttt{C}\quad \texttt{C}\quad \texttt{B}\quad \texttt{A}\quad \texttt{E}\quad \texttt{E}\quad \texttt{D}$$

(17.6)

Key Point #1: The lexicographical order of the suffixes of the coded string t is the same as the order of the group 1 and 2 suffixes of T. Why? Every suffix of t corresponds to some suffix of T (perhaps with some extra letters at the end of it—in this case the extra characters are "EED"). Because the tokens are sorted in the same order as the triples, the sort order of the suffix of t matches that of T. Therefore, we can recursively compute the suffix array for t to get the ordering of the group 1 and group 2 suffixes.

Step 3: recursively compute the suffix array for t. In the example for $\texttt{mississippi\$}$, we obtain the following suffix array A from the recursive call:

$$\begin{array}{ll}
3 & \texttt{AEED} \\
2 & \texttt{BAEED} \\
1 & \texttt{CBAEED} \\
0 & \texttt{CCBAEED} \\
6 & \texttt{D} \\
5 & \texttt{ED} \\
4 & \texttt{EED}
\end{array}$$

(17.7)

and $A = [3, 2, 1, 0, 6, 5, 4]$. Expanding the coding back, we would obtain a *partial* suffix array for T that only includes the suffixes in group 1 and group 2.

Step 4: create the inverse suffix array. For the next steps, we need to know the position of suffix i in the suffix array. This is easy to compute from A: We create a new array S where S_i is the position of i in the suffix array. If A was the full suffix array of T, we could do this with a single scan down A by setting $S_{A[i]} = i$. Because A is actually the partial suffix array of T', we have to do a little extra arithmetic to translate suffix numbers from T' to T and account for the missing suffixes. This can still be done in one pass down A. See Exercise 17.2.

Step 5: sort the group-0 suffixes. Group-0 suffixes are related to group-1 suffixes. Specifically, we can encode a group-0 suffix as the combination of a letter followed by a group-1 suffix. If $i = 0 \mod 3$ then suffix T_i can be represented by

$$(T[i], T[i+1, \ldots]).$$

(17.8)

Here, $T[i+1\ldots]$ is a group-1 suffix. This is a clever insight that we will use in later steps. We therefore can encode group-0 suffixes using

$$(T[i], S_{i+1}),$$

(17.9)

where S_{i+1} is the entry in the inverse suffix array S that we computed in the previous step corresponding to suffix $i + 1$, which is a group-1 suffix.

We can sort the group-0 suffixes using this encoding, again using a radix sort since they have only two digits. This gives us a sorted list L of the group-0 suffixes. This all takes $O(|T|)$ time.

Step 6: merge the group-0 suffixes back in. We have to add the group-0 suffixes into our partial suffix array that contains group-1 and group-2 suffixes. The way to do this is to run a list merge algorithm. You're likely familiar with the list merging done in (say) merge sort. We use that here with our two lists: the list A of group-1 and 2 suffixes and the list L of 0-suffixes, which by the previous steps are each sorted lists. Such a list merge takes $O(|T|)$ if we can compare the items in $O(1)$ time.

The challenge is how to compare an item from the group-0 list L with an item from the group-$\{1, 2\}$ list A. To do this, we use the clever idea about the relationship between the suffixes again.

To compare a group-0 suffix j with a group-1 suffix i, we can test whether

$$\underbrace{(T[i], S_{i+1})}_{\text{group 1 suffix}} < \underbrace{(T[j], S_{j+1})}_{\text{group 0 suffix}}? \tag{17.10}$$

Equation (17.10) is true if and only if the group-1 suffix is lexicographically before the group-0 suffix. To compare a group-0 suffix j with a group-2 suffix i, we can test whether:

$$\underbrace{(T[i], T[i+1], S_{i+2})}_{\text{group 2 suffix}} < \underbrace{(T[j], T[j+1], S_{j+2})}_{\text{group 0 suffix}}? \tag{17.11}$$

The reason for the particular encodings as 2- and 3-tuples is that in each case $S_{i+1}, S_{i+2}, S_{j+1}, S_{j+2}$ are either group-1 or group-2. Suppose $i = 1 \pmod 3$. Then the test we have to do is:

$$(T[i], \underbrace{S_{i+1}}_{i+1 \equiv 2 \bmod 3}) < (T[j], \underbrace{S_{j+1}}_{j+1 \equiv 1 \bmod 3})? \tag{17.12}$$

On the other hand if $i = 2 \pmod 3$, then the test we have to do is:

$$(T[i], T[i+1], \underbrace{S_{i+2}}_{i+2 \equiv 1 \bmod 3}) < (T[j], T[j+1], \underbrace{S_{j+2}}_{j+2 \equiv 2 \bmod 3})? \tag{17.13}$$

Since S_k gives the relative position of suffix k among the group-$\{1, 2\}$ suffixes, we can do the tests by comparing these tuples directly. In either case we are comparing tuples of at most 3 items. Each of these comparisons takes $O(1)$ time, and our list merge to merge A and L takes the total lengths of the lists we are merging which is $O(|T|)$. We now have a complete suffix array containing all the suffixes.

17.2.1 Running time

Theorem 17.1 (Skew algorithm running time). *The Skew algorithm takes $O(|T|)$ to create the suffix array for a string T.*

Proof: For a string of length n, the recurrence for the algorithm is:

$$T(n) = O(n) + T(2n/3), \tag{17.14}$$

where the first term is the time to sort and merge and the second term comes from the fact that the array in the recursive call is 2/3rds the size of the starting array.

So, we have $T(n) \leq cn + T(2n/3)$ for some c. Suppose we "guess" that $T(n) \leq 3cn$. Certainly, this is true when n is 1 for large enough c, so that takes care of the base case. We prove the general statement by induction, assuming it is true for all $i < n$. Then by the induction hypothesis (I.H.) we have:

$$T(n) \leq cn + 3c(2n/3) \qquad \text{by the I.H.} \tag{17.15}$$
$$= cn + 2cn \tag{17.16}$$
$$= 3cn. \tag{17.17}$$

$\square$

17.3   Summary and notes

We've seen a succession of more and more efficient algorithms for suffix array construction, ending up with a linear-time algorithm. The non-naïve algorithms use an encoding of the string that preserves some sorting information plus a linear-time sort algorithm, which is possible since our encodings are all a constant number of digits. The simpler algorithm of Section 17.1 is probably fine for all but the longest strings, since the extra $O(\log n)$ factor is likely not too bad. Puglisi et al. [2007] give a survey and synthesis of various suffix array construction algorithms.

Kasai's algorithm [Kasai et al., 2001] can be used to construct the LCP array for an already constructed suffix array in linear time.

> **Presentation Notes**
> _____
>
> Our presentation of the Skew algorithm follows its original description [Kärkkäinen and Sanders, 2003].

17.4   Exercises

17.1   Let Σ be a constant-sized alphabet. Describe how to sort m length-k strings over Σ in $O(m)$ time, assuming k is a constant. Your answer should not be more than 2 sentences.

17.2   In step 4 of the linear-time suffix array construction algorithm due to Kärkkäinen and Sanders (Section 17.2), the algorithm must be able to access the inverse suffix array. In particular, the algorithm requires a function $f(i_T) \to j_A$ that returns the location j_A in the partial suffix array A computed by the algorithm in this recursion of the suffix corresponding to the suffix starting at index i_T of *the string T that is input to this recursive call*. In order to compute f, you may want to precompute some values.

Give a careful pseudocode implementation of the function f and any pre-computation function *pre*. You may assume *pre* and f have access to any of the data that is available at step 4 of the algorithm. f should run in constant time and *pre* should run in at most $O(|A|)$ time.

17.3   In the Skew algorithm, why do we use 3 letters per block? Could a simpler algorithm work with 2 letters per block?

CHAPTER 18

Burrows-Wheeler Transform

We're going to see a different kind of operation on strings, called the Burrows-Wheeler Transform (BWT). The BWT has been useful in compression—that was its original motivation, and the `b` in `bzip2` stands for this—but it also has formed the basis of a number of modern string algorithms that can operate on text using a small amount of space. The BWT was invented in 1983 and published as a tech report in 1994 by Burrows and Wheeler [Burrows and Wheeler, 1994]. In 2009, the BWT gained popularity because of theoretical advances based on it (i.e., the FM-index; see Chapter 21) and its practical use in genome read aligners Bowtie [Langmead et al., 2009; Langmead and Salzberg, 2012] and BWA [Li and Durbin, 2009, 2010; Li, 2013].

We will see:

- How to compute the BWT of a string,
- How to invert the BWT to recover the original string,
- How to use the BWT to search a transformed string,
- A compression approach that works well with the BWT.

18.1 Definition of the BWT

Let S be a string of length n over an alphabet Σ. We will assume all of our strings end with a special, unique character \$. We assume \$ is "alphabetically" before any character in Σ. We will compute $bwt(S)$, which is another length-n string that is a permutation of the characters of S. We will first define the BWT as the output of a slow algorithm.

A string $y \circ x$ is a *cyclic rotation* of a string S if $S = x \circ y$, where x, y are strings and the $\circ$ operator indicates concatenation. For example, if $S = banana\$$, then $ana\$ban$ is a cyclic rotation of S (with $y = ana\$$ and $x = ban$). To compute the BWT (slowly), list all of the cyclic rotations of S, and then sort them lexicographically. Again, supposing $S = $ "`banana$`":

$$
\begin{array}{lll}
\text{banana\$} & & \text{\$banana} \\
\text{anana\$b} & & \text{a\$banan} \\
\text{nana\$ba} & & \text{ana\$ban} \\
\text{ana\$ban} & \longrightarrow & \text{anana\$b} \\
\text{na\$bana} & & \text{banana\$} \\
\text{a\$banan} & & \text{na\$bana} \\
\text{\$banana} & & \text{nana\$ba} \; .
\end{array} \tag{18.1}
$$

The matrix of characters on the right is the *BWT matrix*. Let M_S be this matrix for a string S. We define $bwt(S)$ to be the string of characters in the last column of M_S from top to bottom. In this example, $bwt(S) = annb\$aa$.

Definition 18.1 (BWT). The *BWT matrix* M_S of a string S has rows consisting of the sorted cyclic shifts of S. The *BWT* of a string S is the last column of the BWT matrix. Formally,

$$bwt(S) = M_S[0, n-1] \circ M_S[1, n-1] \circ \ldots \circ M_S[n-1, n-1]$$

where n is the length of S. ■

18.2   Why is this useful for compression?

Notice in (18.1) that the transformation tended to group equal characters together. There is intuition behind this, given by Burrows and Wheeler in their original paper. The *right suffix* of character s_i is the substring $s_{i+1}s_{i+2}\ldots s_{n-1}$. Consider the word "the." Every place "the" occurs, the right suffix of the "t" will start with $he\ldots$. The rows starting with these right suffixes will be near each other in the BWT matrix because the rows are sorted. For each of these, "t" will be the last character of the row, because the last character of each row comes before the start of its row in S since we are dealing with cyclic rotations. These "t" instances will be near each other in the BWT. Intuitively, strings that have regions of low complexity, with lots of equal characters near each other, are easier to compress (e.g., "aaaaabaaa" can be encoded as 5a1b3a or similar). We'll see (Section 18.5.2) a compression scheme that exploits this.

18.3   Recovering the original string from its BWT

This nice property wouldn't be that useful if we couldn't recover the original string. Luckily, (and sort of amazingly) we can recover the string from just the BWT of the string. We'll first see a computationally slow way to do this, and later see a faster way.

We will recover the string by recovering the BWT matrix. We can then recover the original string as the first row of this matrix (which will start with "$"). First, note that it is easy to recover the first column of M_S. It is simply the letters of S sorted alphabetically so this is the same as the letters of $bwt(S)$ sorted alphabetically. Let's call this first column F. Let's also call the last column (the BWT) L, with the entries of these strings denoted F_i, L_i. Because the matrix M_S consists of all the cyclic shifts, L_i *precedes* F_i in S for each i.

Denote by $L \cdot F$ the "vertical concatenation" of the L and F columns. That is, this creates a two-column matrix LF, with row i equal to $L_i \cdot F_i$. These two-character strings all occur in S. In particular, they each occur at the start of some cyclic rotation. We therefore can sort LF alphabetically to recover now the first *two* columns of M_S. We repeat, computing $sort(L \cdot LF)$ to recover the first 3 columns, and so on. $L = bwt(S)$ remains the correct last column of our growing matrix because we're always keeping the columns we have inferred so far sorted as they are in M_S. After $O(|S|)$ iterations, we will have recovered the entire matrix. Here's an example.

$$
\begin{array}{cccc}
\text{BWT} & & & \\
\$ & \$a & \$a & \$ap \\
a & ap & ap & app \\
e & e\$ & e\$ & e\$a \\
e & ee & ee & ee\$ \\
e \;-\; \substack{\text{Sort these 2}\\\text{columns}} \;\to\; & el \;-\; \substack{\text{Prepend}\\\text{BWT}} \;\to\; & el \;-\; \substack{\text{Sort these 3}\\\text{columns}} \;\to\; & ell \longrightarrow \cdots \\
l & le & le & lee \\
l & ll & ll & lle \\
p & pe & pe & pel \\
p & pp & pp & ppe \\
\uparrow & \uparrow & & \uparrow \\
\text{Sorted BWT} & \text{First 2 columns of matrix} & & \text{First 3 columns of matrix}
\end{array}
\qquad (18.2)
$$

This takes $O(|S|^3 \log|S|)$ time: $O(|S|)$ iterations of vertical concatenations, with $O(|S| \log|S|)$ comparisons to sort, each comparison taking $O(|S|)$ time. Obviously, this is inefficient. But it at least shows that the BWT of a string contains all of the information to reconstruct the string!

18.3.1 The LF mapping

Let's see a way to invert the BWT that is faster. This faster algorithm uses a very important property of the BWT matrix called the LF mapping property. This property gives a relationship between the order of characters in the last and first columns of the BWT matrix.

Theorem 18.2 (LF mapping). *For every character $c \in \Sigma$, the ith occurrence of c in L corresponds to the same character in S as the ith occurrence of c in F, for every i.*

Proof: Define the *right context* of a character F_i to be the $n-1$ remaining columns of the BWT matrix. In F, all occurrences of c are together in a single interval, and they are ordered by their right contexts. In L, the occurrences of c can be scattered in L, but they are also ordered by their right contexts since M_S contains cyclic rotations. Because the instances of c in F and L are sorted according to the same keys, they are in the same order. $\qquad \square$

Here's a figure illustrating the this proof.

$$
\begin{array}{cc}
\$dogwoo & \$dogwoo \\
d\$dogwo & d\$dogwo \\
dogwood & dogwood \\
gwood\$d & gwood\$d \\
od\$dogw & od\$dogw \\
ogwood\$ & ogwood\$ \\
ood\$dog & ood\$dog \\
wood\$do & wood\$do
\end{array}
\qquad (18.3)
$$

The boxes at left show the right contexts of the "o" characters. By the BWT matrix construction, the "o"s in this range are ordered by the strings in these boxes. At right, the boxes show the right contexts of the "o" characters in L. Now, the "right contexts" are at the left because of the cyclic rotation. The order of the "o"s in L are determined by these right contexts. Since the "o"s are sorted by the same "key" in both L and F, they must be in the same order. There are no ties in the ordering because of the $ character.

18.3.2 Faster inversion algorithm

The LF property lets us invert the BWT more quickly. We reconstruct S starting from *the end* of S. S ends with $, which is the first character of F. Which character comes just before $\$$? The first character of the BWT string (aka L). So, now we know S ends with $L_0\$$. This is general: the pairs (F_i, L_i) give predecessor relationships. That is, F_i is the character immediately following L_i in S. This follows again from the fact that we are dealing with cyclic rotations.

Continuing our example: We know the string ends with $L_0 F_0$, and now we want to find the predecessor to L_0. To do this, we need to find the correct occurrence of L_0 in F. If we find the row p that starts with the correct instance of L_0, the predecessor will be the character in the same row in L—that is, L_p. Here's where we use the LF mapping property: If L_0 is the ith occurrence of the character L_0 in L, then the corresponding character in F is the ith occurrence of the character L_0 in F.

How do we know how many instances of the character L_j occur before position j in L? That is, how do we know the i for character L_j? To formalize this question, define:

Definition 18.3 (rank). *rank*(j) is the number of occurrences of the character L_j that occur in positions $\leq j$ in L. ∎

How do we know *rank*(j)? We can precompute these by scanning L once, creating an $|S|$-long array *rank* where *rank*$[j]$ is the number of occurrences of character L_j before or at position j of L. *rank*(j) gives us the i we're looking for in F: we now need to find the ith occurrence of L_j in F.

The row of the ith occurrence of a character c in F is easy to find since F is just the characters of L sorted. We precompute an array C where $C[c]$ is the row at which the range of c characters starts in F. C can be computed by scanning L, counting the number of occurrences of each character. Then, the row for the ith occurrence of any character c in F can be computed as $C[c] + i - 1$.

This leads to the following algorithm to recover S from $bwt(S)$.

Algorithm 18.1. Recover string from BWT L.

Require: C and *rank* arrays.
  function INVERTBWT(L)
     $S \leftarrow$ "$\$$"
     $row \leftarrow 0$
     for $i \leftarrow 1 \ldots |L| - 1$ **do**
         $predecessor \leftarrow L[row]$ ▷ *char to prepend*
         $S \leftarrow predecessor \cdot S$ ▷ *prepend it*
         ▷ *find the right row in F*
         $row \leftarrow C[predecessor] + rank[row] - 1$

The algorithm only implicitly uses F—it doesn't construct the F string directly. This is possible because F has such a simple structure. This inversion algorithm runs in $O(|L|)$ time. The arrays *rank* and C can be computed in $O(|L|)$ time with a single scan of L, for a total running time of $O(|L|)$.

Here's an example in (18.4).

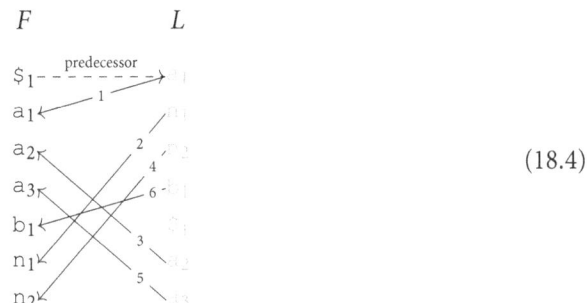

$$ (18.4) $$

The small numbers next to each character give the *rank* of the character. Following the arrows in the order of their numbers will spell out the original string. Each solid arrow connects the equal characters of equal ranks in L and F. For clarity, only the first predecessor arrow is shown (all these arrows go horizontally left to right).

There are some tricks that can be used to speed this up even more. First, it is possible to operate on a *compressed* version of L rather than L directly. Second, it is possible to reduce the size of the rank array via subsampling. These ideas were worked out by Ferragina and Manzini [2000], which we will see in Chapter 21.

18.4   Faster computation of the BWT

Our naïve algorithm to compute the BWT created the entire BWT matrix and took time $O(n^2 \log n)$ to sort n strings (each of the cyclic rotations), each comparison taking $O(n)$ time. In fact, we can compute the BWT from the suffix array of a string in $O(n)$ time. This is because the suffix array and BWT string are very closely related.

The suffix array is created (conceptually) by sorting all the suffixes of S. Look at the BWT matrix M_S but with, in each row, the characters after the "$" deleted. Each row in this new matrix represents a suffix of S, and they are still in sorted order because $ comes before every letter. The suffix array of S is the indices of the suffixes in the order that they appear in M_S.

Consider row i of M_S. The suffix number in this row is given by $SA[i]$, where SA is the suffix array of string S. Because of the cyclic shifts, the ith letter of the BWT is the one that comes just before this suffix: $S[SA[i]-1]$. So $bwt(S)$ can be computed by the following algorithm (assuming 1-indexing for everything).

Algorithm 18.2. Creating the BWT from the suffix array.

> **function** BWTFROMSA(S, SA)
> $bwt \leftarrow$ array of length $|S|$
>   **for** $i \leftarrow 1 \ldots |S|$ **do**
> $bwt[i] \leftarrow S[SA[i]-1]$ ▷ *use BWT $\leftrightarrow$ SA correspondence*
>   **return** bwt

Handling the $ character is a special case. Since the $ character appears at the end of the first suffix, when we look at the character "before" this suffix, we really mean the last character of the string (aka $). This is a minor change to the pseudocode in Algorithm 18.2. Here's an example in (18.5) (again assuming 1-indexing).

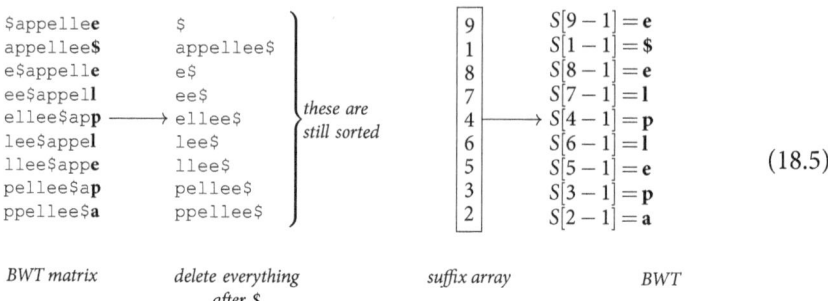

$$
\begin{array}{cccc}
\text{\$appelle\textbf{e}} & \text{\$} & 9 & S[9-1]=\textbf{e} \\
\text{appellee\textbf{\$}} & \text{appellee\$} & 1 & S[1-1]=\textbf{\$} \\
\text{e\$appell\textbf{e}} & \text{e\$} & 8 & S[8-1]=\textbf{e} \\
\text{ee\$appel\textbf{l}} & \text{ee\$} & 7 & S[7-1]=\textbf{l} \\
\text{ellee\$ap\textbf{p}} \longrightarrow & \text{ellee\$} & 4 \longrightarrow & S[4-1]=\textbf{p} \\
\text{lee\$appe\textbf{l}} & \text{lee\$} & 6 & S[6-1]=\textbf{l} \\
\text{llee\$app\textbf{e}} & \text{llee\$} & 5 & S[5-1]=\textbf{e} \\
\text{pellee\$a\textbf{p}} & \text{pellee\$} & 3 & S[3-1]=\textbf{p} \\
\text{ppellee\$\textbf{a}} & \text{ppellee\$} & 2 & S[2-1]=\textbf{a} \\
\end{array}
\qquad (18.5)
$$

BWT matrix   *delete everything*   *suffix array*   *BWT*
after $

This algorithm is just $O(|S|)$ array lookups once the suffix array is created. Our $O(|S|)$-time algorithm to create the suffix array (Chapter 17) gives an $O(|S|)$-time algorithm to create the BWT.

18.5 Applications of the BWT

18.5.1 Search

The BWT can be directly searched to answer the question, "does a string q occur in S?" This will require expanding our *rank* array to make the search process efficient, but first let's see the main idea.

The first idea is to search for $q = q_1, \ldots, q_m$ right-to-left. That is, we will first match the last character of q, then q_{m-1}, and so on. The reason to do this is the same reason that we invert $bwt(S)$ from last character to first character: because (F_i, L_i) gives a predecessor relationship. As with the inversion algorithm, we are going to be conceptually alternating between looking at L and F.

To start, we're looking for occurrences of q_m. We can find the range of these in F in the same way as before using the C array. Let $[u \ldots d]$ be the range of rows starting with character q_m. We then look at the same range in L. If q occurs in S, q_{m-1} occurs at least once in this range.

Suppose the rank of the first occurrence of q_{m-1} in this range is u' and the rank of the last occurrence of q_{m-1} in the range is d'. Look now at the range for q_{m-1} in F. The only instances of the characters that we care about in this range are those with rank between u' and d'. This is true again by the LF-mapping property: since the instances of q_{m-1} in this range are all those within an interval, they are also exactly the instances within some range in F. This range is easy to find: it's $[C[q_{m-1}] + u' - 1 \ldots C[q_{m-1}] + d' - 1]$.

We update u and d to be this range, and we just repeat, now looking for q_{m-2} in the new range in L, and so on. If we are able to continue this way through all of q without the range $[u, d]$ becoming empty, we know q occurs in S. If $[u, d]$ becomes empty, we report that q does not exist.

Here's an example searching for "abb" in the BWT of "abbbaba".

$$(18.6)$$

We do not have an easy way to find *where* q occurs. To do that, one must store extra data structures (such as the suffix array). We can however count the number of occurrences: it is the number of positions inside the final $[u, d]$ range.

One last detail: how do we efficiently find u' and d' in each iteration? We could linearly scan the *rank* array, since u' and d' are just the ranks of the first and last occurrence of the character within the current range. This is too slow, however. Instead, we expand our *rank* array to have $n \times |\Sigma|$ entries, where

$$Rank(i, c) = \text{\# occurrences of } c \text{ at positions} \leq i \text{ in } L.$$

Now, at a step when we are looking for character c, we have

$$u' = Rank(u - 1, c) + 1,$$
$$d' = Rank(d, c).$$

This new *Rank* array is somewhat larger than our old *rank*. Though if $|\Sigma| = O(1)$, it is still linear in the size of L. Using it, we can find the new range in $O(1)$ time, and thus our search procedure takes $O(|q|)$ time.

18.5.2 Compression

The BWT was originally motivated by compression, and Burrows and Wheeler give in their paper a specific implementation of an algorithm to compress a string that takes advantage of the BWT's reordering properties. Their algorithm is based on a general "move-to-front" (MTF) compression scheme, which was introduced by Bentley et al. [1986]. It turns out that the BWT is particularly suited to this kind of scheme, because MTF is adaptive: as the frequency of letters change, the code adapts, using fewer bits to encode more frequent characters.

Let's see this compression scheme. The best way to explain it is to think of the compressor C sending a string S to the decompressor D over a lossless, perfect channel. This channel might be a file on disk and the write and read steps might be separated in time indefinitely. In the MTF scheme, each of C and D maintain separate lists L_C and L_D of characters that they will update to keep in sync. To send a character x, C and D communicate in the following way:

Algorithm 18.3. Move-to-front compression *Send(x)*.

1. C finds x in the list L_C. If x is not in the list, append it to the end. Let i_x be the position of x in the list (1-based).
2. C sends i_x to D (using a variable number of bits—see (18.7) and the text following it). If C just added x to the list, C also sends x.
3. D receives i_x. If $i_x > |L_D|$, D reads a character x from the channel and appends it to L_D.
4. D prints the character at position i_x of L_D.
5. Both C and D move x to the front of their lists.

Since C and D add characters to the end of their lists at the same time, and reshuffle their lists in the same way, L_C and L_D are the same list of characters throughout the communication. This means that D will decode the message correctly.

Here's an example.

$$
\begin{array}{ll}
\Sigma & \\
 & \text{do\$oodwg} \\
\text{\$dgow}\dotsb\!\!>1 & \\
\text{d\$gow} & 13 \\
\text{od\$gw} & 132 \\
\text{\$odgw} & 1321 \\
\text{o\$dgw} & 13210 \\
\text{o\$dgw} & 132102 \\
\text{do\$gw} & 1321024 \\
\text{wdo\$g} & 13210244 = \text{MTF(``do\$oodwg'')}
\end{array}
\tag{18.7}
$$

In the example (18.7), the protocol is modified a little: when Σ is small, both C and D can start with the letters of Σ in some predetermined order.

What does this have to do with compression? Frequently used characters will end up near the front of the lists since each time they are used they are moved to the front. (See e.g., Borodin and El-Yaniv [1998] for more information about algorithms like MTF.) So, i_x in step 2 will generally be a small number for characters we have seen a lot recently. We take advantage of that by using a variable number of bits to send i_x: common characters will have small numbers that can be encoded in few bits. Rarer characters will have larger codes, but that's ok because they are rarer.

A specific way to do this is with a "prefix code" (an idea we will return to in Chapter 32). To send $i \geq 1$, we send $\lceil \log_2 i \rceil$ 0s, followed by the binary encoding of i using $\lfloor \log_2 i \rfloor + 1$ bits. This gives a codeword for i of length $1 + 2\lceil \log i \rceil$. The point of this is that the encoding is related to the magnitude of i: when i is small, this will use fewer bits than when i is big.

Theorem 18.4. *If b occurs n_b times in string S of length n, then the average cost to transmit b using the MTF procedure is no more than*

$$1 + 2\log(n/n_b). \tag{18.8}$$

Proof: Let p_i be the position of b in L_C when b is transmitted for the ith time ($i = 1, \ldots, n_b$). The average cost of sending b is then (ignoring floors and ceilings for convenience):

$$avgcost(b) = \frac{1}{n_b} \sum_{i=1}^{n_b} (2\log p_i + 1) = 1 + 2\sum_{i=1}^{n_b} \frac{1}{n_b} \log p_i. \tag{18.9}$$

Since log is a concave function, we have $\sum_{i=1}^{n_b} \frac{1}{n_b} \log p_i \le \log \sum_{i=1}^{n_b} \frac{p_i}{n_b}$. In addition, $\sum_{i=1}^{n_b} p_i \le n$ because to increment the position of b, some other character must be output and moved to the front. Combining with (18.9), we have

$$avgcost(b) \le 1 + 2\log \sum_{i=1}^{n_b} \frac{p_i}{n_b} \le 1 + 2\log(n/n_b). \tag{18.10}$$

$\square$

It also turns out that transmitting with the *optimum*, static prefix code must use at least $\log(n/n_b)$ bits for each character (which we will not prove). This implies:

Theorem 18.5. *Let $M(S)$ be the average number of bits per character used to compress S with MTF, and let $OPT(S)$ be the average number of bits per character used in an optimal, static prefix code. Then $M(S) \le 2OPT(S) + 1$.*

18.6   Summary and notes

The BWT is a very important transformation of the string. It is efficiently invertible, efficiently searchable without inverting it, and useful for compression. Fast, practical, memory-efficient approaches for BWT construction [e.g, Kärkkäinen, 2007] are important for genomics applications. BWT continues to be important for compression, and theoretical bounds on compression are known for some BWT-based schemes [Manzini, 2001]. The BWT is the basis of compressed, full-text indices that allow you to search a compressed version of the string without decompression. We'll see that the FM-index is one approach for that in Chapter 21.

The ideas in the FM-index and the BWT have been essential for achieving fast read mapping for biological sequences. Tools such as BWA [Li, 2013] and Bowtie [Langmead et al., 2009] use the BWT to achieve practical performance for mapping millions of short sequences onto a long, fixed reference. A long line of follow up tools extends these ideas.

Presentation Notes

Due to their importance, the BWT and related concepts are covered in many sources. One particularly approachable and clear source are Ben Langmead's lecture notes and teaching materials [Langmead, 2024, 2013]. Another resource that covers BWT and many extensions is Mäkinen et al. [2015].

18.7   Exercises

18.1   Compute the BWT for the following string: ceaselessness.

18.2   Give two distinct 9-letter words over the alphabet $\Sigma = \{a, b\}$ that are not cyclic shifts of each other but that have the same first 3 columns of the BWT matrix.

Hint: consider paths in the graph $G = (V, E)$ where $V = \Sigma^3$ and there is an edge from "xyz" to "yzw" for any $x, y, z, w \in \Sigma$.

18.3   The string "aaaaa$" is its own BWT if you ignore the $. Give a scheme, for any k, to form a string with this property that uses k different symbols.

CHAPTER 19

RRR Compressed Bit Vectors

We will talk now about bit vectors, which we can think of as strings over an alphabet of $\{0, 1\}$. Of course, bit vectors have a huge number of uses, including marking pages or items in memory that are "dirty" (have been modified since last saved), and so on. Bit vectors of length n can be thought of as equivalent to sets of items, where the universe of possible items is the set of numbers from $1 \ldots n$, and the 1 bits indicate which elements are in the set.

When treating a bit vector as a string, all of the algorithms that we have already discussed can apply. With bit vectors, a few additional operations are often useful. Bit vectors are often used when space is at a premium, so supporting the required operations while storing a vector of n logical bits in fewer than n bits of actual memory is often very helpful. In this chapter, we will see a specific data structure, introduced by Raman et al. [2007], that stores bit vectors in small space.

19.1 Rank and select operations

Which operations should we support for bit vectors? Obviously, we need to support bit access $S[i]$, meaning returning the value of the ith bit in bit vector S. We are going to consider *read-only* bit vectors, so we will not need to support the assignment operation $S[i] \leftarrow b$.

Rather than talk specifically about $S[i]$, we instead work with two more general kinds of operations:

- $\text{rank}_1(S, i) = $ the number of 1 bits at or before position i in S.
- $\text{select}_1(S, j) = $ the position of the jth 1 bit in S.
- $\text{rank}_0(S, i)$ and $\text{select}_0(S, j)$ are defined analogously for 0-value bits.

With these operations, one can compute $S[i]$, the value of the ith bit using

$$S[i] = \text{rank}_1(S, i) - \text{rank}_1(S, i - 1). \tag{19.1}$$

The two terms in (19.1) will be equal if $S[i]$ is 0; otherwise, the first will be 1 more than the second. The rank and select operations are inverses of each other:

$$\text{rank}_1(S, \text{select}_1(S, j)) = j. \tag{19.2}$$

Our main focus is going to be how to implement the rank operation—we'll see a bit about select, but it turns out to be a little more complex.

19.2   RRR bit vector data structure

The RRR [Raman et al., 2007] bit vector data structure stores a read-only bit vector S of length n by breaking it into blocks of $u = O(\log n)$ bits and storing several pieces of information for each block.

Basic block info. For each block i, we are going to conceptually store:

- $w_i =$ the number of 1-bits in block i. This is the "weight."
- $p_i =$ a pointer into block value tables (to be described next).
- $s_i =$ the number of bits used to represent p_i.

Block value tables. The RRR exploits the fact that there can only be so many different patterns of bits in each of the blocks. It further groups these patterns by how many 1 bits there are in each block (w_i).

The block value tables consist of a table for each possible value of w_i, which can range between 0 and u. Each row of the table corresponds to a particular pattern of w_i bits set to 1. Each row contain values of $\text{rank}_1(i)$ for the corresponding patterns of bits. A figure will help make this clear.

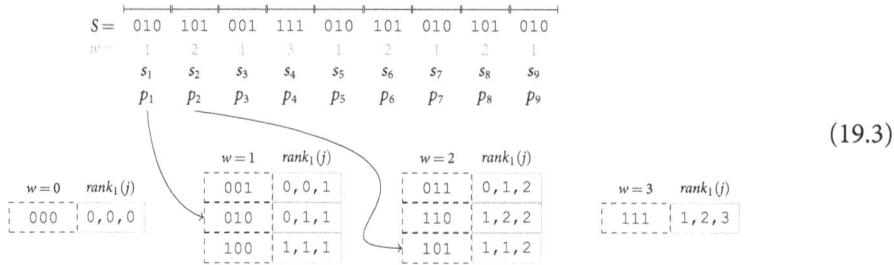

$$(19.3)$$

Here, $u = 3$. The dashed boxes correspond to possible bit patterns for the number of 1-bits given by the value of w. The solid boxes give the value of the rank for each bit for the bit vector in the adjacent dashed box. Let the table for w_i be called $T[w_i]$.

Using just the information we've described so far, one can already compute rank_1 for the *first* block: use w_1 to find the right table, use p_1 to find the right row in that table, and look at the proper entry in the list of ranks for that row. For the next set of values let's compute the rank_1 for subsequent blocks.

Prefix sums. We also store, for each block, the prefix sums of the w_i and s_i. That is

- $f_i =$ the sum of the w_j for $j < i$,
- $q_i =$ the sum of the s_j for $j < i$.

Why do we store f_i? Since we're going to want to compute rank_1, the f values give us the blockwise rank_1 already. The problem that the q values solve is that if we use a variable number of bits for each p_i, we need to figure out where each p_i starts within the stream of bits that represents the p values. The q values let us find the starting bit for each p_i.

Note. We will not actually store all these values for each block as (say) a list of integers. Rather, the RRR structure will store them more compactly. But for now, we should think about these values being available as if they were in arrays of integers.

19.3 Computing $\mathrm{rank}_1\,(S, i)$

Using the RRR data structure, we can compute $\mathrm{rank}_1(S, i)$ by finding the block $x = \lfloor i/u \rfloor$ that contains position i. Then

$$\mathrm{rank}_1(S, i) = f_x + T[w_x][p_x][i - xu], \tag{19.4}$$

where the notation $T[w_x][p_x][i - xu]$ indicates that we look at the p_x row in the $T[w_x]$ table (the table for weight w_x); in the rank vector in that row, we look at the entry at position $i - xu$. This gives the correct rank since the f_x gives the rank up to just before block x, and the rank vector in entry $T[w_x][p_x]$ gives the rank for each position in the block. Rank_0 can be computed simply by $\mathrm{rank}_0(S, i) = i - \mathrm{rank}_1(S, i)$.

19.4 Compactly encoding the RRR information

We will now see how to encode these values (w_i, p_i, s_i, f_i, q_i for each block i) compactly.

19.4.1 Prefix-sum data structure

One of the main tools we'll use is the following theorem (due to Tarjan and Yao [1979]; Pagh [1999]) and its corresponding data structure to store vectors in the RRR data structure. It shows that an ordered list of bounded integers that don't change very fast as you walk down the list can be compactly stored.

Theorem 19.1 (Tarjan and Yao; Pagh; simplified). *Let $z_1, \ldots, z_k$ be integers such that $|z_i| = n^{O(1)}$ and $|z_i - z_{i-1}| = O(\log n)$ for some positive integer n, then the ordered list $z_1, \ldots, z_k$ can be represented in*

$$O(k \log \log n) \tag{19.5}$$

bits allowing for constant-time access to any z_i.

Proof: Use the following representation, where we break the list of integers into blocks of size $O(\log n)$.

The first integer in each block is explicitly written—these are the "key frames." The other integers in each block are encoded as deltas from the first.

The "key frame" integers take at most $O(\frac{k}{\log n} \log n) = O(k)$ total bits for all of them since there are $O(k/\log n)$ of them, and each of their magnitudes is at most $n^{O(1)}$. An integer of value at most n^c for some c can be written using $O(\log n^c) = O(c \log n) = O(\log n)$ bits.

The integers encoded by the deltas, by the assumption of the theorem, all differ by at most $O(\log n)$ from the preceding integer. Therefore, the last integer in a block differs by at most $O((\log n) \times (\log n)) = O((\log n)^2)$ from the key frame integer, and this is the largest possible delta. Hence, we can encode each of the deltas in $O(\log (\log n)^2) = O(2 \log \log n)$ bits. There are approximately k deltas stored, so the total space to store the deltas is $O(k \log \log n)$ bits.

Since $k < k \log \log n$, our total space usage for both the deltas and the key frame integers is $O(k \log \log n)$.

Access time is $O(1)$ since for any position i, the key frame integer z_j can be found in $O(1)$-time, as can the value $z_i + z_j$. ☐

19.4.2   Space to store the vectors of integers

The RRR vector uses this representation to store the f_i values, where each f_i is the number of 1 bits before block i. Theorem 19.1 applies because, for all i:

- **Condition 1:** $f_i \leq n$ since there can be at most n 1-bits.
- **Condition 2:** $|f_{i+1} - f_i| = O(\log n)$ since $u = O(\log n)$.

In this setting, k (from the theorem) equals the number of blocks, which is $n/u = O(n/\log n)$, and therefore the prefix sum vector f can be represented in $O((n/\log n) \log \log n)$ bits, just plugging $k = O(n/\log n)$ into the theorem.

Let's look at the space usage for various parts of the RRR structure so far.

The f values: The f vector takes $O((n/\log n) \log \log n)$ space, as just described.

The w values: The w_i values are all $\leq u$ because there are at most u 1-bits in a block of length u. Therefore, each w_i takes $O(\log u) = O(\log \log n)$ bits. There are $O(n/\log n)$ of them, so the total space for listing out the w_i values would be $O((n/\log n) \log \log n)$, which is the same as we are using to store the f values. But, in fact, we don't need to store the w_i values themselves, since we can retrieve $w_i = f_{i+1} - f_i$.

The p values: The p_i values point into tables of various sizes. Each p_i uses s_i bits by definition, which differs depending on which table p_i points into. How big is each s_i? The table that p_i points into has a row for each u-long vector that has exactly w_i 1 bits. There are

$$\binom{u}{w_i} \tag{19.7}$$

such vectors possible. Therefore, p_i can specify the row in this table using $\left\lceil \log_2 \binom{u}{w_i} \right\rceil$ bits. Define

$$\mathcal{B}(w_i, u) = \left\lceil \log_2 \binom{u}{w_i} \right\rceil, \tag{19.8}$$

which is the number of bits needed to select a subset of w_i elements from a universe of u elements. Each p_i takes $\mathcal{B}(w_i, u)$ bits; that is $s_i = \mathcal{B}(w_i, u)$. Therefore, the total space for the p values is

$$\sum_i \mathcal{B}(w_i, u). \tag{19.9}$$

We will analyze this estimate in more detail soon.

The s and q values: Recall that the q_i values are the prefix sums of the s_i values. Each $s_i = \mathcal{B}(w_i, u)$ by the previous discussion. This is $\leq u$ since clearly one can specify a u-long bit vector using u bits. Also $q_i \leq n$ since it is at most the number of bits in the blocks up through block i. We also have $|q_i - q_{i-1}| = O(\log n)$, since the q prefix sum is increased by at most $|s_i| \leq u = O(\log n)$ after each block. Hence, both Condition 1 and 2 (See 19.4.2) of Theorem 19.1 are satisfied, and we can store the q values in $(n/\log n) \log \log n$ space.

We don't need to actually store the s_i values, since we store the q_i values. We can retrieve s_i via $q_{i+1} - q_i$. (Though, again, the s_i values also satisfy both of the conditions of Theorem 19.1; see Exercise 19.1.)

So far, we've used $O(n \log \log n / \log n)$ space to store the f and q values plus the space for the tables and the p values, that leads to the following lemma:

Lemma 19.2. *To store a bit vector of n bits, the RRR data structure that supports rank$_1$, takes $O(n \log \log n / \log n + \sum_i \mathcal{B}(w_i, \log n) + tables)$ bits, where tables is the number of bits to store the tables.*

Proof: The $n \log \log n / \log n$ term accounts for f and q. The term $\sum_i \mathcal{B}(w_i, \log n)$ accounts for the p values, and *tables* accounts for the last piece. $\qquad\square$

We now dive deeper into the table storage and the $\sum_i \mathcal{B}(w_i, \log n)$ terms to refine Lemma 19.2.

19.4.3 Space for the tables

The tables seem like they could really take a lot of space. But, in fact, they are small. Let's see how small they are.

First, we need only to store the solid-outlined parts of the tables in (19.3). The dashed parts of the table are never used. We have a table for each possible value of w (which ranges between 0 and u). The table for a particular value w has $\binom{u}{w}$ rows. Each row has u entries of rank values. Each rank value is a number that is at most w and so can be encoded in $\log w$ bits (ignoring ceilings). That means that the total number bits needed for all of the tables is:

$$\sum_{w=0}^{u} \binom{u}{w} u \log w \leq \sum_{w=0}^{u} \binom{u}{w} u \log u \quad \text{because } w \leq u \tag{19.10}$$

$$= u \log u \sum_{w=0}^{u} \binom{u}{w} \tag{19.11}$$

$$= u(\log u)2^u \quad \text{by the binomial theorem.} \tag{19.12}$$

Or another way to see this directly is that we have a row in *some* table for every one of the 2^u possible u-bit patterns. Each of those rows stores u numbers using at most $\log u$ bits per number.

We now refine our choice of u a bit. Since we want u to be $O(\log n)$, let's choose $u = \epsilon \log n$ for some $\epsilon < 1$. Then 2^u becomes n^ϵ. If, for example, we take $\epsilon = 1/2$, then the space is $O(\sqrt{n} \log n \log \log n)$ which is asymptotically less than $O(n \log \log n / \log n)$, so it "fits" in the first term of our goal of $O(n \log \log n / \log n + \sum_i \mathcal{B}(w_i, \log n) + tables)$ from Lemma 19.2. Therefore, the *tables* term disappears.

19.4.4 Space for the p values

Next, let's look at the $\sum_i \mathcal{B}(w_i, \log n)$ term. The space to store the p values is:

$$\sum_{i=1}^{m} \left\lceil \log_2 \binom{u}{w_i} \right\rceil, \tag{19.13}$$

where $m = n/u$ is the number of blocks. We can upper bound this by eliminating the ceiling:

$$\sum_{i=1}^{m} \left\lceil \log_2 \binom{u}{w_i} \right\rceil \leq m + \sum_{i=1}^{m} \log_2 \binom{u}{w_i}. \tag{19.14}$$

Using the fact that the sum of logs is equal to the log of the product (that is, $\sum_{i=1}^{m} \log_2 \binom{u}{w_i} = \log_2 \prod_{i=1}^{m} \binom{u}{w_i}$), we have:

$$\text{space for } ps \leq m + \sum_{i=1}^{m} \log_2 \binom{u}{w_i} = m + \log_2 \prod_{i=1}^{m} \binom{u}{w_i}. \tag{19.15}$$

We can rewrite the righthand side in (19.15) as something a little simpler using the following lemma.

Lemma 19.3. *We have the following inequality using m, u, w_i just defined:*

$$\prod_{i=1}^{m} \binom{u}{w_i} \leq \binom{\sum_{i=1}^{m} u}{\sum_{i=1}^{m} w_i} = \binom{n}{w}, \tag{19.16}$$

where n is the vector length and w is the number of 1 bits in the vector.

Proof: In the righthand side of the inequality, note that $\sum_{i=1}^{m} u = n$.

- The righthand side of the inequality is the number of ways to pick $\sum_i w_i = w$ objects from a universe of n things.
- The lefthand side is the number of ways to pick $\sum_i w_i$ objects from a universe of n things *where we must take w_i from each block.*

In pictures, the righthand side is the number of ways of picking w locations on this line of length n:

$$
\begin{array}{ccccc}
u & u & u & u & u \\
\end{array}
$$
$$
\begin{array}{ccccc}
w_0 & w_2 & w_4 & w_6 & w_8
\end{array}
\tag{19.17}
$$

while the lefthand side is the number of ways of picking w locations on this line while ensuring that exactly w_i items are picked in each block. The lefthand side is more constrained than the righthand side, so the number of ways of choosing objects counted on the lefthand side cannot be more than the number of ways of choosing the objects counted on the righthand side. $\qquad\square$

Using Lemma 19.3, we can continue (19.15):

$$
m + \log_2 \prod_{i=1}^{m} \binom{u}{w_i} \leq m + \log_2 \binom{n}{w} \qquad \text{by Lemma 19.3} \tag{19.18}
$$

$$
\leq m + \left\lceil \log_2 \binom{n}{w} \right\rceil = m + \mathcal{B}(w, n). \tag{19.19}
$$

This also uses the fact that the log function is monotonically nondecreasing.

Since $m = n/u = n/\log n$, we end up with the following:

Lemma 19.4. *The space to store the p values is $O(\mathcal{B}(w, n) + n/\log n)$, where $\mathcal{B}(w, n)$ is as defined earlier and here means the number of bits needed to specify which of the n-vectors containing w 1-bits is being stored.*

19.5 Supporting the select operation*

19.5.1 The select data structures

To support select_1 queries, we will need to add some additional data structures. The first set of which is to expand our tables of bit vectors (19.3) with an additional column that gives the value of select_1 for each position in every possible u-long bit pattern. In other words, we add a vector to each entry in this table, so that $T_{\text{select}}[w_x][p_x][j]$ is the position of the jth 1-bit in the weight w_x table for row p_x. This at most doubles the size of our tables.

To answer select queries, the challenge is to find the block that contains j. To do this, we first add a new array that lets us segment our blocks in an intelligent way. Let $m = n/u$

be the number of blocks, and define $v = \lceil (\log m)^2 \rceil$. We create an array C where $C[k]$ stores a pointer to the block that contains the $\text{select}_1(kv)$-th bit for every k.

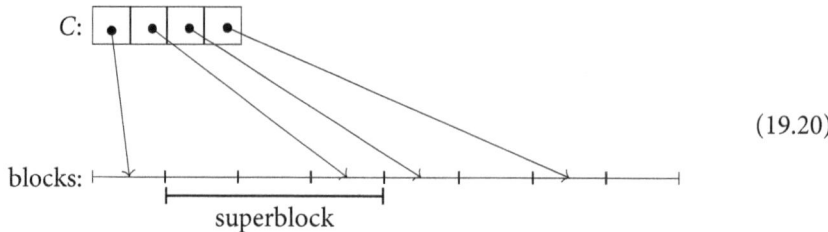

$$(19.20)$$

The entries of C partition the blocks into superblocks ($\vdash\!\!-\!\!\dashv$) of non-uniform number of bits: the ith superblock goes from block $C[i-1]+1$ to block $C[i]$. We call the number of bits covered by a superblock its *span*. Even though their spans may be different, each of these superblocks contains about $(\log m)^2$ 1 bits, by the definition of select_1 and how we constructed C (since each successive entry of C points to the block containing the subsequent $(\log m)^2$ 1 bit).

We will store some additional information for each superblock. The information we store depends on how big the superblock is. If the superblock is $\leq (\log m)^4$ bits, we call the block *dense*, since its run of $\leq (\log m)^4$ bits contains approximately $(\log m)^2$ 1 bits. Otherwise, we call the superblock *sparse*.

For sparse superblocks, we store in a simple array, in sorted order, the positions of the 1 bits contained in the span of this superblock. These are just the answers for the select_1 queries for this subset of 1 bits. Each of these positions takes $O(\log n)$ bits to store, and there are $(\log m)^2$ 1 bits in a sparse superblock, so the total space for this list is $O((\log m)^2 \log n)$ bits. But there are at most $O(n/(\log m)^4)$ sparse superblocks (since each of the sparse superblocks is at least $(\log m)^4$-bits long). Therefore, the total space to store all these arrays for all the sparse superblocks is $O(n/\log n)$ (see Exercise 19.4).

For each dense superblock, we have to do something more involved. We create a complete tree with branching factor $\sqrt{\log m}$ with a leaf for each block in the segment. The branching factor of $\sqrt{\log m}$ means that every internal node has $\sqrt{\log m}$ children. Since dense superblocks only span at most $(\log m)^4$ blocks, this tree has constant depth (approximately depth 8). At each internal node v, we store an array I_v of length $\sqrt{\log m}$ where the ith entry is the number of 1 bits in the leaves of child i's subtree.

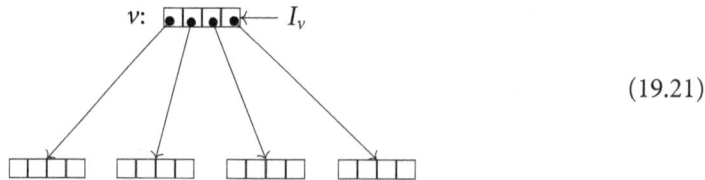

$$(19.21)$$

The count in each entry of the array is at most $(\log m)^2$, so each entry of the array takes $O(\log \log m)$ bits and the whole array takes $\sqrt{\log m} \log \log m$ bits at each node. Suppose the span of superblock i is m_i bits. There are $O(m_i/\sqrt{\log m})$ nodes in the level just above the leaves, so there are $O(m_i/\sqrt{\log m})$ internal nodes (with arrays) overall. Each of these internal nodes contains an array of length $\sqrt{\log m} \log \log m$.

Therefore, the total space for the internal nodes for this tree is $m_i \log \log m$ which is $\leq m_i \log \log n$.

But there are at most $O(n/\log n)$ dense superblocks, so across all the trees, the total number of leaves is at most $O(n/\log n)$. That is

$$\sum_i m_i \log \log m \leq \sum_i m_i \log \log n \tag{19.22}$$

$$= \log \log n \left(\sum_i m_i \right) \tag{19.23}$$

$$= O(n \log \log n / \log n). \tag{19.24}$$

So overall, adding support for select doesn't increase our asymptotic space usage.

19.5.2 Answering select queries

To answer $\text{select}_1(j)$, we use the C array to find the superblock that contains j (by looking at entry j/v). What we do now depends on whether the superblock is dense or sparse.

If it's sparse, we compute $s = \text{rank}_1$ for the start of the superblock. That tells us how many of the j 1 bits are "taken care of" by the blocks before this superblock. We then look at entry $j - s$ of the list of positions that are explicitly stored in the list of positions in a sparse superblock. This takes $O(1)$-time since we can already compute rank_1 on our bit vector in $O(1)$ time.

If the superblock is dense, we traverse the tree associated with this superblock, looking for the leaf that contains the jth bit using the I_v array. At each node v, the array I_v tells us which child to advance to. Once we find the leaf that points to the block b containing j, we can again use rank_1 to find the number s of 1 bits before that block b and return the $j - s$ entry of select_1 column in the T_{select} tables in the row corresponding to b. Traversing the tree takes $O(1)$ time for two reasons: first, since the depth of the tree is constant, we only have to walk down a constant number of levels until we find a leaf. Second, the array stored in the node is of size $O(\sqrt{\log m} \log \log m)$ which is much smaller than $O(\log n)$—that is smaller than a pointer to a position into your bit string, so searching for the correct child can be done quickly with some small tables.

There are some implementation and technical details that we have omitted (see Raman et al. [2007] for more information). But overall, this gives the idea about how to support select_1 on our vectors without adding to the asymptotic space, or the runtime.

19.6 Summary and notes

Theorem 19.5 (RRR space usage). *The RRR data structure takes*

$$O\left(\mathcal{B}(w, n) + n \log \log n / \log n\right) \tag{19.25}$$

bits, where $\mathcal{B}(w, n)$ is the number of bits needed to specify an n-bit vector with w 1-bits.

Proof: This follows from Lemma 19.2 by plugging in values derived from the our discussion, including Lemma 19.4. To do this, we note that $n/\log n$ is smaller than $n \log \log n/\log n$. $\quad\square$

This is good space usage! $\mathcal{B}(w, n)$ is approximately the information theoretic minimum needed to store the vector (assuming any vector of w 1-bits is possible) and $\log \log n/\log n$ is much smaller than 1 (and goes to 0 as n grows). And it's even better than that: not only do we use fewer bits than just writing out the string, we can answer rank (and select) in $O(1)$ time!

What big ideas did we use? The main idea is breaking the vector into blocks, which is a common approach to string and vector algorithms. The second big idea is that we use the fact that vectors with few or lots of 1-bits can be encoded more compactly than vectors with approximately a 50/50 split between 0 and 1 bits.

> ### Presentation Notes
>
> Our presentation of RRR is an expanded form of the original argument used in Raman et al. [2007]. Alex Bowe has a nice shorter description as well [Bowe, 2011].

19.7   Exercises

19.1   Show that the s_i values satisfy the conditions of Theorem 19.1.

19.2   Extend the proof of Theorem 19.1 to explicitly handle the case when the z_i values might be positive or negative.

19.3   Let S be a subset of $\{0, \ldots, m-1\}$ for some m, represented as a bit vector B, where $B[i] = 1$ if $i \in S$. Show how to answer predecessor$(x) = \max_{y \in S}\{y \in S \mid y < x\}$ in $O(1)$ time using rank_1 and select_1.

19.4   Prove that $\frac{n \log n}{\log^2 \frac{n}{\log n}}$ is $O(n/\log n)$. Hint: show that

$$\frac{n \log n}{\log^2 \frac{n}{\log n}} \leq \frac{2n}{\log n}$$

for large enough n.

19.5   In the RRR data structure, the w_i values immediately give us the s_i values. (a) Explain why that is so. (b) We can get the w_i values using the difference between adjacent f values. Explain why we still need to store the q_i values.

19.6   Give some practical ways that the space usage of RRR vectors could be reduced without changing how they operate. These don't need to reduce the asymptotic space usage.

19.7   Let $L(T)$ be the BWT of a string T (Chapter 18). We saw how to count the number of occurrences of a pattern P in T using L and the LF mapping, but not to find the *position* in T of the pattern.

(a) If we also kept around the suffix array $A(T)$ we could find the positions of P using the correspondence between BWTs and the suffix arrays. Describe (in about 1 paragraph) how to report 1 position of P using a constant amount of additional work after doing the BWT search. (Of course, if you have $A(T)$, you could just ignore the BWT, but answering this question will help with (b).)

(b) Having to keep $A(T)$ around defeats the purpose of the space-savings of the BWT. Suppose instead we only keep suffix array entries with $A(T)[i] = rk$ for some fixed r and every $k = 0, 1, \ldots$. This corresponds to sampling the suffix array entries for every rth suffix in T. We store these entries in a shorter array $A'(T)$ in the same order as they appear in $A(T)$. Additionally, let's keep an RRR bit vector B of length $|A(T)|$ that says which entries we kept: $B[i] = 1$ if entry i of $A(T)$ is present in $A'(T)$. (Khan et al. [2009]; Tomohiro et al. [2014].)

Show how to report a position of P in T using the BWT search and the subsampled $A'(T)$ (and B). The extra work you have to do should take $O(r)$ time with a constant-sized alphabet.

Hint: Use *rank* on B to map between positions in A' and A and use some additional operations after the BWT search to relate the BWT search result to some entry in A'.

CHAPTER 20

Wavelet Trees

We want to generalize the rank and select operations to strings of larger alphabets. This will be useful of course for regular string data, but having this ability will let the rank and select operations be used in many different contexts where you would not expect them to be useful. To do this, we will use *wavelet trees* [Grossi et al., 2003; Ferragina et al., 2009a; Navarro, 2014], which are a versatile data structure for sequence data. We will see some surprising uses for wavelet trees, including efficiently storing graphs.

Wavelet trees support the operations:

- $\text{rank}_c(S, i) :=$ the number of occurrences of character c at or before position i in S.
- $\text{select}_c(S, j) :=$ the position of the jth occurrence of c in S.
- $S[i] =$ "access character i."

These are analogous to the rank_1 and select_1 queries we saw in Chapter 19, but now c could be any letter from our larger alphabet Σ. As usual, our goal will be to support rank, select, and access quickly while using small space. We assume our string S is fixed and read-only, so we do not need to support assignment ($S[i] \leftarrow c$).

To build a wavelet tree that stores a string S and supports these operations, we will depend on similar operations defined on bit vectors, which to remind you are:

- $\text{rank}_1(S, i) :=$ the number of 1 bits at or before position i in S.
- $\text{select}_1(S, j) :=$ the position of the jth 1 bit in S.
- $\text{rank}_0(S, i)$ and $\text{select}_0(S, j)$ are defined analogously.

We saw how to support these in Chapter 19 on RRR compressed bit vectors.

20.1 The wavelet tree data structure

A wavelet tree is a tree with a binary vector $B(v)$ stored at each node v. We will describe the wavelet tree as if we are storing uncompressed bit vectors, but of course the natural thing to do is to use the RRR compressed bit vector (or other compressed bit vector data structure) to store each of these $B(v)$ bit vectors.

The root r has a bit vector of $|S|$ bits (conceptually). Entry i of $B(r)$ is 0 if the letter $S[i]$ is in the first half of the alphabet Σ. Entry $B(r)[i]$ is 1 if $S[i]$ is in the second half of the alphabet.

The root (and every other internal node) has 2 children. The left child corresponds to the subsequence of the string that uses letters from the first half of Σ. That is, the string represented at the left child consists of the string with all letters from the second half of Σ

removed. The right child corresponds to the subsequence of the string using letters from the second half of Σ.

This structure is repeated recursively to form the full tree. At each level the size of the relevant alphabet is halved. The leaves are the nodes with an alphabet size of 1.

Example 1. Here's an example where S is a string over the alphabet of $\{A, C, G, T\}$.

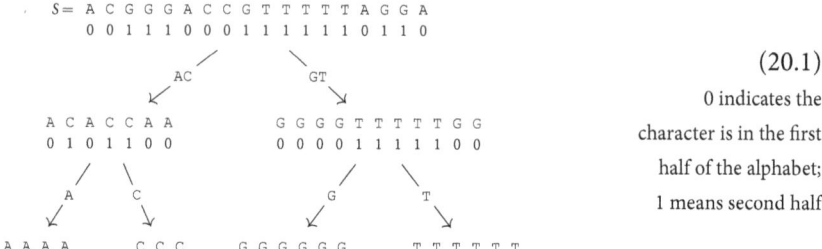

(20.1)

0 indicates the
character is in the first
half of the alphabet;
1 means second half

In this case, the initial "first half" of the alphabet consists of $\{A, C\}$ while the second half consists of $\{G, T\}$. Note that we do not store the string at each internal node—they are in the figure for clarity—we just store the bit vector. The "string" at the leaves will consist of a single character, repeated. At the leaves we just store this character.

Example 2. Here's another example with $\Sigma = \{1, \ldots, 7\}$.

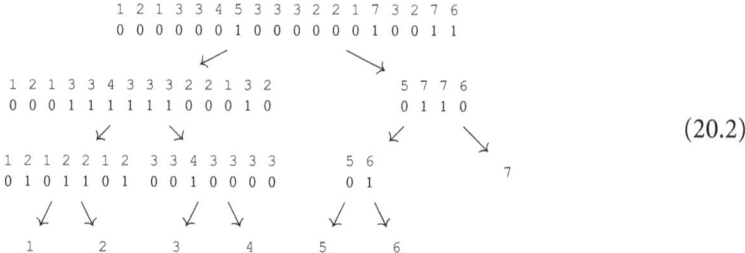

(20.2)

20.1.1   The access $S[i]$ operation

To find the character at position i in S, we traverse the wavelet tree starting at the root. Note that the bits that logically correspond to the character i occur in the bit vectors in a path from the root to a leaf. We find this path by proceeding to the left at each node u, if the bit corresponding to location i in $B(u)$ is 0 and right if that bit is 1.

The big question is how to find "the bit corresponding to location i" at each node. For the root, this is easy: it's the ith bit. But for the next node u that we visit (a child of the root), the bit corresponding to logical index i won't be at position i in $B(u)$ since we've thrown some characters out of the string represented at node u.

Don't worry! We can find the correct index using $\text{rank}_{0/1}(i)$. Specifically, if we are at node u, and the current bit is 0 (meaning we're heading to the left), the location corresponding to the ith position of u in the child c_{left} is

$$i_{c_{\text{left}}} = \text{rank}_0(u, i).$$ (20.3)

If the bit is 1, and we're heading right, then the location in the child corresponding to the bit is

$$i_{c_{\text{right}}} = \text{rank}_1(u, i). \tag{20.4}$$

Here's what this looks like.

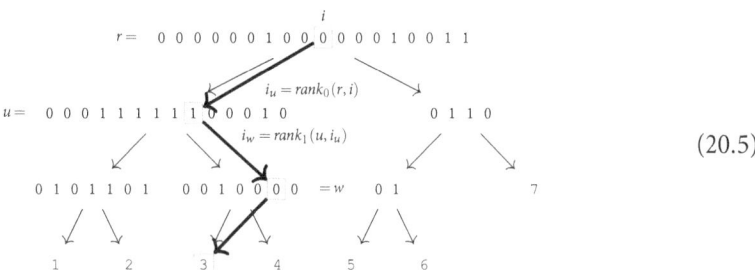

$$(20.5)$$

Eventually, this process will end at a leaf. Each of the leaves corresponds to a single character in the alphabet, which we return as the result of $S[i]$. This also tells us a bit more: in the (20.5) example, if we compute $rank_0(w, i_w)$ that will be 5, telling us not only that the ith character is "3", but that it is the 5th "3" in the string. This idea generalizes to compute $rank_c(i)$ for any character c, as we will see next.

20.1.2   The $\text{rank}_c(S, i)$ operation

Recall that the $\text{rank}_c(S, i)$ means the number of occurrences of character c at or before index i. Again, we start at the root. We use $\text{rank}_{0/1}$ to find the corresponding locations as we move down the tree just as with the $S[i]$ operation: we use rank_0 if we move left and rank_1 if we move right.

Let c be the character we are searching for in the rank_c operation. The difference from the $S[i]$ operation is that we go to the left child if c is in the first half of the remaining alphabet, and we move to the right if c is in the second half of the remaining alphabet. In other words, we're following the path from the root to the leaf that contains c.

Eventually, we reach a leaf which is logically encoding the string "$ccccccccc\ldots$" (repeated the number of times c occurs in the original string). The index into this leaf string gives the rank. Here's an example for $rank_5(i)$.

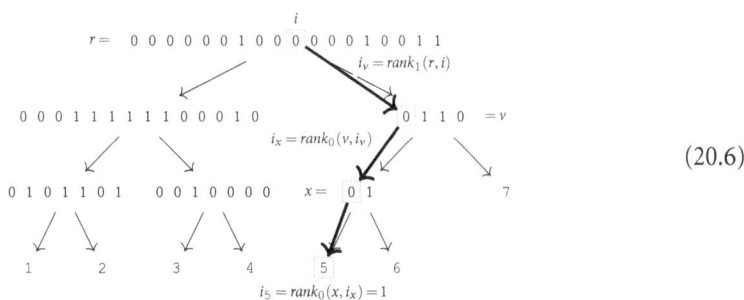

$$(20.6)$$

20.1.3 The $select_c(S, j)$ operation

Finally, consider the $select_c(S, j)$ operation. This is the inverse of the $rank_c(S, i)$ operation, so we compute it by following the inverse path in the tree. We start at the *leaf* corresponding to character c and position j in the string represented by this leaf. This is the jth occurrence of character c in the original string S.

We then walk up toward the root, using $select_{0/1}$ to find the corresponding position in the parent. If we're coming from a left child, we use $select_0$; if we're coming from a right child, we use $select_1$. The index we land on at the root is the index of the jth occurrence of c in S.

That is, we repeat "$i \leftarrow select_b(p, i)$" where p is the parent of the current node, and $b = 0$ if the current node is the left child of p, and $b = 1$ if it's the right child.

Here's an example looking for the 4th "3" in the string.

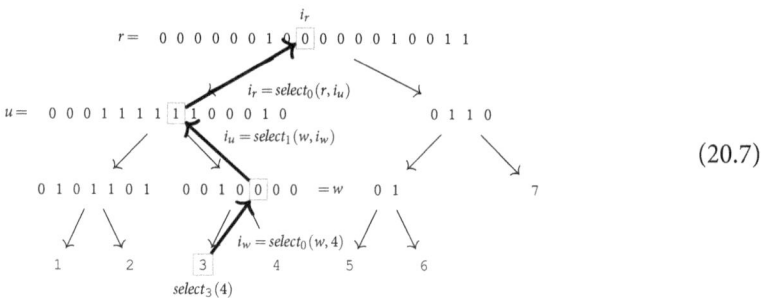

$$(20.7)$$

20.1.4 Running times

The tree height is $\log |\Sigma|$, where Σ is the alphabet. The rank, select, and access operations each follow a root-leaf path in the tree, taking $O(1)$ time at each node, since we can compute $rank_{0/1}$ in constant time. Therefore, rank, select, and access take $O(\log |\Sigma|)$ to run. If the alphabet size is constant, this is $O(1)$. We'll see some applications where we have non-constant-sized alphabets.

In fact, we don't have to use a balanced tree shape. We can instead use a tree shape that puts more common characters closer to the root (say, a Huffman code tree from Chapter 32). This would make accesses to frequent characters faster and would give good space usage even without compressed bit vectors.

What about space? If you were going to write out a string over alphabet Σ in the normal way, you would use $O(\log |\Sigma|)$ bits per character. That's the same space as writing down, for each character separately, the bits encountered along the path through this tree followed by the access $S[i]$ operation. The total number of bits in the tree is no more than you would use to write the string uncompressed. But if we use compressed bit vectors at each node, we get the space savings of (say) RRR compressed vectors, so the space will be less than the original string.

20.2 Applications of wavelet trees

20.2.1 Inverted indices

The traditional inverted index represents a document by an array I of lists, where each entry in the array $I[w]$ corresponds to a word w and contains a list of locations of that word. For example, (20.8).

and	18, 47, 55
internet	39
search	17, 32
strings	11, 20, 38
the	12, 26

Our focus in this book are algorithms that operate on strings. The ability to process, store, search, and manipulate strings with computational efficiency has changed the world in many ways. Web search engines regularly process terabytes of strings, internet shopping sites deal with many product descriptions, and social networks handle large collections of comments and posts.

$$(20.8)$$

This is very good to identify all the places where any word occurs—if the array is sorted alphabetically, it's easy to find the list of positions for a word. This representation is a lot harder to use to identify the ith word. That is, it no longer easily represents the original string.

Wavelet trees enable both traditional inverted index operations and efficient representation of the original list of words [Claude and Navarro, 2008]. Let S be the original text, represented as a string of word identifiers. In other words, take the alphabet Σ to be the set of words. Store this word-based string in a wavelet tree.

$S[i]$ represents the ith word. This can be computed in $O(\log|\Sigma|)$ time by the wavelet access operation. Now $\Sigma = O(|S|)$, where $|S|$ is calculated in words. Hence, with an $O(\log|S|)$ slowdown (compared with storing the words in an array), we can still access the original list of words in order.

In addition, $\text{select}_w(S, j)$ now gives the position of the jth occurrence of word w in the same $O(\log|S|)$ time. This is the information that is normally easily available from an inverted index. To recover the full location list, we compute $\text{select}_w(S, j)$ for $j = 1 \ldots$.

20.2.2 Document retrieval

Let $D_1, \ldots, D_m$ be a collection of documents. We want to quickly answer the question:

- In which documents does a given word w appear?

To solve this [Arroyuelo et al., 2010], we again represent each document D_i as a string of word identifiers. We concatenate the documents together, separated by a word $ that doesn't occur elsewhere in any document. Now, we have a long string S over the word alphabet where the query word w (represented by $\times$) occurs in some locations (maybe many times per document).

$$(20.9)$$

We store S in a wavelet tree.

The following code, where $\text{RANK}(w, i)$ and $\text{SELECT}(w, j)$ compute the rank and select on S for word w, prints the documents that contain w:

Algorithm 20.1. Find documents containing word w.

Require: RANK and SELECT operate on S defined in (20.9).
   function FINDDOCS(w)
      $i \leftarrow 1$
      repeat
         $pi \leftarrow \text{SELECT}(w, i)$ ▷ *location of ith occurrence*
         $di \leftarrow \text{RANK}(\$, pi) + 1$ ▷ *document containing it*
         output di
         $p' \leftarrow \text{SELECT}(\$, di)$ ▷ *end of that document*
         $i \leftarrow \text{RANK}(w, p') + 1$ ▷ *find 1st occurrence after end of doc*
      until i is past the end of concatenated string

How does this work? pi finds the location of the ith occurrence of the word (at the start, $i = 1$ so the first time through this is finding the first occurrence). di then counts the number of \$ markers before that position. That gives the index of the document that contains the ith occurrence, which is then printed. p' is then computed, using select on the \$ separator, to be the end of the dith document. i is then reset to be 1 more than the number of occurrences of w before the end of the dith document, skipping the other instances of w in the dith document. The loop proceeds, now looking for the ith occurrence of w, which is the first after the current document.

Each time through the loop prints out a document. The work in each iteration consists of two select and two rank operations, each taking $O(\log N)$ time, where N is the total length of the documents. This leads to a runtime of $O(k \log N)$ time, where k is the number of documents containing the word w.

20.2.3 Storing graphs

Wavelet trees are useful for things that do not look like strings or arrays. For example, given a directed graph G, where every vertex has at least one in-edge and one out-edge, we want to answer the following queries quickly:

- $successor(u, i) :=$ the ith vertex v such that edge (u, v) exists.
- $predecessor(u, j) :=$ the jth vertex v such that edge (v, u) exists.

These operations allow for efficient graph traversals.

Let V be the vertices of G. We assume, without loss of generality, that V is the set $\{1, \dots, |V|\}$. We represent G as a concatenation of adjacency lists, in some order. That is, if A_u is the list of vertices that are successors of u, then $S = A_{u_1} A_{u_2} \dots$ for $u_i \in V$. This gives us a "string" S over the alphabet V, which we store in a wavelet tree. We also store a bit vector B where an entry $B[i] = 1$ if i is the start of the adjacency list of a vertex. We store B in a data structure that supports rank and select in $O(1)$ time (such as an RRR vector). In pictures we have:

$$\text{Adjacency lists:}\quad \boxed{2\,|\,3\,|\,5\,|\,4\,|\,3\,|\,5\,|\,6\,|\,2\,|\,6\,|\,5\,|\,4\,|\,1\,|\,2}$$
$$B = \quad 1\ \ 0\ \ 0\ \ 1\ \ 0\ \ 0\ \ 1\ \ 1\ \ 0\ \ 0\ \ 1\ \ 1\ \ 0 \tag{20.10}$$

and now:

- $successor(u,i)$: $p = \text{select}_1(B,u)$; return $S[p+i-1]$.

Here, p is the location of the start of the adjacency list of node u, and $S[p+i-1]$ is the ith successor of u in that adjacency list. The select_1 operation takes $O(1)$ time and the access operation $S[p+i-1]$ takes $O(\log|\Sigma|) = O(\log|V|)$ time. (This operation doesn't really make use of the wavelet tree directly—if we only need $successor$, and aren't worried about space, we could store S in a regular array.)

In addition:

- $predecessor(u,j)$: $p = \text{select}_u(S,j)$; return $\text{rank}_1(B,p)$.

Here, p is the location of the jth occurrence of node u in the list. The rank_1 operation identified the index of the node that contains index p. This also takes $O(\log|V|)$ time. The advantage of this graph representation is that it is very space efficient.

20.3 Summary and notes

Wavelet trees compactly store strings. They allow for access almost as fast as for a plain array. They also allow for fast rank and select queries. They have many applications and extensions, often storing things not normally thought of as strings. Significant work on the use, construction, and properties of wavelet trees has been undertaken [e.g., Gagie et al., 2012; Claude et al., 2011].

> ### Presentation Notes
>
> The applications that we describe here were pointed out by Navarro [2014], which gives many other applications and benefits for the data structure (as well as more details about variants). Makris [2012] provides another nice survey.

20.4 Exercises

20.1 You are given an ordered collection of documents $D_1, \ldots, D_t$ that represent some main document as it changed over time ($D_{t'}$ is the document at time t'). You can treat each of the documents as a list of words.

Define `span(w)` to be (i,j) where i is the first document in which w appears and j is the last. Show how to preprocess $D_1, \ldots, D_t$ into a wavelet tree so that you can find span(w) for any w in $O(\log n)$ time, where n is the total length (number of words) of all the documents.

20.2   Define the following operation on a string S:

$$CountRange(S, i, j, \ell, u) := |\{k : i \le k \le j \text{ and } \ell \le S[k] \le u\}|. \qquad (20.11)$$

In other words, $CountRange(S, i, j, \ell, u)$ is the number of characters in substring $S[i \ldots j]$ that fall between characters ℓ and u in the alphabet Σ. Use wavelet trees to give an $O(\log|\Sigma|)$-time algorithm to compute $CountRange$ (assume you can do operations on $O(\log n)$-bit numbers in $O(1)$ time).

20.3   Suppose you have a wavelet tree T that represents an ordered sequence $S = x_1, \ldots, x_n$ of integers $x_i \in [1 \ldots m]$. Give a $O(\log m)$-time algorithm to answer the following type of queries:

- $IthInRange(T, j, k, i) :=$ the ith smallest integer in the range $x_j, \ldots, x_k$.

20.4   We define the kth circular rotation of a length n string S by $S^k[i] = S[(i + k) \bmod n]$, assuming zero indexing. Given a wavelet tree T for string S of length n over alphabet Σ, describe how to implement rank and select for S^k without using any extra space and still running in $O(\log(|\Sigma|))$ time.

Hint: Draw a picture to see where each part of the string is sent.

20.5   In the application of wavelet trees to graphs in Section 20.2.3: (a) explain what breaks down in the approach if some node has 0 in-degree and if some node has 0 out-degree; (b) explain how to modify the approach to support graphs that have 0 in- or out-degree nodes.

CHAPTER 21

The FM Index for Compressed Searching

In Chapter 18, we saw how the BWT can be used to search a string T for occurrences of substrings and also how the BWT can be used to improve compression. That naturally raises the question about whether we can do both at the same time: search a *compressed* BWT string efficiently. The answer is yes, and this was first pointed out via the design of the FM-index [Ferragina and Manzini, 2000, 2005]. In this chapter, we'll see how to do this.

21.1 BWT and wavelet trees

First, recall the algorithm for counting the number of occurrences of a pattern P using a BWT string (Section 18.5.1). This algorithm uses the LF mapping to repeatedly jump (conceptually) between the BWT string L and the sorted input string F. It uses an array $C[x]$, where $C[x]$ is the count in T of all the characters that come before x alphabetically. Using the notation of Ferragina and Manzini [2000], we have:

Algorithm 21.1. Use BWT to count occurrences of a pattern in a string.

Require: RANK operates on the *BWT*, and array C has been computed.
    function BWTCOUNT(BWT, P)
        $c \leftarrow P[|P|]$ $\triangleright$ *c is the current character*
        $sp = C[c] + 1$ $\triangleright$ *Start of range*
        $ep = C[c + 1]$ $\triangleright$ *End of range*
        $i \leftarrow |P|$ $\triangleright$ *Current position in P*
        while $(sp \leq ep)$ **and** $(i \geq 2)$ **do**
            $c \leftarrow P[i - 1]$
            $sp = C[c] + \text{RANK}(c, sp - 1) + 1$
            $ep = C[c] + \text{RANK}(c, ep)$
            $i \leftarrow i - 1$
        if $ep < sp$ **then**
            return "not found"
        else
            return $ep - sp + 1$ $\triangleright$ *Size of final range*

The RANK(c,p) value is the number of occurrences of the character c that occur in positions $\leq p$ in the BWT string $L = BWT(T)$. In the notation we have used in recent chapters,

$$\text{RANK}(c,p) = rank_c(L,p). \tag{21.1}$$

What data structure could we use to efficiently support $rank_c(L,p)$ queries? The wavelet tree is designed for this! We can simply create a data structure

$$\langle Wavelet(BWT(T)), C \rangle \tag{21.2}$$

that consists of storing the BWTed string in a wavelet tree, plus the count array C. That's all we need to be able to do the BWT count occurrences operation. If our wavelet tree uses either compressed RRR vectors or is unbalanced so that common characters have shorter paths, then we've got a compressed representation of the BWT that can be queried. This approach doesn't really use the property that BWT makes the text more compressible. Despite this, there is some theoretical and practical evidence that this direct approach is the right approach for many applications [Mäkinen and Navarro, 2007b; Ferragina et al., 2004].

21.2   A more compression-focused approach

The FM-index is a data structure that applies several compression steps to the BWT before creating an index on top of those compressed representations.

The scheme Ferragina and Manzini [2000] propose to compress T, computes $BWT(T)$, then uses the move-to-front (MTF) encoding (as in Section 18.5.2) to adaptively encode common characters with fewer bits. It then does run-length encoding (RLE) to replace runs of 0s with the count of the number of 0s. Finally, they use a Huffman code (see Chapter 32) to encode the final stream. So, the proposed compressed version $\tilde{T}$ of a string T is:

$$\tilde{T} = Huffman(RLE(MTF(BWT(T)))), \tag{21.3}$$

which seems like it might be pretty hard to search. But from the previous discussion, we only need the array C (small and easy to keep) and the ability to compute $rank_c(BWT(T), i)$ for any $c \in \Sigma$ and any position i.

In the rest of this chapter, we will see the scheme for computing $rank_c(BWT(T), i)$ without having to keep T around. Instead, we can keep the compressed string $\tilde{T}$ and some additional (small) data structures.

21.2.1   Computing rank on the compressed string

Our RRR discussion (Chapter 19) was focused on computing the *rank* operation on compressed bit vectors. Here, we have a similar problem: computing *rank*s on compressed strings. So, naturally, the solutions are going to look somewhat similar. The big difference is that with RRR we had a "simple" compression scheme based on lookup tables; but here we have a "complex" compression scheme involving BWT, MTF, RLE, and Huffman encoding, so we need an approach that works with such complex transformations.

The first idea is the same as an idea in RRR: break our string up into substrings of some length u (to be determined later). The blocks are not compressed individually. We compress the entire string, and then we break up the *compressed* string into blocks corresponding to length-u substrings of $BWT(T)$. This is possible since the compression scheme encodes each character one at a time (though see Exercise 21.5). This looks like (21.4).

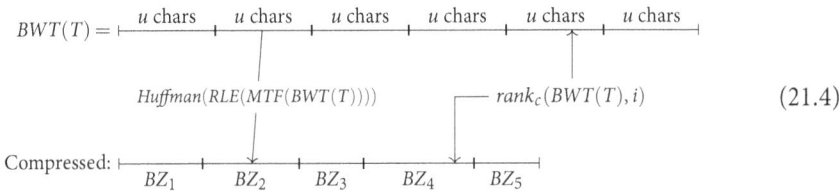

$$(21.4)$$

Block BZ_i is the compressed bit string that represents u characters of the original string. The number of bits for each block likely varies.

Our job is to be able to compute the number of occurrences of any character c up through any position p (where p is in the original uncompressed coordinates, of course). We will store some extra information for each block and some groups of blocks to compute this quickly.

We conceptually group the blocks into superblocks of u blocks.

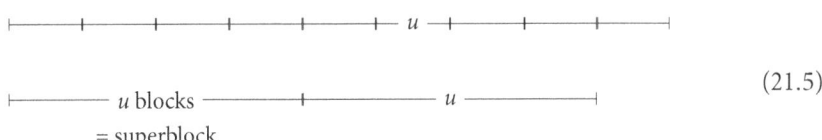

$$(21.5)$$

We store the following count arrays:

- for each superblock j: store a $|\Sigma|$-long array giving the number of occurrences of each character up through and including superblock j.
- for each block i: store a $|\Sigma|$-long array giving the number of occurrences of each character up through and including this block i *since the end of the last superblock*.

21.2.2   Space usage for count arrays

How much space is this? Let n be the length of our string. Choose $u = \Theta(\log n)$. We have n/u blocks. Each block has a $|\Sigma|$-long array of counts. Since the counts are only from the end of the last superblock, each count is at most u^2, so it takes $O(\log u)$ bits at most for a count, and each block array takes $O(|\Sigma| \log u)$ bits.

Taking (as usual) $|\Sigma|$ to be a constant, we use

$$\frac{n}{u} \log u = \frac{n \log \log n}{\log n} \tag{21.6}$$

space for all the block tables, which is $< n$ asymptotically.

We have $(n/u)/u = n/u^2$ superblocks, each with a $(|\Sigma| \log n)$-bit array to store the counts. Each count could be as much as n because the counts are cumulative. This leads to:

$$\frac{n}{(\log n)^2} \log n = \frac{n}{\log n} \tag{21.7}$$

bits in total for the superblocks, which is smaller than n and smaller than (21.6).

21.2.3 Computing occurrence counts

With these tables, to compute the number of occurrences of a character c at or before a position p, we first find which block b_p and superblock sb_p it is in—this is easy since the block sizes are based on breaking up the $BWT(T)$ string. We then compute:

$$
\boxed{\begin{array}{l}\text{Count of } c \\ \text{up through} \\ \text{superblock} \\ sb_p - 1\end{array}} +
\boxed{\begin{array}{l}\text{Count of } c \text{ up} \\ \text{though block} \\ b_p - 1 \text{ since} \\ \text{the end of the} \\ \text{last superblock}\end{array}} +
\boxed{\begin{array}{l}\text{Count of } c \text{ up} \\ \text{to } p \text{ from the} \\ \text{start of block} \\ b_p\end{array}}. \tag{21.8}
$$

The first two counts are directly read from the $|\Sigma|$-long arrays stored with each block and each superblock. But how do we obtain the count of the occurrences of c within p's block (the last block of (21.8))? That's the last piece of the puzzle.

21.2.4 Computing counts within a block

We now deploy the same trick we did with RRR (and with the Four Russians speed up): precomputation. Since we chose our block sizes to be small (exponentially smaller than the length of the string) we can store the answers for every *type* of block.

The type of block i is determined by the bits in the block BZ_i plus the state of the MTF array when the compression of the block was started. That is, given MTF_i and BZ_i we can uncompress this block—or more to the point, the pair (BZ_i, MTF_i) completely determines the text of the block. For each block, we can store the MTF_i list in $|\Sigma| \log |\Sigma|$ space, which is a constant number of bits for each block.

We store a lookup table M for each of these block types:

- $M[BZ_i, MTF_i, c, p]$ = the number of occurrences of character c through position p within a block of type (BZ_i, MTF_i).

The size for these tables is:

$$O\left(\underbrace{|\Sigma|^u}_{\text{\# BZ bit patterns}} \times \underbrace{|\Sigma|!}_{MTF} \times \underbrace{|\Sigma|}_{\text{for each character } c} \times \underbrace{u}_{\text{posn in block}} \times \underbrace{\log u}_{\text{count}}\right). \tag{21.9}$$

Taking $|\Sigma|$ to be constant, this becomes:

$$O(u|\Sigma|^u \log u) = O(n^\epsilon \log n \log \log n) \qquad (21.10)$$

if we take $u = \epsilon \log_{|\Sigma|} n$ for some ϵ, using the same trick as we used in Section 19.4.3 for RRR. We can choose this ϵ to be small enough to make the tables take less space than our string. (See Exercise 21.3.)

21.3 Summary and notes

You'll notice some ideas that are related to ideas that we saw with RRR bit vectors and with the "Four Russians" speed up: break the string into chunks for which we can pre-compute and store the answers to queries. In the "Four Russians" approach (Chapter 9), this was used for a speed improvement. In RRR and the FM index (Chapters 19 and 21), this idea is used for space savings.

There are a lot of hidden constants in the analysis in this chapter that need to be considered when implementing this in practice. It's nice that we can say something theoretically about the space usage, but for any particular application it's not always clear whether we will get space performance that is worth this added complexity.

A line of work has improved the practical space efficiency to create a data structure called the r-index [Gagie et al., 2020]. The space used by the r-index is related to the number of runs r of identical characters in the BWT [Mäkinen and Navarro, 2007a], which grows slowly if there is a lot of repetition in the indexed string. Construction and applications of the r-index have also been demonstrated [e.g., Boucher et al., 2019; Kuhnle et al., 2020; Mun et al., 2020; Rossi et al., 2022].

> **Presentation Notes**
>
> The bulk of this chapter is inspired by the various original papers that introduced the FM-index.

21.4 Exercises

21.1   Explain why the number of BZ bit patterns is $O(|\Sigma|^u)$.

21.2   (a) Show that for any n, there is an ϵ such that

$$n^\epsilon (\log n)(\log \log n) < n. \qquad (21.11)$$

(b) Further show that for sufficiently large n, there is a constant that does not depend on n such that $n^\epsilon (\log n)(\log \log n) < n$.

21.3   In order to use the precomputed table M, we must be able to retrieve the bit pattern of the ith compressed block. The challenge is that since each BZ_i may be of a different length, it's not immediately clear where block i begins in the bitstream that encodes $\tilde{T}$. Explain how to add some additional data tables so that the bit pattern BZ_i can be found given i. Show that they do not increase the asymptotic space usage.

21.4   Although the $|\Sigma|!$ in (21.9) is constant if $|\Sigma|$ is constant, the number of permutations can be huge for even a modest-sized alphabet. Give an alternative argument that does not involve $|\Sigma|!$ and that bounds the space of the tables by $n \log \log n$ (which is bigger than n, but not by too much), even when $|\Sigma|$ is arbitrarily large (but still a constant).

21.5   We used the fact that each block of the compressed string corresponds to a block of the $BWT(T)$ string. But the run-length encoding (RLE) breaks this assumption if a run of 0s spans a block boundary. Explain how to handle this case and fix the argument.

CHAPTER 22

Storing Text for Editing

You probably most directly interact with string data in text editors and word processors. These programs allow a user to manipulate possibly very long strings by inserting, changing, and deleting text. They must find a way to store the text of the document so these kinds of edits are efficient to make.

The simplistic approach can be very inefficient: suppose we just store our document as one big string in an array in contiguous memory. This has very good space usage: it only uses space to directly store the string; but edits could be very slow. To insert a character in the middle of a length-n string would take $O(n)$ time to shift the characters after the insertion point down to make room for the new character. If n is large, doing this every time the user types or deletes a character is going to be slow.

One solution that has been used by some editors is a *line array*. This is an array L of pointers to strings, each of which represents a line (a substring between newline characters) in the document:

(22.1)

This can work ok if the text is expected to have relatively few lines, each of reasonable length. This is usually true of programming source code, for example. But if the file has at least one very long line, then edits to that line could be very slow (since it devolves to the simplistic slow approach). This approach also devolves to the simplistic approach if the file has many very short, say constant-sized, lines, since then each line can be thought of as a character in a big alphabet and inserting or deleting lines could take $O(n)$ time.

Next we'll see some data structures that attempt to fix some of these problems.

22.1 Gap buffers

In a text editor, nearly all of the edits are going to happen where the cursor is at the moment. The problem with the simplistic approach is that if the cursor is at a position in the middle of the document, and the user types a character to insert it, we don't have any place to put it until we move all the existing characters after the insertion point down in the array.

Gap buffers are a way to avoid that. If we have n characters currently in our text T, we put them in an array G of size $N > n$, and we use the extra space to leave a gap where the cursor currently is. Specifically, if the cursor is after character i, then $T[1, \ldots, i]$ is placed in the first i entries of the array G. $T[i+1, \ldots, n]$ is placed in the *last* $n - i$ entries of G. Since $|G| = N > n$, this leaves a gap of unused entires, with the insertion point pointing to the start of the gap.

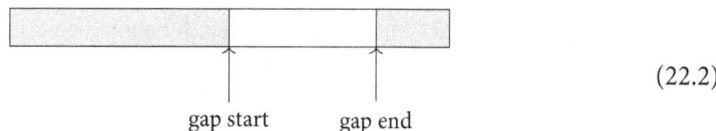

$$(22.2)$$

gap start   gap end

When we insert a character now, it just goes into the gap and the gap gets smaller by 1 entry. When we move the cursor forward from position i to $i+1$, we move $T[i+1]$ to $G[i+1]$ and shift the endpoints of the gap toward the end of the string by 1 position. When we move the cursor backwards, we do the opposite. Deletion (e.g., via backspace) grows the gap.

While this approach does move the characters in the string around a lot—on every cursor move for example, it does only a constant amount of work for every edit or move operation. This is much better than the possibly $O(n)$ work for modifying a non-gapped string.

This approach has two drawbacks: (1) N needs to be sufficiently larger than n to handle insertions for a while, and this uses extra memory (and we need some way to estimate what N should be), and (2) once we run out of gap space, we have to copy everything over to a new bigger array, which is time-consuming and could lead to bursty slowdowns.

22.2 Piece tables

Instead of moving characters around a gap, an alternative approach is embodied by the *piece table*, where the text is broken into immutable pieces. A piece table has two strings: F and M. F represents the original text in the file at the point we start editing. F is never changed. M holds other text, as described in (22.3), and M is only ever appended to—we never change any of the characters in M, we only add new characters to its end.

We represent the current string T as a list of *pieces* that refer to F or M. Each piece is a triple (s, i, ℓ), where $s \in \{F, M\}$, i is a position in s, and ℓ is a length. In this way, a piece p represents some string $str(p)$. The string T is specified by a list $p_1, \ldots, p_k$ of pieces that spell out $T = str(p_1) \circ \cdots \circ str(p_k)$.

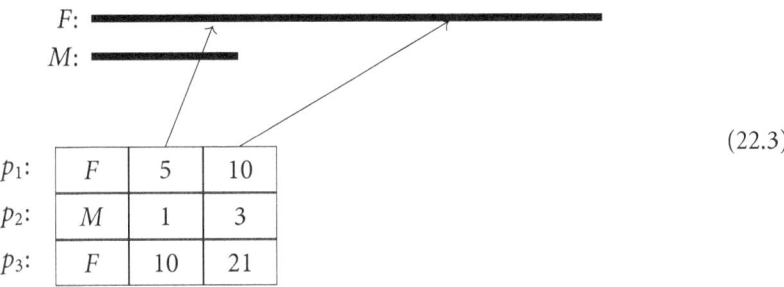

$$(22.3)$$

p_1:	F	5	10
p_2:	M	1	3
p_3:	F	10	21

In the example (22.3), the original F string has had its first 4 characters deleted, and after the original 15th position, 3 characters have been inserted.

To insert a string α at position i, we append α to M, create a piece $p = (M, |M| - |\alpha|, |\alpha|)$ that points to this new substring at the end of M, and add p to to the list of pieces at text position i. If position i occurs in the middle of an existing piece, we split that piece. For example, inserting a string of length 4 at position 15 into the example (22.3) would look like (22.4).

p_1:	F	5	10
p_2:	M	1	3
p_3:	F	10	2
p_4:	M	4	4
p_5:	F	11	21

$$(22.4)$$

Deletion works by removing pieces (or partial pieces) from the list.

If we use a linked list to represent the list of pieces, deletion takes $O(1)$ time, and insertion takes $O(1)$ time plus the time it takes to append α to M. If we further use a linked list of strings to represent M instead of a simple array, appending α to M takes time $O(|\alpha|)$, which is amortized constant time for every inserted character.

This also has the advantage that since we keep all of the text ever entered, even if it is eventually deleted, supporting undo operations can be easier.

Piece tables do have some tradeoffs, however. Since M only grows, if you make a lot of edits (especially intermixed insertions and deletions), the memory usage could grow considerably. Instead of being proportional to the size of the represented string, it will be proportional to the size of the original string plus the total size of all the edits. Also, finding character $T[i]$ takes time proportional to the number of pieces before position i, which could be $O(|T|)$, a very bad access time.

22.3 Ropes

Ropes [Boehm et al., 1995] avoid this worst case access time and growth of space usage based on the history of edits by building on balanced binary trees. A rope is a balanced binary tree where:

1. every leaf contains a substring of the text T we want to represent;
2. an in-order traversal of the tree (visiting left subtrees before right subtrees) will visit the leaves in the order that they appear in T; and
3. every node u records it's LENGTH(u), which is the number of characters represented by the leaves of its subtree.

This looks like (22.5).

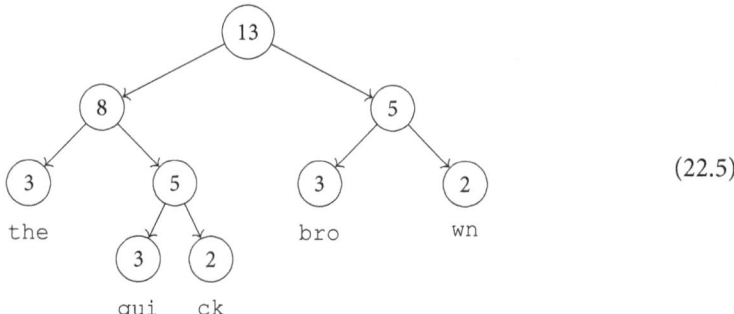

(22.5)

We can view the internal nodes as concatenating the substrings represented by their children.

Ropes support several operations with good guaranteed runtimes. For example, to reconstruct the entire string, one need only do an in-order depth-first traversal of the tree, visiting left subtrees before right subtrees. By property 2, the leaves will be encountered in the order they should be appended to construct T. We can view the rope as a binary search tree ordered by the leaf substrings' position in the full string. Since every internal node has 2 children, the total number of nodes is $O(n)$, assuming there are no empty strings in the leaves.

Accessing by character position is also fast. To find $T[i]$, we set $j \leftarrow i$ and start at the root. When we're at any internal node u, we compare j to the length of the string stored at subtree LEFT(u). If $j \leq$ LENGTH(LEFT(u)) then we move to the left child (and leave j unchanged). If $j >$ LENGTH(LEFT(u)), we set $j \leftarrow j -$ LENGTH(LEFT(u)) and move to the right child. The rationale behind this is that, since $j >$ LENGTH(LEFT(u)), position j must be in the second string concatenated by node u (the right child), and when we move to the right child it's like we're "throwing away" the "before" part of the string, so we update the index j to account for that. When we get to a leaf, we return the jth character of the string stored there.

Because we assume that the tree is balanced, it will have depth $O(\log|T|)$, and since we do constant work at each node walking down the tree, accessing the ith character takes $O(\log|T|)$ time. This is slower than the constant access that we would get using a simple array-based string, but it's much faster than the $O(|T|)$ time for piece tables.

Two ropes can be concatenated in $O(1)$ time simply by creating a new root node and having its left and right children point to the two ropes to be concatenated.

Insertion works just how you'd expect. To insert α after position i, traverse the tree to find the node that contains position i, split that node into two nodes: one that contains the string β before i and one that contains the string γ after i. Create a node that contains α. Finally, create two new internal nodes to hook up the β, α, γ nodes in order, like (22.6).

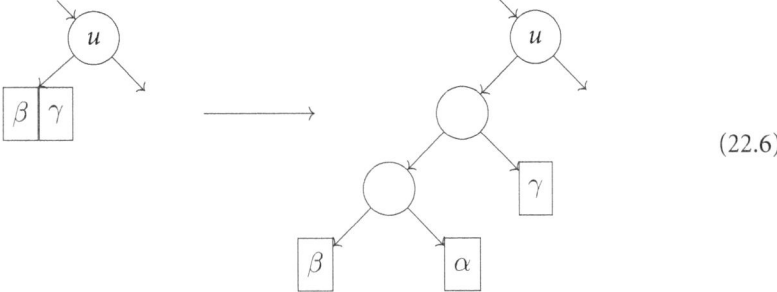

$$(22.6)$$

(There is a special case if γ is empty; see Exercise 22.1.) In reality, since we build the rope on top of some balanced tree structure, we split the $\beta\gamma$ node and then do whatever insert operation is supported by that balanced tree data structure.

There's one more thing we have to do when inserting: we have to update the lengths of the trees stored there. But the only nodes we need to change are those along the path from the root to the place where we're doing the insertion (since those are the only ancestors of the newly inserted substring). We can add $|\alpha|$ to the length of each node along this path as we're walking down the tree.

We can also split a rope into two. The operation SPLIT(r, i), where r is a rope and i is a leaf node of r, returns two ropes: one that represents the string before i, and one that represents the string after i.

To achieve that, first, if i is not at the beginning of a node, divide the node it's in into two nodes, adding a new parent, so that position i is at the start of a node u. (If i is already at the start of a node, we can skip this splitting step). Let P be the path from the root to the u that starts at position i.

Conceptually, what we need to do is break off u and the right subtrees hanging off of P that don't contain u and concatenate them into a new rope. For SPLIT$(r, 7)$, this looks like (22.7).

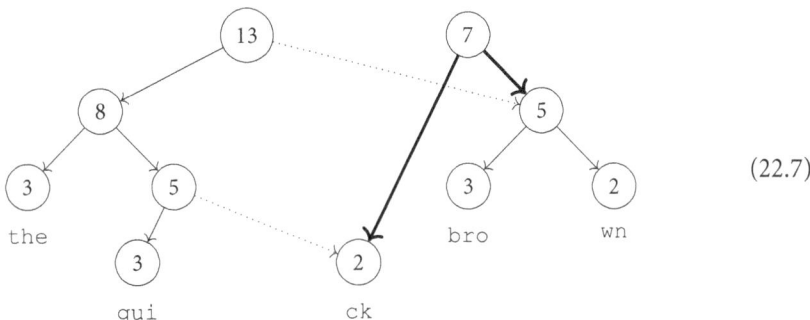

$$(22.7)$$

The dotted lines represent old links, and the thick edges are the newly added part of the second rope. Again, we also have to update the lengths of the affected nodes along P. Depending on the balanced tree chosen for the implementation, we also may need to (or want to) clean up nodes that only have 1 child or to rebalance after the split.

We can implement the split via the following recursive procedure, which walks down the tree (just like the $T[i]$ operation).

Algorithm 22.1. Split rope r into two ropes, dividing the string at position i.

Require: Assumes i is at the start of a node.

```
1: function SPLIT(r, i)
2:     if i < LENGTH(LEFT(r)) then                        ▷ i is in the left subtree
3:         r₁, r₂ ← SPLIT(LEFT(u), i)
4:         return ⟨r₁, CONCAT(r₂, RIGHT(r))⟩
5:     else if i > LENGTH(LEFT(r)) then                   ▷ i is in the right subtree
6:         r₁, r₂ ← SPLIT(RIGHT(u), i − LENGTH(LEFT(u)))
7:         return ⟨CONCAT(left(u), r₁), r₂⟩
8:     else                                   ▷ Assumes position i starts a node
9:         return ⟨LEFT(r), RIGHT(r)⟩
```

Lines 2–4 handle the case when we are at node r and the split position i is to the left: we recursively split the string represented under the left child. This will give us three sub-ropes:

- The rope (r_1) representing the string before i,
- The rope (r_2) representing the string between i and the start of the right child of r, and
- The rope ($\text{RIGHT}(r)$) representing everything after that.

r_1 is the first rope we want to return from the split, and the concatenation of r_2 and $\text{RIGHT}(r)$ is the second. The case when the split point to the right is similar.

One interesting point is the "else" condition in line 8. Because the split position i is assumed to be at the start of a node, there must be some node in the tree that represents exactly a substring starting at i, and this node must be on the path from the root to the node starting at position i. The recursion will terminate when this node is reached.

Since the tree is balanced, SPLIT takes $O(\log n)$ time.

The operation $\text{DELETE}(r, i, j)$ deletes the characters between i and j. We can use SPLIT to implement it.

Algorithm 22.2. Delete characters between i and j.

Require: Assumes i is at the start of a node.

```
1: function DELETE(r, i, j)
2:     r₁,ᵢ₋₁, rᵢ,ₙ ← SPLIT(r, i)
3:     rᵢ,ⱼ, rⱼ₊₁,ₙ ← SPLIT(rᵢ,ₙ, j − i + 1)
4:     return CONCAT(r₁,ᵢ₋₁, rⱼ₊₁,ₙ)
```

This takes $O(\log n)$ time since that is what SPLIT takes.

To summarize, with ropes, we achieve $O(\log n)$ access, insertion, deletion, split, and $O(1)$ concatenation—assuming the tree is balanced. While access and split are slower than with a simple array, insertion and deletion are much faster. To keep the tree balanced, we may have to rebalance it after the operations. This will affect the running time in a way that depends on what type of balanced tree we maintain. The rebalancing operation also needs to be aware that we are tracking subtree lengths, and must keep those up-to-date as well.

22.4 Summary and notes

The simplistic array-based representation of strings is about the best we can do in terms of space without compressing the string. But it makes frequent edit operations slow. The data structures in this chapter represent some of the approaches that text editors use to store strings that are expected to undergo many edits. They represent various tradeoffs—from easy to implement and practically fast in most cases, to good guaranteed runtimes (and practically fast in most cases), but harder to implement. The particular choice of data structure requires empirical tests to compare these options.

The ideas behind the piece table are not limited to only 2 string buffers (F and M). A natural extension is to use arbitrary many append-only buffers that the piece table can refer to. Another natural extension is to reuse parts of these buffers as needed (i.e., rather than appending an "a" if that is typed, find a substring that matches "a" in some existing buffer).

Piece trees [VS Code, 2018] are another representation that merges the ideas of piece tables and ropes. They represent the string as a balanced binary search tree of substrings, ordered by position of the substring in the full string. Instead of storing the strings directly at each leaf, substrings are stored in a set of append-only buffers, and nodes in the tree refer to those substrings.

Presentation Notes

A good reference for text editor data structures is Crowley [1998]. A classic reference on text editing data structures is Finseth [1999]. The descriptions of the data structures here are informed by the original papers, or various online technical discussions of text editor implementations. J.S. Moore seems to have invented piece tables while he and colleagues were implementing the 77-Editor text editor [Moore et al., 1973].

22.5 Exercises

22.1   Describe how to handle the case for γ being empty in the insertion for ropes.

22.2   Write out pseudocode to the insertion operation for ropes.

22.3   Give a slightly modified implementation of DELETE for ropes that is better for parallelism.

22.4   Write the preprocessing step for SPLIT that converts a rope into a form where the insertion point is at the start of a node.

22.5   Give an alternative implementation of INSERT for ropes that uses calls to SPLIT and CONCAT instead of traversing the tree directly.

22.6   We often want to find ranges of a string by lines—substrings separated by newline characters. This is important for a text editor that needs to show some number of lines of the current string. Give an extension to ropes that enables efficient search for the ith line of the file.

22.7   Implement gap buffers and ropes and compare the performance over various distributions of random edits.

Part IV

Sketching

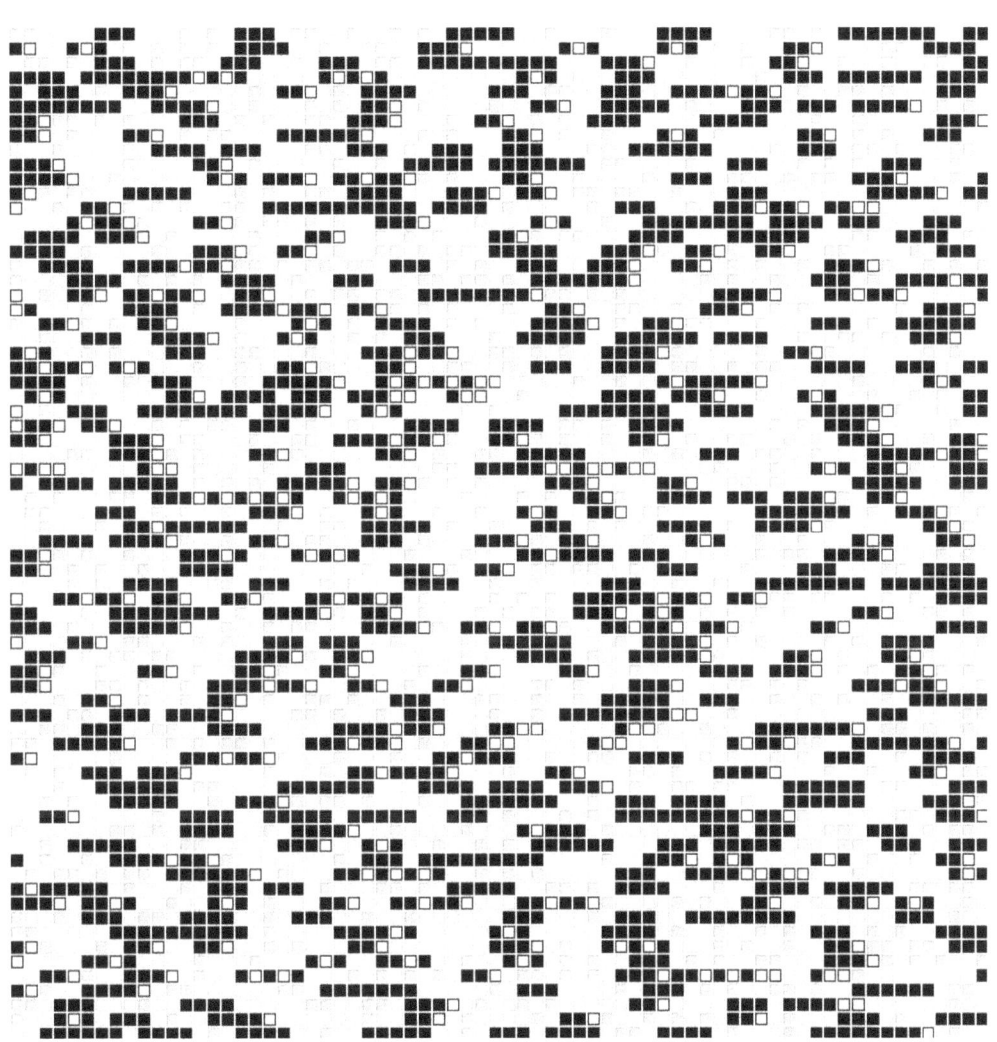

CHAPTER 23

Locality Sensitive Hashing

We will continue talking about algorithms that work in very small space, but with two twists: (1) the data is too large even to store a compressed representation, and (2) we allow some errors (with low probability) in the answers to our queries. Hash functions are a technique that helps with these criteria. In this chapter, we will see a use of hash functions for storing summaries, or *sketches*, of documents so that similarity between documents can be estimated. In the next chapter, we'll see a use of hash functions for storing sets of items.

Computing the similarity between documents, strings, or sets is a very common and fundamental operation. If you have a large collection of documents (i.e., web pages, genomic sequencing reads), you might want to cluster these documents, remove near duplicates, or find the most similar document to a query document. All of these tasks rely on being able to estimate pairwise similarities between the items in your collection based on some similarity measure $\text{sim}(X, Y)$ between X and Y. This similarity measure must be between 0 and 1.

The appropriate similarity function depends on the application and on the type of data. When comparing natural language text, the similarity measure could compare the sets of words that are present. If items are vectors, a natural measure of similarity is the cosine of the angle between the vectors. When comparing genomic sequences, the edit similarity may make the most sense. We will see examples of each of these in this chapter.

To quickly estimate the similarity between items, we will design a "sketch" of each of the items so that we can store these sketches in a reasonable amount of space, and we choose a $\text{sim}(X, Y)$ function that works directly on the sketches. That way we avoid ever looking at the full documents after computing their sketches.

23.1   Estimating similarity between strings

One class of such sketches is based on *locality sensitive hash functions* introduced in Indyk and Motwani [1998]:

Definition 23.1 (A Locality Sensitive Hash (LSH) Family). A set F of hash functions is a *locality sensitive hash family* for similarity measure sim if

$$\Pr_{h \in F}[h(X) = h(Y)] = \text{sim}(X, Y) \tag{23.1}$$

for all X, Y in the domain of the functions in F. Note that the probability is taken over the choice of hash function, drawn from the hash family F. ∎

Suppose such an LSH family existed. How could we use it to estimate the similarity $\text{sim}(X, Y)$ between X and Y? The idea is that we could draw t hash functions from this family and compute:

$$\text{sketch}(X) = \langle h_1(X), h_2(X), h_3(X), \ldots, h_t(X) \rangle, \tag{23.2}$$

$$\text{sketch}(Y) = \langle h_1(Y), h_2(Y), h_3(Y), \ldots, h_t(Y) \rangle. \tag{23.3}$$

Define the random variable B_i to be 1 if $h_i(X) = h_i(Y)$, and define $\widehat{\text{sim}}(X, Y) = \frac{1}{t} \sum_{i=1}^{t} B_i$. In other words, $\widehat{\text{sim}}$ is the empirically estimated probability of a hash collision between the entries of the sketches of X and Y. We can show that $\mathbb{E}[\widehat{\text{sim}}(X, Y)] = \text{sim}(X, Y)$:

$$\mathbb{E}[\widehat{\text{sim}}(X, Y)] = \mathbb{E}\left[\frac{1}{t} \sum_{i=1}^{t} B_i \right] \tag{23.4}$$

$$= \frac{1}{t} \sum_{i=1}^{t} \mathbb{E}[B_i] \tag{23.5}$$

$$= \frac{1}{t} \sum_{i=1}^{t} \Pr[h_i(X) = h_i(Y)] \tag{23.6}$$

$$= \frac{1}{t} \sum_{i=1}^{t} \text{sim}(X, Y) = \text{sim}(X, Y), \tag{23.7}$$

where (23.5) follows from linearity of expectation, and (23.7) follows from Definition 23.1. In other words, the expected value of the empirical estimate of collisions equals our similarity. Usually, collisions in hashing are *bad*, but here they have been repurposed to be something good!

The benefit of this is that instead of storing the collection of documents directly, we store only $\text{sketch}(X)$ for each document X, and maybe we can make $\text{sketch}(X)$ small by using a family of hash functions with a small range and a small t. In fact, if we're a little flexible with our similarity measure, we can always have the hash functions output a single bit in $\{0, 1\}$. That way a sketch can be stored as simply a bit vector of length t. To see this, we need to introduce another (key) definition related to hash functions:

Definition 23.2 (Universal Hash Function Family). A set $\mathcal{H}$ of hash functions, each from the universe U to the range $\{1, \ldots, m\}$, is called *universal* if

$$\Pr_{h \leftarrow \mathcal{H}}[h(X) = h(Y)] = \frac{1}{m} \tag{23.8}$$

for all $X \neq Y \in U$. In other words, we can treat the probability of collision the same as if the hash function distributes items into the m bins uniformly. ■

Given any U and m, there exists a universal hash function family [Cormen et al., 2009]. We can use such universal hash functions to restrict our LSH function to families of binary hash functions.

Lemma 23.3. *Given an LSH family F for similarity sim(X, Y), then there is an LSH family F'* *for similarity* $sim'(X, Y) = \frac{1+sim(X,Y)}{2}$ *where each* $h' \in F'$ *outputs a single bit.*

Proof: Suppose the hash functions in F go to some range R. Let B be a set of universal hash functions such that $b \in B$ goes from R to $\{0, 1\}$. Construct $F' = \{b \circ h \mid$ for all $b \in B$ and $h \in F\}$; that is for $h' \in F'$, we have $h'(X) = b(h(X))$. Then there is a collision between X and Y using h' when either there is a collision using h or when there is not a collision using h but there is a collision using b:

$$\Pr_{b \circ h \in F'}[b(h(X)) = b(h(Y))] = sim(X, Y) + (1 - sim(X, Y))(1/2) \qquad (23.9)$$

$$= \frac{1 + sim(X, Y)}{2}. \qquad (23.10)$$

$\square$

23.2 An LSH family for the Jaccard similarity

We will now see that an LSH family can be constructed for the Jaccard similarity [Broder, 1997]. The Jaccard similarity measure is a widely used similarity measure between two sets of items. If X and Y are sets, then

$$\text{Jaccard}(X, Y) = \frac{|X \cap Y|}{|X \cup Y|}. \qquad (23.11)$$

The Jaccard is often used to compare documents of text, for example, where X and Y are taken to be the set of words (or k-mers) contained each document. The Jaccard measures what fraction of the words that appear in either document are shared between the two documents. We will see an LSH family F_{Jaccard} for Jaccard introduced by Broder [1997].

A *permutation* of a set X is a function π from elements in X to the numbers $\{1, \ldots, |X|\}$ such that π doesn't map two elements of X to the same number. For example, if $X = \{a, b, c, d, e\}$, one permutation π is:

$$\begin{array}{cccccc} X = & a & b & c & d & e \\ \pi = & 5 & 2 & 3 & 4 & 1. \end{array} \qquad (23.12)$$

Because each of the items in X are assigned to a unique number, this is equivalent to specifying an ordering of the elements of X. In the case of (23.12), π gives the order *ebcda*.

To define F_{Jaccard}, we will use the concept of a *minwise independent permutation family*:

Definition 23.4 (Minwise independent permutation family). A set P of permutations on U is *minwise independent* if, for each set $X \subset U$ and any $e \in X$, we have:

$$\Pr_{\pi \in P}\left[\min_{x \in X}\{\pi(x)\} = \pi(e)\right] = \frac{1}{|X|}. \qquad (23.13)$$

■

This definition says that all elements $e \in X$ have equal chance to become the minimum when shuffling according to a randomly chosen permutation π in P. Small minwise independent families exist [Indyk, 2001], but at least the set of all permutations is minwise independent since any element is equally likely to become the minimum.

We now define:

$$F_{\text{Jaccard}} = \left\{ h_\pi(X) = \min_{x \in X}\{\pi(x)\} \text{ for all } \pi \text{ on } U \right\}. \tag{23.14}$$

That is, each member of our LSH family of hash functions maps a set to its minimum element under a randomly chosen permutation of the elements of our universe.

Theorem 23.5. *Let h be a randomly chosen function from F_{Jaccard}. Then*

$$\Pr_h[h(X) = h(Y)] = \Pr_h\left[\min_{x \in X}\pi(x) = \min_{y \in Y}\pi(y)\right] \tag{23.15}$$

$$= \frac{|X \cap Y|}{|X \cup Y|} = Jaccard(X, Y). \tag{23.16}$$

Proof: Let $\alpha = \min_{a \in X \cup Y}\{\pi(a)\}$. Then

$$\min_{x \in X}\{\pi(x)\} = \min_{y \in Y}\{\pi(y)\} = \alpha \tag{23.17}$$

if and only if $\alpha \in \pi(X \cap Y)$. To see this, remember that each element in the entire universe is assigned a unique position by π, and π is a permutation of the universe, not just of the items in one of the sets. So we have the following picture.

$$\tag{23.18}$$

The permutation π shuffles the elements of U (the universe). This is like shuffling the rectangle "cards." Then $\min_{x \in X}\{\pi(x)\}$ is the shuffled location of the leftmost card with a ⊘ and $\min_{y \in Y}\{\pi(y)\}$ is the shuffled location of the leftmost card with a ◉. Viewed this way, we can see that:

$$\alpha = \min_{x \in X}\{\pi(x)\} = \min_{y \in Y}\{\pi(y)\} \Longleftrightarrow$$

$$\pi^{-1}(\alpha) \in X \cap Y = \{\text{things in both } X \text{ and } Y\}. \tag{23.19}$$

This gives:

$$\Pr[\min_{x \in X}\{\pi(x)\} = \min_{y \in Y}\{\pi(y)\}] = \Pr[\alpha \in \pi(X \cap Y)] \tag{23.20}$$

$$= \sum_{e \in X \cap Y} \Pr\left[\min_{a \in X \cup Y}\{\pi(a)\} = \pi(e)\right]. \tag{23.21}$$

In (23.21), e ranges over the "cards" for which it is possible to have the same minimum: the different possibilities for $\pi^{-1}(\alpha)$. The equality in the $[]$ is the event that the e "card" was put first among all the relevant cards. Continuing, we can simplify (23.21) with

$$= \sum_{e \in X \cap Y} \frac{1}{|X \cup Y|} \tag{23.22}$$

$$= \frac{|X \cap Y|}{|X \cup Y|}, \tag{23.23}$$

where (23.22) follows from minwise independence.   □

The MinHash algorithm. We can put these ideas together into what is called the MinHash algorithm: Choose some number t of hash functions $h_1, \ldots, h_t$ from F_{Jaccard}. This proof shows that each $h \in F_{\text{Jaccard}}$ has a collision with probability equal to the Jaccard similarity. The sketch(X) of each document X is the vector of the application of these t hash functions to X. To estimate Jaccard(X, Y), use Equal(sketch(X), sketch$(Y))/t$, where the Equal function is the number of dimensions in which two vectors are equal.

How many samples do we need? The last question is: how large must t be for the error to be small. Intuitively, the error decreases rapidly with t (the estimate when $t = 1$ is horrible, but at $t = 100$ the number of collisions is probably pretty close to the true probability of collision). To get a better handle on this, we can use Chernoff bounds.

Theorem 23.6 (Chernoff bounds, simple form). *Let B be the sum of independent 0/1 random variables B_i. Let $\mu = \mathbb{E}B$. Then*

$$\Pr[B \geq (1+\delta)\mu] \leq \exp\left(\frac{-\delta^2\mu}{3}\right) \tag{23.24}$$

$$\Pr[B \leq (1-\delta)\mu] \leq \exp\left(\frac{-\delta^2\mu}{2}\right) \tag{23.25}$$

for any $0 < \delta < 1$.

In our case, let

$$B_i = \begin{cases} 1 & \text{if } h_i(X) = h_i(Y) \\ 0 & \text{otherwise} \end{cases} \tag{23.26}$$

and let $B = \sum_i B_i$. Our estimate of the similarity is B/t and $\mu = t \cdot \text{Jaccard}(X, Y)$. Substitute these values into (23.25) and we get:

$$\Pr\left[B \leq (1 - \delta)t \cdot \text{Jaccard}(X, Y)\right] \leq \exp\left(-\frac{\delta^2 \text{Jaccard}(X, Y)}{2}t\right). \tag{23.27}$$

Setting $t = 2/(\delta^2 \text{Jaccard}(X, Y)) \ln(1/\epsilon) = O(\log \frac{1}{\epsilon})$ gives the probability of error to be $\leq \epsilon$. The case of overestimating (23.24) is almost exactly the same.

This choice of t depends on the size of the Jaccard distance. This makes sense: very small Jaccard distances are going to need a large number of samples to get within a small multiplicative factor. In practice, one can often assume a reasonable lower-bound on the Jaccard distances you really care about.

23.3 An LSH family for the weighted Jaccard

Using the Jaccard similarity to compare sequences using the k-mers they contain does not account for k-mers occurring multiple times in the sequence. K-mers occurring more than once is the typical situation when the k-mers are much shorter (as usual) than the sequences being compared. If we don't account for k-mer multiplicities, we're throwing away information about how similar two sequences are: "AA" and "AAAAA" contain the same set of k-mers, and have Jaccard similarity 1, but of course are pretty different sequences.

To improve our estimate, we could use the *weighted* Jaccard over multisets. A multiset A is a set where each element $x \in A$ has a count $\chi_A(x)$ of occurrences in the set. Then, the weighted Jaccard is:

Definition 23.7 (Weighted Jaccard). The *weighted Jaccard* similarity between two multisets (A, χ_A) and (B, χ_B) is

$$J^w(A, B) = \frac{\sum_x \min\{\chi_A(x), \chi_B(x)\}}{\sum_x \max\{\chi_A(x), \chi_B(x)\}}. \tag{23.28}$$

■

When A and B are sets, χ_A and χ_B are each always either 0 or 1, and (23.28) exactly equals the unweighted Jaccard similarity (see Exercise 23.5).

To obtain an LSH for the weighted Jaccard, we make each of the instances of each k-mer unique in the sequence by pairing it with its occurrence number. If k-mer x occurs i times, then we have

$$(x, 0), (x, 1), \ldots, (x, i - 1) \tag{23.29}$$

in our set. Specifically, if M is the multiset of k-mers in a string S, the representation of S is:

$$R(S) = \{(x, 0), \ldots, (x, \chi_M(x) - 1) \mid x \in S\}. \tag{23.30}$$

We've turned a multiset M into a normal set $R(S)$, but of pairs instead of k-mers. We can now apply the MinHash algorithm to obtain an LSH for the weighted Jaccard. Our universe of elements has expanded. Instead of strings in Σ^k, we now have pairs that are subsets of $\Sigma^k \times \{0, \ldots, |S| - 1\}$ for a string S. This means we should choose random permutations over this expanded universe. That is, if our strings are all of length $\leq n$, we will choose π as a random permutation over the set of all permutations of $\Sigma^k \times \{0, \ldots, n-1\}$.

We used the occurrence number to create a regular set, but when comparing two sequences S_1 and S_2, if $(k, i) \in R(S_1)$ and $(k, j) \in R(S_2)$ these are still the "same" k-mer, even if $i \neq j$. Our hash function, after doing the permutation, should throw away the occurrence number. Specifically, we define the set of hash functions as:

$$H = \left\{ h_\pi(R) = \left(\arg\min_{(x,i) \in R} \pi(x, i) \right) [0] \right\}, \tag{23.31}$$

where the $[0]$ notation means take the first item of the minimum (x, i) pair. In other words, the hash function uses a random permutation to permute the universe of unique pairs, finds the minimum pair under the permutation that exists in our representation, and returns the k-mer (only) that is associated with the minimum element.

Applying the same argument as with MinHash, we can show:

Theorem 23.8 (Weighted Jaccard LSH). *H defined in (23.31) is an LSH for a weighted Jaccard.*

23.4 An LSH family for a cosine-related similarity

Often documents are not represented by sets but by vectors. For example, a webpage could be represented by a vector v that has an entry for each keyword and contains 1 in that entry if the keyword is present and 0 if not. A good measure of the distance (dissimilarity) between two documents u, v is then the *angle* $\theta(u, v)$ between the vectors that represent them, or some function of that angle. This distance is frequently used in information retrieval applications. We will see an LSH family F_{AngleSim} for the measure:

$$\text{AngleSim}(u, v) = 1 - \frac{\theta(u, v)}{\pi} \tag{23.32}$$

between two vectors u and v. This LSH family was introduced by Charikar [2002].

The main idea is to choose a random hyperplane r and define a hash function as:

$$h_r(x) = \begin{cases} 1 & r \cdot x \geq 0 \\ 0 & r \cdot x < 0. \end{cases} \tag{23.33}$$

We then take

$$F_{\text{AngleSim}} = \{h_r(x) \text{ for all choices of random hyperplanes } r\}. \tag{23.34}$$

Theorem 23.9. *When h is drawn randomly from $F_{AngleSim}$, we have:*

$$\Pr_h[h(u) = h(v)] = 1 - \frac{\theta(u, v)}{\pi} = AngleSim(u, v) \tag{23.35}$$

for all u, v.

Proof: The probability that a random hyperplane splits u and v is $2\frac{\theta(u,v)}{2\pi}$, where the factor of 2 comes from the fact that we can specify the hyperplane in two different ways.

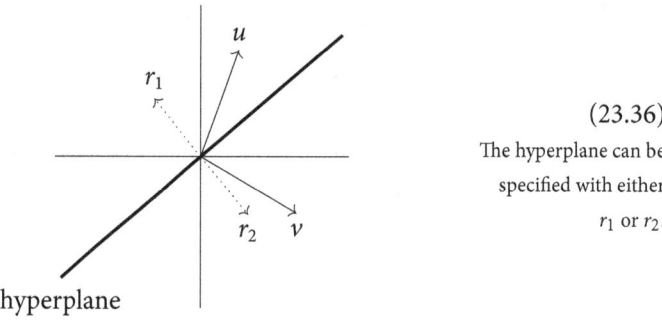

(23.36)

The hyperplane can be specified with either r_1 or r_2.

□

The arguments about how many samples you need for Jaccard work just as well for AngleSim.

23.5 A gapped LSH family for edit distance

One would like an LSH for the edit distance between strings, instead of merely settling for the Jaccard of the string's constituent k-mers. To work toward this, we will generalize our definition of LSH to *gapped LSH*:

Definition 23.10 (Gapped LSH). A distribution of hash functions from a family H is (s_1, s_2, p_1, p_2)-*sensitive* for similarity measure sim if, for all $x \neq y$,

$$\text{sim}(x, y) \geq s_1 \implies \Pr_{h \leftarrow H}[h(x) = h(y)] \geq p_1, \text{ and} \tag{23.37}$$

$$\text{sim}(x, y) \leq s_2 \implies \Pr_{h \leftarrow H}[h(x) = h(y)] \leq p_2, \tag{23.38}$$

where $s_1 \geq s_2$ and $p_1 \geq p_2$. Such a distribution of hash functions is called a *gapped LSH* for sim. ∎

This definition says that if two items x and y are similar enough (sim $\geq s_1$) then the probability that a hash function makes them collide is high enough ($\geq p_1$), and conversely, if they are not too similar (sim $\leq s_2$) then the probability of a collision is low ($\leq p_2$). For items of intermediate similarity, the definition places no restrictions on what the hash functions do. This is why we call it "gapped."

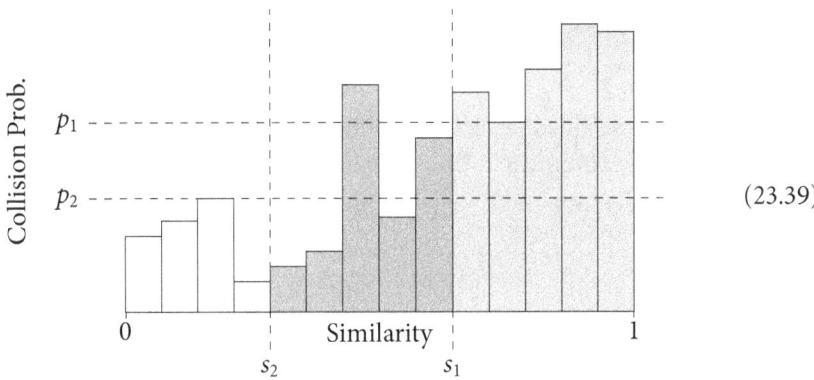

$$(23.39)$$

A family is a traditional LSH if it is (r, r, r, r)-sensitive for any $r \in (0, 1)$ (see Exercise 23.4). A gapped LSH that isn't a traditional LSH doesn't give us an estimate of the distance directly. Rather, it only distinguishes between high- and low-similarity items. This is often what we need in practice anyway: if we want to find all the similar pairs, we can first filter by those that collide: if the items are similar, that probability of collision is high, so we'll likely catch the similar pairs (and also likely not include the very dissimilar pairs).

Using the weighted Jaccard instead of the Jaccard when comparing strings preserves information about their multiplicities, but one can find (Exercise 23.3) two strings with the same k-mers and the same multiplicities, with a very large edit distance. What is missing is accounting for the *order* of the k-mers.

We can do this by extending our approach for weighted Jaccard to keep more k-mers, preserving their order. We introduce the technique with an example.

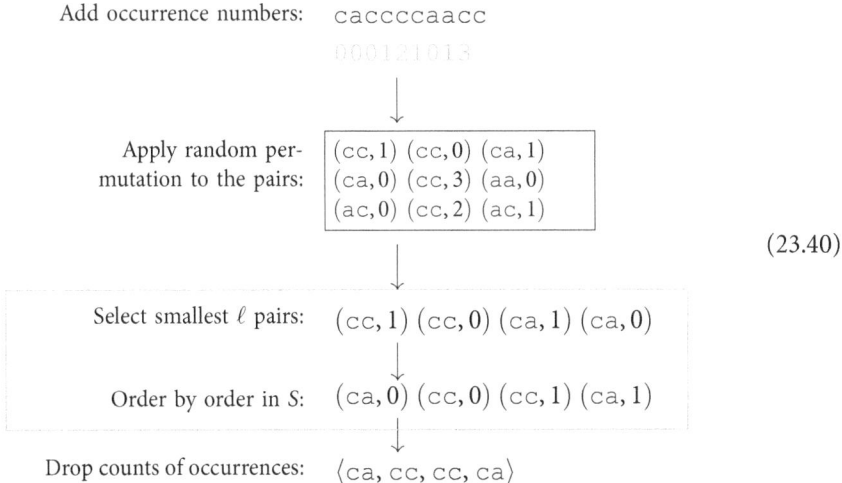

$$(23.40)$$

The part in the final box is the difference from the weighted Jaccard approach in Section 23.3. First, instead of keeping only the one arg min, we keep the ℓ smallest pairs according to the random permutation. Second, we order these pairs based on their order in the original sequence. In general, $h_\pi^\ell(S)$ is defined to be a hash function, using a random permutation on $\Sigma^k \times \{0 \ldots n-1\}$, that selects the ℓ smallest items, orders them according to their order in the sequence, and the output of the hash is a vector of the selected k-mers in this order.

This approach is called *Order Min Hash* (OMH) [Marçais et al., 2019a]. When $\ell = 1$, this is exactly the weighted Jaccard LSH defined in (23.31). For longer ℓ, more information about the ordering of k-mers in the sequence is preserved. In fact, one can prove that this hash family is a gapped LSH for edit distance:

Theorem 23.11. *Let n be the maximum length of a sequence and k be the choice of k-mer lengths. For any $\ell \in [2, n-k]$, and any choice of gapped LSH boundaries $0 < s_2 \leq s_1 < 1$, there are values p_1 and p_2 such that the Order Min Hash is (s_1, s_2, p_1, p_2)-sensitive for the edit dissimilarity.*

The edit dissimilarity is the traditional edit distance between strings of length n that counts the number of substitutions, insertions, and deletions, all divided by n. Unfortunately, we don't have space here to prove Theorem 23.11, the current simplest version of which is somewhat involved; the full proof is found in Marçais et al. [2019a].

23.6 Summary and notes

Hashing is a powerful technique. We've seen it used in string algorithms in Wu-Manber, Rabin-Karp, and now for the perhaps unexpected use of estimating similarities for various types of measures. More background regarding sketching and its use in genomics can be found in Marçais et al. [2019b]. Other sketches for edit distance that take a two-step approach of first embedding the edit similarity into a simpler metric can be found in Bar-Yossef et al. [2004] and Ostrovsky and Rabani [2007].

> **Presentation Notes**
>
> ---
>
> Some of the material in this chapter is derived from CMU 15-451 course lecture notes.

23.7 Exercises

23.1   Let F be a minwise independent set of permutations over a universe U. Let $n = |U|$ and assume n is large. Prove that $|F|$ is $\Omega(LCM(\{2, \ldots, n\}))$, where $LCM(S)$ is the least common multiple of the integers in set S.

23.2   To compare two strings x and y each of length n, one often uses the Jaccard similarity between the k-mers of x and y. Let $edit(x, y)$ be the edit distance between

x and y with the cost of mismatches and indels $= 1$. Define the edit similarity:

$$editsim(x, y) = 1 - edit(x, y)/n \qquad (23.41)$$

Give a construction where the Jaccard similarity between the sets of k-mers for x and y is 1, but the edit similarity is $\leq 2k/n$.

23.3   Extend Exercise 23.2 to find an example where the multiset of k-mers is the same, but the edit distance is large.

23.4   Prove that if a hash function family H is (r, r, r, r)-sensitive for *sim* for any $r \in (0, 1)$, then $\Pr_{h \leftarrow H}[h(x) = h(y)] = sim(x, y)$ for all x, y.

23.5   Prove that when A and B are (non-multi) sets, the weighted Jaccard similarity (Definition 23.7) equals the Jaccard similarity (Eq. 23.11).

CHAPTER 24

Bloom Filters

Let's change gears to a completely different problem, but one where (a) hashing still provides an elegant solution and (b) space is at a premium. Suppose we want to maintain a set S that is a subset of a very large universe U of possible items. We want our storage to be small, and we want to support the following two operations:

1. `add`(S, x): put item x into our set S,
2. `query`(S, x): return **true** if x has been added to S in the past.

We do not need to support deletion of elements. It's possible to solve this problem using a hash table, but that would use $\log |U|$ bits per item (plus the overhead for any unused entries in the hash table). That's too large for some applications, and we can get a huge improvement if we allow some error.

Specifically, if $x \in S$ we will be *required* to return **true** on `query`(S, x). However, we allow some false positives: if $x \notin S$ we can wrongly return **true** when S is queried for x. Of course, we want to make this false positive rate (FPR) as small as possible.

What is an application of this data structure? Suppose we have a very long list of websites that host malicious code. Whenever a user visits a website we would like to quickly check whether that website is one of the malicious ones. Over time, new malicious sites will be discovered and need to be added to our list. We can take S to be the set of bad websites, and when a new one is discovered we can `add` it to S. At every visit to a site, we use `query` to check whether the site is malicious. We can tolerate some false positives, since these will just result in a warning to the user that they can ignore (or we can use a more expensive check to see if the site really is malicious). We can't tolerate false negatives, since saying a website is safe when it is known not to be is very bad.

The Bloom filter is one solution to this problem [Bloom, 1970]. It has been widely used for network applications and for problems in genomics. Apparently, some versions of the Chrome browser used a Bloom filter for exactly this malicious website application. This allows it to ship a small data structure that stores a long list of bad websites. Bloom filters also have the advantage that it is not possible to easily extract the actual items contained within them—this allows the list of bad websites to be kept somewhat secret.

24.1 How do Bloom filters work?

A Bloom filter is simply an array B of m bits, plus t hash functions $h_1, \ldots, h_t$ where each $h_i : U \to \{0, \ldots, m-1\}$. We'll see how to choose m and t shortly. B is initially set to all zeros. The operations can be implemented as follows.

1. $\text{add}(S, x)$—set all bits given by the hash functions to 1:

    for $i \leftarrow 1 \ldots t$ **do**
    $\quad B[h_i(x)] \leftarrow 1$

2. $\text{query}(S, x)$—return **true** if all the bits given by the hash functions are 1:

    for $i \leftarrow 1 \ldots t$ **do**
    $\quad$ **if** $B[h_i(x)] = 0$ **then**
    $\quad\quad$ **return false**
    return true

For example, in the case (24.1), we know that x is probably in S (and we would report that it was in the set). We know for sure that y is not in the set, since otherwise all the bits the hash functions point to would be set to 1.

$$h_1(x) \qquad h_2(x) \qquad h_3(x)$$

| 1 | 1 | 0 | 0 | 1 | 0 | 0 | 1 | 1 | 1 | 0 | 0 | 0 | 1 | 0 | 1 |

$$h_3(y) \qquad h_1(y) \qquad\qquad h_2(y)$$

(24.1)

24.2 How should we set the filter parameters?

It's clear that if we've added x to S, then $\text{query}(S, x)$ will always be true, so we have no false negatives. But it might be the case that, due to hash collisions, all the bits for x were set by other elements even though x was never added. This will lead to a false positive. So the question is: how often does that happen? That depends on m and t: increasing the space of the Bloom filter (m) decreases the chance of a collision and hence decreases the FP rate. Increasing t will mean more bits are set to 1 for each item. We now see how to choose these parameters to minimize the error rate.

Theorem 24.1. *The probability of a false positive in a Bloom filter is minimized by choosing* $t = (m/n) \ln 2$, *where* n *is the number of items added to the filter,* t *is the number of hash functions, and* m *is the length of the filter.*

Proof: After a single item is added, the probability that a particular bit is 0 is $\left(1 - \frac{1}{m}\right)^t$. Since the hash functions can be viewed as "randomly" flipping bits, we can assume that the bits being flipped are independent. After n items have been added, the probability that a particular bit is set to 1 is:

$$1 - \left(1 - \frac{1}{m}\right)^{tn}. \tag{24.2}$$

We get a false positive when all t bits are set for a given element x. The probability of this is:

$$\Pr[\text{FP}] = \left(1 - \left(1 - \frac{1}{m}\right)^{tn}\right)^{t} \tag{24.3}$$

$$= \left(1 - \left[\left(1 - \frac{1}{m}\right)^{m}\right]^{tn/m}\right)^{t} \tag{24.4}$$

$$\approx \left(1 - e^{-tn/m}\right)^{t}, \tag{24.5}$$

where we've used the fact that $(1 - 1/m)^m$ is about e^{-1}.

This expression is minimized by choosing $t = \frac{m}{n} \ln 2$. We can see this by taking the derivative of the natural log of the expression with respect to t—minimizing the log of an expression is equivalent to minimizing the expression since the log function is monotonically increasing:

$$\frac{\partial \ln \Pr[\text{FP}]}{\partial t} = \frac{\partial}{\partial t} \left[t \ln(1 - e^{-tn/m})\right] \tag{24.6}$$

$$= t \frac{\partial}{\partial t} \left[\ln(1 - e^{-tn/m})\right] + 1 \ln(1 - e^{-tn/m}) \tag{24.7}$$

$$= t \left[\frac{-e^{-tn/m}\left(-\frac{n}{m}\right)}{1 - e^{-tn/m}}\right] + \ln(1 - e^{-tn/m}), \tag{24.8}$$

where (24.7) uses the product rule. Setting to 0 lets us find the minimum:

$$t \left[\frac{e^{-tn/m}\left(\frac{n}{m}\right)}{1 - e^{-tn/m}}\right] + \ln(1 - e^{-tn/m}) = 0 \tag{24.9}$$

$$\ln(1 - e^{-tn/m}) = -t \left[\frac{e^{-tn/m}\left(\frac{n}{m}\right)}{1 - e^{-tn/m}}\right] \tag{24.10}$$

$$(1 - e^{-tn/m}) \ln(1 - e^{-tn/m}) = e^{-tn/m}\left(\frac{-tn}{m}\right) \tag{24.11}$$

$$(1 - e^{-tn/m}) \ln(1 - e^{-tn/m}) = e^{-tn/m} \ln(e^{-tn/m}), \tag{24.12}$$

where the last line uses the fact that $x = \ln e^x$. Equation (24.12) is of the form $(1 - x) \ln (1 - x) = x \ln x$. This means that $(1 - x) = x$, so we have:

$$(1 - e^{-tn/m}) = e^{-tn/m}. \tag{24.13}$$

Solving this for t using straightforward manipulations gives us $t = \frac{m}{n} \ln 2$. □

Using these optimal parameters, a Bloom filter uses a small number of bits per item, as the next theorem shows.

Theorem 24.2. *Let ρ be a given false positive probability. A Bloom filter with FPR ρ containing n items can be achieved with $O(n \log \frac{1}{\rho})$ bits.*

Proof: We will use the optimal number of hash functions from the previous theorem. Substituting that into (24.5), we have:

$$\rho \approx \left(1 - e^{-((m/n)\ln 2)n/m}\right)^{(m/n)\ln 2} \tag{24.14}$$

$$= \left(1 - e^{-\ln 2}\right)^{(m/n)\ln 2} \tag{24.15}$$

$$= \left(\frac{1}{2}\right)^{(m/n)\ln 2}. \tag{24.16}$$

Solving for ρ:

$$\ln \rho = \ln 2^{-(m/n)\ln 2} = -(m/n)(\ln 2)^2. \tag{24.17}$$

Therefore, a given ρ error probability can be achieved by setting $m = \frac{n\ln(1/\rho)}{(\ln 2)^2}$. □

Thus, we need only $O(\log \frac{1}{\rho})$ bits on average per item if we are willing to tolerate ρ chance of a false positive.

24.3 Notes on and extensions to Bloom filters

Where are the items? A Bloom filter stores a set of items *without ever storing the items*. The access time is always $O(tH)$, where H is the time to compute the hash value, independent of the number of items in the filter.

Computing the union of two sets. Suppose we have Bloom filters B_1 and B_2 of the same length, using the same set of hash functions. We can compute a Bloom filter $B_1 \cup B_2$ that represents the union of the sets that B_1 and B_2 encode by computing the bitwise OR of B_1 and B_2. The resulting vector has 1 bit everywhere that would have been set if the union of the items were directly inserted into it.

Deletion. It's not possible to delete an element from a Bloom filter without possibly introducing false negatives (missed elements). This is because if two elements x and y both set bit i to 1, setting it to 0 when deleting x will also delete y. One partial solution to this is to maintain two Bloom filters: the regular filter that contains all the items ever added to S and a deletion filter D that contains all the items ever deleted from S. Then `query(S, x)` returns **true** if x is in B but not in D, but now false positives in D correspond to false negatives.

Counting Bloom filters. If instead of a single bit in each entry, we store a small number of bits, we can support limited counts of the items [Fan et al., 2000]. Adding an item increments the entry by 1 unless it is already all 1s, in which case no increment happens (and the count is saturated). Deletion decrements the counts. To estimate the number of items that have been added, one takes the minimum value encoded on each of the locations to which it hashes. So long as the count of any item is less than or equal to a number that can be represented by each entry, this data structure supports insertion, deletion, and approximate counting queries, though not in optimal space (since we keep extra bits around even for entries that don't need them).

Changing the size of a Bloom filter. In general, it is not possible to resize a Bloom filter after items have been added to it since this would move the targets of the hash functions around. However, if the filter length is a power of 2, its size can be (repeatedly) reduced by a factor of 2. To do this, we "fold" the left half of the filter onto the right half, "OR"ing them together. We then modify our hash functions to throw away their high-order bit, also reducing their range by a factor of 2. If it is difficult to guess how many items will be in the filter, one can start out with a large filter and halve it repeatedly as it becomes clear that fewer items will be stored. Unfortunately, *doubling* the size of a filter is not as easy.

24.4 Cascading Bloom filters

Chaining together Bloom filters has been a common technique, either to reduce false positives, or to aid in storing additional information with each item. We will now see one (non-optimal) technique for this [Chazelle et al., 2004].

Suppose you want to store a set S, and each item in the set is assigned a number $f(x) \in \{1, 2\}$. That is, we want to store both S, and a two-coloring of S (an item is either color 1 or color 2). To do this, we create *two* filters to store S: one filter A_0 holds the set $\{x \in S \mid f(x) = 1\}$. The other filter B_0 holds the set $\{x \in S \mid f(x) = 2\}$. Assume that these filters have a false positive error of ϵ. To test if an item is in S and retrieve its value, we ask both $x \in A_0$? and $x \in B_0$? Look at the possible answers.

$x \in A_0$?	$x \in B_0$?	implication
no	no	$x \notin S$
yes	no	$x \in S$ with probability $(1 - \epsilon)$ and $f(x) = 1$.
no	yes	$x \in S$ with probability $(1 - \epsilon)$ and $f(x) = 2$.
yes	yes	Problem!

The last case is a problem: we don't know which of the two answers is a false positive (or perhaps they both are!). We are not able to tell whether $f(x)$ is 1 or 2. Here's where chaining together more Bloom filters helps. We create two Bloom filters A_1 and B_1. A_1 stores the items in A for which "$x \in B_0$?" falsely says "yes". B_1 stores the items in B for which "$x \in A_0$?" falsely says "yes". We continue this until the sets are small enough that they can be stored explicitly.

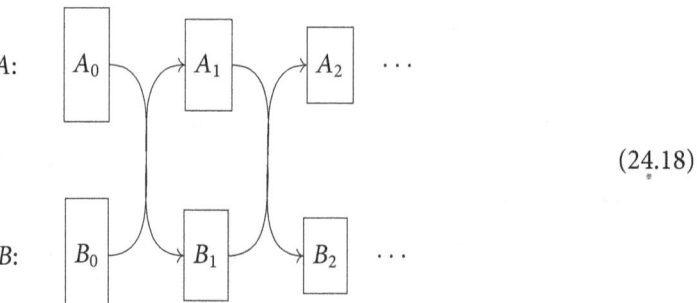

$$(24.18)$$

Since the filters down the line need to store smaller and smaller sets, we make them smaller as we go down the chain. To be more specific, suppose we have t hash functions in our filters, and let $n_i = \max\{|A_i|, |B_i|\}$—that is n_i is the maximum number of elements that one of the filters at level i has to store. We know n_i when we're constructing this chain of filters, since at construction time we have the set S and the values of function $f(x)$ for each $x \in S$. Make the filter at level i have size:

$$2^{t^i} t n_i \qquad (24.19)$$

bits. This size is chosen so that the expected total size of all the filters will be small. Even though the $2^{t^i} t$ factor is growing quickly in terms of i, we expect the n_i term to shrink fast, since we have to encounter a rare false positive for an item to be passed to the next level. Let's look at the chance for an item x that is in A_i to be passed on to A_{i+1}. For this to happen, x must be a false positive in filter B_i. There are at most $t|B_i|$ bits set to 1 in filter B_i (since each item stored it in sets at most t bits). Filter B_i is, by construction, $2^{t^i} t n_i$ bits long. So $t|B_i|/(2^{t^i} t n_i)$ fraction of the bits are set to 1 in B_i. For there to have been a false positive, each of the t hash functions must have hit one of these set bits. That happens with probability:

$$\left(\frac{t|B_i|}{2^{t^i} t n_i} \right)^t. \qquad (24.20)$$

The two ts cancel out, and $|B_i| \le n_i$ by definition of n_i. So we have the probability of an item going from level i to level $i + 1$:

$$\left(\frac{t|B_i|}{2^{t^i} t n_i} \right)^t \le \left(\frac{1}{2^{t^i}} \right)^t = 2^{-t^{i+1}}. \qquad (24.21)$$

At each level, there is quite a small probability that an item gets passed along to the next level.

How many items end up in level i? For that to happen, they have to pass through all the previous levels, which happens with probability:

$$\prod_{j=1}^{i} 2^{-t^j} = 2^{-\sum_{j=1}^{i} t^j}. \qquad (24.22)$$

The $\sum_{j=1}^{i} t^j$ term in the exponent is almost a geometric series, except we start at $j=1$ instead of $j=0$, so we reindex:

$$\sum_{j=1}^{i} t^j = \sum_{j=0}^{i-1} t^{j+1} = t\sum_{j=0}^{i-1} t^j. \tag{24.23}$$

Now we can use the identity $\sum_{b=0}^{n} c^b = \frac{c^{n+1}-1}{c-1}$:

$$t\sum_{j=0}^{i-1} t^j = t\frac{t^i-1}{t-1} = \frac{t^{i+1}-t}{t-1}. \tag{24.24}$$

Putting this value for the exponent back into (24.22), we have that the probability that any item ends up at level i is less than or equal to:

$$2^{-(t^{i+1}-t)/(t-1)} \tag{24.25}$$

and the expected number of elements at level i is

$$\mathbb{E}n_i = |S|2^{-(t^{i+1}-t)/(t-1)}. \tag{24.26}$$

What is the total expected size then of all our cascade of filters? We plug in our estimate of the number of elements at level i into our filter size (24.19) and sum them up. Suppose we have α total levels:

$$\mathbb{E}\sum_{i=0}^{\alpha} 2^{t^i} tn_i = t\sum_{i=0}^{\alpha} 2^{t^i}\mathbb{E}n_i \tag{24.27}$$

$$= t\sum_{i=0}^{\alpha} 2^{t^i}|S|2^{-(t^{i+1}-t)/(t-1)} \tag{24.28}$$

$$= t|S|\sum_{i=0}^{\alpha} 2^{t^i-(t^{i+1}-t)/(t-1)} \tag{24.29}$$

$$= t|S|\sum_{i=0}^{\alpha} 2^{(t^i(t-1)-(t^{i+1}-t))/(t-1)}. \tag{24.30}$$

Distributing and canceling the terms in the exponent gives us:

$$\sum_{i=0}^{\alpha} 2^{(t^i(t-1)-(t^{i+1}-t))/(t-1)} = \sum_{i=0}^{\alpha} 2^{-(t^i-t)/(t-1)}, \tag{24.31}$$

which is bounded above by a constant. So our total expected space usage from (24.30) is:

$$O(t|S|). \tag{24.32}$$

We therefore expect to use, on average, t bits for every item in our set.

24.5 Summary and notes

There has been an impressive line of work improving on data structures for solving "approximate membership queries," which are the type of queries that Bloom filters support (one-sided, bounded error answers to $x \in S$?). Improvements include near optimal space usage, supporting both insertion and deletion, removing the assumptions about the hash functions that are used, among other advancements.

One such technique are *single hash filters*. To store a set S, these draw a random universal hash function $h : U \to \{0, \ldots, |S|/\epsilon\}$ and store the set $S_h = \{h(x) \mid x \in S\}$. If set S_h can be stored compactly, then this gives us the properties we want: if $x \in S$ then $h(x) \in S_h$. If $x \notin S$, then

$$\Pr_h[h(x) \in S_h] \leq \sum_{y \in S} \Pr_h[h(x) = h(y)] \tag{24.33}$$

$$= \sum_{y \in S} (1/(|S|/\epsilon)) = |S|/(|S|/\epsilon) = \epsilon, \tag{24.34}$$

where $\Pr_h[h(x) = h(y)] = 1/(|S|/\epsilon)$ since we assume h is a universal hash function. Implementing this scheme comes down to finding efficient ways to store set S_h, and there has been considerable work on this [Pagh et al., 2005].

Trees of related Bloom filters have been used to enable large-scale search of biological sequences [Solomon and Kingsford, 2016, 2017; Sun et al., 2018].

Presentation Notes

A good reference for analysis and use of Bloom filters is Broder and Mitzenmacher [2005], and most of the derivation here of the false positive rate and optimal number of hash functions is from there. The derivation for finding the minimum in Theorem 24.1 was communicated to me by Ke Chen.

24.6 Exercises

24.1   A strategy to partially implement deletion in Bloom filters is to construct a hybrid data structure A that maintains two Bloom filters B and D. B contains the items that have been added, and D contains the items that have been deleted. The add operation adds an item to A as usual. The (new) delete operation adds the item to D, and the query operation now returns true if and only if the traditional Bloom filter query returns true on A and false on D. If the false positive rates of B and D are f_B and f_D respectively, what is the false *negative* rate of the hybrid A?

24.2   Suppose you want to create a Bloom filter with an FPR of ρ to store $n > 0$ items. Give the (a) number of hash functions and (b) the length of the Bloom filter that will achieve this.

CHAPTER 25

Sketching with Minimizers

Suppose we want to align a short string r to a long string S. We do not expect r to always match exactly, so we'd like to permit edits such as insertions, deletions, and substitutions. One approach to do this is to use local or semi-global alignment (Section 7.5). But this is costly if S is very long, and even more costly if we want to align many $\{r_1, \ldots\}$ to the same S. This is the situation we are in if we have many reads r_i from a genomic sequencing experiment and we want to align them to a reference genome string S.

To speed up such alignments, we can compute a set of small sequences from S and their locations in such a way that we can easily check if an r_i contains any of those sequences. These "seed" sequences can then anchor an edit distance computation around the location of each matching seed in S.

The *minimizers* method is used to quickly infer that two strings, or a set of strings, are likely to have a match, or rule out the possibility that they have a match. Even though data structures such as suffix trees/arrays (Chapters 13 and 16) and the FM-index (Chapter 21) can answer exact match queries relatively fast (even optimally asymptotically speaking), these queries are still slow in some practical situations. We can speed things up by quickly separating strings that may have a match from those than cannot have a match before using techniques such as suffix array or FM-index. The minimizer method does this by defining a rule for selecting short substrings in a deterministic way.

The minimizers method was introduced in 2004 [Roberts et al., 2004] for the problem of computing read overlaps. This method is identical to the *winnowing* method introduced in 2003 for plagiarism detection [Schleimer et al., 2003]. The method did not see any other use than the original overlap computation until 2014. Since then, it has been adopted by many bioinformatics software programs (e.g., read aligner `minimap` [Li, 2016; Li and Birol, 2018], metagenomic classification with `kraken` [Wood and Salzberg, 2014; Wood et al., 2019], k-mer counter KMC [Deorowicz et al., 2015], etc.).

In addition to its practical use, the minimizer method is also quite interesting from a purely theoretical perspective, since it is tightly connected with properties of the de Bruijn graph, which is of independent interest.

In this chapter, we will see:

- an example of minimizers use,
- the basic properties of minimizers,
- the expected density to measure the quality of particular minimizer schemes,

- how to extract minimizers efficiently, and
- the link between minimizers and the de Bruijn graph.

25.1 Accelerating read overlap computation

Suppose we have n strings $\{r_1, \ldots, r_n\}$ on an alphabet Σ of size $\sigma = |\Sigma|$, and we want to find which sequences align to each other. That is, we want to compute all pairwise alignments and find those pairs with a significant alignment. An example of such a setup is in the context of genome or transcriptome assembly, where the strings r_i are sequencing reads, and we want to find the pairs of reads that have an overlap.

The naïve algorithm of doing all pairwise $O(n^2)$ alignments is prohibitively expensive, and most of it is wasteful as often only a small number of pairs will have a significant align-ment. It is possible to create a generalized suffix tree T (Section 13.4.3) or similar data structure with all the sequences and to find the exact matches between each r_i and T, or to use the suffix-prefix overlap algorithm of Section 13.4.4 if you are interested only in those types of overlaps. However, in practice, with hundreds of millions of strings, these are still costly computations (both in time and memory). One use of minimizers is to reduce the number of pairwise comparisons that must be done.

25.2 Definition of minimizers

A minimizer scheme is a scheme for selecting k-mers from a sequence that has 3 parameters: k is the length of the k-mers, w is the length of the overlapping windows in the sequence to consider, and $\mathcal{O}$ is an order on the k-mers.

The order $\mathcal{O}$ is a permutation of the set of all k-mers Σ^k. For example, with $k = 2$ and the DNA alphabet $\Sigma = \{A, C, G, T\}$, one such order, the lexicographic order, is

$$\mathcal{O} = [AA, AC, AG, AT, CA, CC, CG, CT, GA, GC, GG, GT, TA, TC, TG, TT]. \tag{25.1}$$

Formally, $\mathcal{O} : \Sigma^k \to [0, \sigma^k - 1]$ is a function that gives to each k-mer its ranking value in the order (i.e., $m_1 <_{\mathcal{O}} m_2$ if and only if $\mathcal{O}(m_1) < \mathcal{O}(m_2)$).

Given a string S, the minimizer breaks down the sequence into overlapping windows of w consecutive k-mers, and in each window it selects the smallest (lowest rank) k-mer according to the order $\mathcal{O}$. This is the origin of the name "minimizer": the selected k-mers are the ones in each window that are minimum according to the order $\mathcal{O}$. The result is either (depending on the application) the set of selected positions or the set of selected k-mers with their selected positions.

Here in (25.2) is an example of selecting minimizers from a sequence S using the lexicographic order.

$$\mathcal{O}: \quad \text{AAA} < \text{AAC} < \text{AAG} \ldots$$

$$S: \quad \text{CACTGCTGTACCTCTTCT}$$

```
       CACTGCT-----------
       -ACTGCTG----------
       --CTGCTGT---------
       ---TGCTGTA--------
       ----GCTGTAC-------
       -----CTGTACC------
       ------TGTACCT-----
       -------GTACCTC----
       --------TACCTCT---
       ---------ACCTCTT--
       ----------CCTCTTC-
       -----------CTCTTCT
```

(25.2)

minimizers: $\{\text{ACT}:1, \text{CTG}:2:5, \text{ACC}:9, \text{CCT}:10, \text{CTC}:11\}$

Formally, let $f_{\mathcal{O}} : \Sigma^{w+k-1} \to [0, w-1]$ be the function that takes the sequence of a window (w k-mers is $w + k - 1$ characters) and returns the position of the smallest k-mer in the window: $f_{\mathcal{O}}(\omega) = \arg\min_{0 \leq i < w} \mathcal{O}(\omega[i : i + k - 1])$. In case of ties (because the same, lowest-rank k-mer appears multiple times in the window), the leftmost of the smallest k-mers is selected. The set of selected positions is

$$\mathcal{S}_{k,w,\mathcal{O}}(S) = \{i + f_{\mathcal{O}}(S[i : i + w + k - 1]) \mid 0 \leq i < |S| - w - k\}. \tag{25.3}$$

25.3 Properties of minimizers

Fact 25.1 (Minimizer properties). *The selected positions and selected k-mers of a minimizer satisfy these two properties:*

Small gaps *The distance between two selected positions is $\leq w$. This means that there are no large gaps between selected positions and no part of the sequence is "ignored."*

Consistency *If two sequences have an exact match of at least $w + k - 1$ characters (the window size in characters), then the minimizer method selects at least one k-mer common between the two sequences.*

To compute all pairwise alignments, one selects the parameters w and k using the pigeonhole principle. Suppose we are interested in finding all pairs of sequences with an alignment of at least m characters with an identity (fraction of matching characters) of at least I. Then, two strings that satisfy this criteria must have an exact match of at least $L = (1 - I)m$ characters. One can choose w and k such that $w + k - 1 \leq L$. For example, if we require $m = 100$ and $I = 95\%$, then $L = 20$ is the length of an exact match, and we can choose $k = 10$ and $w = 11$.

Two strings that do not have a selected k-mer in common do not have an exact match of length L and therefore cannot meet our criteria. The small gaps and the consistency properties guarantee that no significant alignment will be missed. On the other hand, strings that share a common selected minimizer may have an alignment matching our criteria, which we can check using a more costly algorithm. Each string is put into the buckets labeled by its selected minimizers.

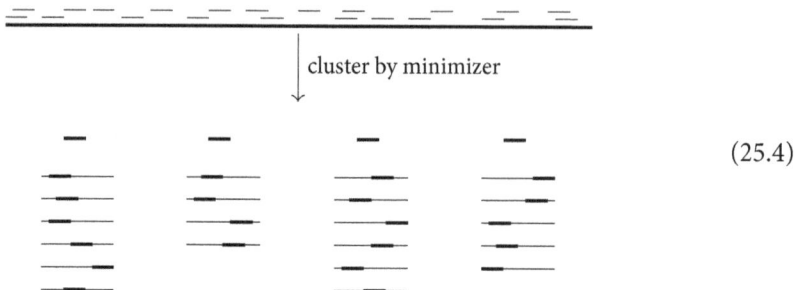

$$(25.4)$$

Strings within the same buckets are likely to have a significant alignment, while reads in different buckets cannot have a good alignment. We need only compare sequences in the same bin. In each bucket, the all pairs alignment problem is (hopefully) feasible, and less wasteful.

25.4 Expected density

A minimizer is a sampling process selecting positions from a sequence. The *density* measures how sparsely the positions are selected.

Definition 25.2 (Density). The *particular density* for a sequence S is the number of selected positions over the number of k-mers in S:

$$d_{k,w,\mathcal{O}}(S) = \frac{|S_{k,w,\mathcal{O}}(S)|}{|S| - k + 1}.$$

$$(25.5)$$

The *expected density* is obtained at the limit of the density for a randomly chosen sequence (each character is independently, identically distributed) as the sequence length goes to ∞. ∎

The density d is always $1/w \leq d \leq 1$, where the upper bound corresponds to selecting every k-mer in the sequence, while the lower bound corresponds to selecting exactly one k-mer every w bases.

A low density is generally desirable. In the pairwise alignment problem, a lower density means that each string is placed in fewer bins, hence the average size of each bin is smaller. Smaller bins mean that the problem of pairwise alignment in each bin is smaller and faster to solve.

In the following, we assume that the parameters k and w are fixed by the problem we are trying to solve (as in Section 25.3). The choice of the order $\mathcal{O}$ is not constrained by the problem, and regardless of the choice of $\mathcal{O}$, the properties of small gap and consistency are satisfied. Different choices of orders give different densities for fixed k and w, and this motivates the problem of finding the best order.

Problem 25.3 (Minimizer expected density problem). *For fixed k and w, what is the order $\mathcal{O}$ with the smallest expected density?* ◆

This problem is still open, although there has been a line of work [e.g., Marçais et al., 2018; Kille et al., 2025; Zheng et al., 2021c] finding schemes with better and better density, and tighter lower bounds. We will focus on the density achieved by particular types of minimizers.

25.5 Random minimizers

Definition 25.4 (Random minimizer). A random minimizer scheme is one where the order $\mathcal{O}$ is chosen at random uniformly among the permutations of Σ^k. ∎

One mathematical challenge is to estimate the expected density of a random minimizer. A difficulty with this is that there are two sources of randomness to consider: (1) the randomness of the sequence and (2) the randomness of the choice of the order $\mathcal{O}$. To get a handle on this, we look at *charged contexts*:

Lemma 25.5. *A context is two consecutive windows (equivalently, $w + 1$ consecutive k-mers or a substring of length $w + k$).*

1. *The positions selected by a minimizer scheme form a non-decreasing sequence.*
2. *The selected k-mer is different in the first and second window of a context if and only if the smallest k-mer of the context is the first or last one. Such a context is called charged.*
3. *The number of selected positions is equal to the number of charged contexts.*

Proof: Consider the following context, which is made of 2 consecutive windows, and the 3 possible cases for the position of the (leftmost) smallest k-mer among these $w + 1$ k-mers: first, last, or neither.

$$
\begin{array}{l}
\text{Smallest:} \quad \underline{\text{first}} \qquad \underline{\text{neither}} \qquad \underline{\text{last}} \\[2mm]
\text{Window 2:} \quad \rule{6cm}{0.4pt} \\[2mm]
\text{Window 1:} \quad \rule{6cm}{0.4pt} \\[2mm]
\text{Context:} \quad \rule{6cm}{0.4pt}
\end{array}
\tag{25.6}
$$

In a context, it is not possible for the selected k-mer in window 2 to be before the selected k-mer in window 1. If that were the case, the selected k-mer in window 2 would also be the smallest k-mer in window 1 and to the left of the selected k-mer in window 1, which contradicts the definition of $f_{\mathcal{O}}$. This proves part 1.

If the smallest k-mer in the context is the first one, then it is in window 1 and not in window 2, hence a new position must be selected in window 2. In the case where it is the last one, it is in window 2 and not in window 1, hence it could not have been selected in window 1 and a new position was selected in window 2. Otherwise, when it is neither the first nor the last, it belongs to both window 1 and 2, and is selected in both windows. This proves part 2.

Part 3 holds because if a context is not charged, then no new position is selected. The number of selected positions is therefore equal to the number of charged contexts.   □

Estimating the expected density is therefore equivalent to estimating the number of charged contexts in a long random sequence. It turns out that the case where there are repeated k-mers in a context is important in the analysis. The point of the next lemma is that when k is large enough compared to w (a reasonable assumption in practice), then having repeated k-mers in a context is a rare event that we can ignore in the analysis.

Lemma 25.6. *If $k \geq (3 + \epsilon) \log_\sigma(w)$, then the probability of having two identical k-mers in a context chosen at random is $o(1/w)$.*

Proof: Let $E_{i,j}$ be the event that the k-mers starting at positions i and j are identical. Regardless of whether these two k-mers share any sequence, the number of "free" bases to choose outside of these two k-mers is $w + k - k = w$. Hence, $\Pr[E_{i,j}] = \sigma^w / \sigma^{w+k} = \sigma^{-k} = 1/w^{3+\epsilon} = o(1/w^3)$. This looks like (25.7).

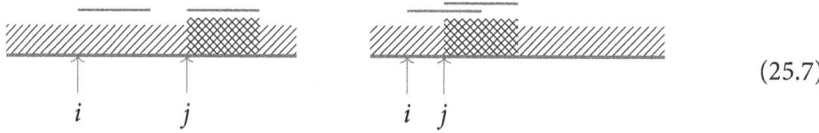

$$(25.7)$$

The probability that at least one of the events $E_{i,j}$ occurs is upper-bounded by $\binom{w+1}{2} o(1/w^3) = o(1/w)$.   □

We can now compute the expected density of random minimizers:

Theorem 25.7. *If $k \geq (3 + \epsilon) \log_\sigma(w)$, then the expected density of a random minimizer is $\leq 2/(w + 1) + o(1/w)$. In other words it is expected that slightly over 2 positions per context will be selected on average.*

Proof: Let $c \in \Sigma^{w+k}$ be a context, and let $I(c)$ be the event that context c has two identical k-mers. By the previous lemma, over the randomness of the choice of the context, $\Pr_c[I(c)] = o(1/w)$.

Provided that $I(c)$ does not hold (denoted as the event $\overline{I(c)}$), then over the randomness of $\mathcal{O}$, the probability that the context is charged is $\Pr_\mathcal{O}[c \text{ is charged}] = 2/(w + 1)$. This holds because every k-mer in the context is distinct, and they each have the same probability of being the smallest one.

The context is charged if either the first or last k-mer in the context is smallest (Lemma 25.5), these events are disjoint and happen with probability $1/(w + 1)$. Therefore:

$$\Pr_{c,\mathcal{O}}[c \text{ is charged}]$$

$$= \Pr_c[\overline{I(c)}] \Pr_{\mathcal{O}}[c \text{ is charged} \mid \overline{I(c)}] + \Pr_c[I(c)] \Pr_{\mathcal{O}}[c \text{ is charged} \mid I(c)] \qquad (25.8)$$

$$\leq \Pr_{\mathcal{O}}[c \text{ is charged} \mid \overline{I(c)}] + \Pr_c[I(c)] \qquad (25.9)$$

$$= \frac{2}{w+1} + o\left(\frac{1}{w}\right). \qquad (25.10)$$

$\square$

This theorem was originally proved in the winnowing [Schleimer et al., 2003] and minimizers papers [Roberts et al., 2004], although both proofs were missing the necessary hypothesis that $k \geq (3 + \epsilon) \log_\sigma(w)$. The next theorem makes it clear that this hypothesis is necessary: if k is small compared to $\log_\sigma(w)$, then a density of $O(1/w)$ is not possible.

Theorem 25.8. *If* $\log_\sigma(w) - k \xrightarrow[w\to\infty]{} \infty$, *then no minimizer method has a density of* $O(1/w)$.

Proof: Let $\mu_{\mathcal{O}}$ be the absolute smallest k-mer according to $\mathcal{O}$. Any context starting with $\mu_{\mathcal{O}}$ is charged, and the proportion of such context is σ^{-k}. The expected density is also at least σ^{-k}. Because $w\sigma^{-k} = \sigma^{\log_\sigma(w)-k} \xrightarrow[w\to\infty]{} \infty$, the density cannot be $O(1/w)$. $\square$

25.6 Selecting minimizers

Once we choose an order $\mathcal{O}$, we can process a sequence S to select the minimizer k-mers quickly.

Algorithm 25.1. Select minimizers from sequence.

```
 1: procedure MINIMIZERS(S, k, w, O)
 2:     Q ← queue initialized via Exercise 25.6
 3:     i ← w + 1
 4:     while i < |S| − k + 1 do
 5:         m ← S[i, i + k − 1]                    ▷ m is k-mer at position i
 6:         ▷ Remove big k-mers from queue:
 7:         while m <_O k-mer at tail of Q do
 8:             Pop tail of Q
 9:         Append (m, i) to tail of Q
10:         ▷ Previous minimizer out of window?
11:         if k-mer position at head of Q is < i − w then
12:             Pop head of Q
13:             output head of Q as next minimizer
14:         else if |Q| = 1 then                   ▷ New k-mer is smallest?
15:             output head of Q as next minimizer
16:         i ← i + 1
```

If the comparison between k-mers using the order $\mathcal{O}$ takes $O(k)$ time, then Algorithm 25.1 runs in $O(k|S|)$, or equivalently amortized $O(k)$ time for each k-mer in S. The total number of iterations of the loop at line 7 over the entire algorithm is $O(|S|)$ since a k-mer can be removed at most once. Therefore, the total running time is $O(k|S| + |S|) = O(k|S|)$ since each of the $O(|S|)$ iterations of the outer loop (line 4) takes $O(k)$ time, and the total cost of the removals is $O(|S|)$. This is optimal and less than the naïve implementation—loop over the w k-mers in each window—which takes $O(kw|S|)$ worst case.

The interesting part of this algorithm is the loop at line 7. Thanks to this loop the current minimizer is always at the head of the queue and there is never a need to search the queue for the smallest element. The reason this is true is the following:

Fact 25.9. *In Algorithm 25.1, when a k-mer is added to Q, it must be bigger (according to $\mathcal{O}$) than the final element of Q after the loop at line 7, and therefore Q contains k-mers in increasing order of $\mathcal{O}$.*

Proof: The loop at line 7 removes any k-mers lower than the added k-mer, and that is the only place k-mers are added to Q, and so Q must contain k-mers in increasing order. □

Fact 25.10. *After line 12, Q only contains k-mers from the current length-w window.*

Proof: At most 1 k-mer leaves the window at every iteration of the loop at line 4, and it will be removed in line 12. □

Fact 25.11. *The minimum k-mer in a window is always in Q.*

Proof: Every k-mer is added at some point in line 9, so in particular, the minimum k-mer was added. K-mers are removed only when a smaller k-mer enters the window from the right (loop at line 7) or when they leave the window (line 12). Once a k-mer leaves the window, it's not the minimum in the window, so this cannot affect the minimum in the window. Since no k-mer smaller than the minimum can enter the window, the minimum k-mer is not removed at line 7. □

Combining Facts 25.9, 25.10, and 25.11, the head of Q will always contain the lowest k-mer in the window. This shows that the algorithm will always output the minimizer in the window.

25.7 De Bruijn sequences and de Bruijn graphs

A de Bruijn sequence is either a circular sequence of length σ^k or a linear sequence of length $\sigma^k + k - 1$ that contains every k-mer exactly once. For example, 0011 is a circular de Bruijn sequence for $\sigma = 2$ and $k = 2$; 00110 is a linear one for the same σ and k.

That such a sequence exists is surprising, but it turns out that for every k there is a very large number of them.

To see this, we need to introduce the de Bruijn graph (and the general concept of a line graph):

Definition 25.12 (Line graph). Given a directed graph $G = (V, E)$, the *line graph* of G is $L(G) = (E, E')$ where $(e_1, e_2) \in E'$ if in G the tail of edge e_1 is equal to the head of edge e_2. Note that the edges of G become the nodes of $L(G)$. ∎

We can now define the class of de Bruijn graphs. There is one such graph for each k and σ:

Definition 25.13 (de Bruijn graph). The de Bruijn graph $B_{0,\sigma}$ of order 0 on the alphabet of size σ is a single node with σ self-loops. The de Bruiijn graphs of higher order are obtained by repeatedly taking the line graph: $B_{k+1,\sigma} = L(B_{k,\sigma})$. ∎

Here are the de Bruijn graphs of orders $0, 1, 2, 3$ over the binary alphabet.

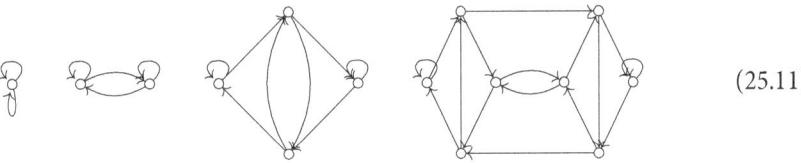

$$(25.11)$$

There is another equivalent definition of de Bruijn graphs.

Definition 25.14 (string-based de Bruijn graph). The de Bruijn graph $B_{k,\sigma} = (V, E)$ has $V = \Sigma^k$, where Σ is an alphabet of size σ. Edge $(u, v) \in E$ if and only if the last $k - 1$ characters of node u equal the first $k - 1$ characters of node v. ∎

Exercise 25.2 asks you to prove that this definition is equivalent to our previous one.

These graphs are quite rich in their structure. They have become widely used in genome assembly, and we'll see a bit more about them in Chapter 29 on genome graphs. Some of their properties are:

Fact 25.15. *De Bruijn graphs have these properties:*

1. $B_{k,\sigma}$ *is regular with in-degree and out-degree equal to* σ.
2. $B_{k,\sigma}$ *has* σ^k *vertices and* σ^{k+1} *edges.*
3. $B_{k,\sigma}$ *is Eulerian and Hamiltonian.*
4. *The vertices of* $B_{k,\sigma}$ *is the set of k-mers,* Σ^k.
5. *Each string* $S \in \Sigma^*$ *corresponds to a unique walk in* $B_{k,\sigma}$ *of length* $|S| - k + 1$, *and conversely each walk of length L corresponds to a unique string of length* $L + k - 1$.
6. *Each Hamiltonian tour of* $B_{k,\sigma}$ *generates a de Bruijn sequence.*

Proof: A graph is *regular* if all the nodes have the same in-degree and the same out-degree.

$B_{0,\sigma}$ is regular with in-degree and out-degree equal to σ. By induction, this property is preserved by taking the line graph.

A graph with the same in-degree and out-degree for every vertex is Eulerian. An Eulerian tour of $B_{k,\sigma}$ becomes a Hamiltonian tour of $B_{k+1,\sigma}$.

A Hamiltonian tour of $B_{k,\sigma}$ visits every k-mer exactly once.   □

Using these properties (and some additional logic), one can prove the following, which was shown independently by de Bruijn [1946] and Flye Sainte-Marie [1894].

Theorem 25.16. *The number of de Bruijn sequences of order k on alphabet of size σ is:*

$$\frac{(\sigma!)^{\sigma^{k-1}}}{\sigma^k}. \tag{25.12}$$

This number is staggering and surprisingly high given that a priori the existence of a single de Bruijn sequence was not obvious. On the DNA alphabet of size 4, the number of de Bruijn sequences is shown in (25.13).

k	Number de Bruijn sequence
1	12
2	66 355
3	10^{21}
4	10^{87}
5	10^{351}
6	10^{1411}
7	10^{5650}
8	$10^{22\,610}$
9	$10^{90\,449}$
10	$10^{361\,809}$

$$\tag{25.13}$$

De Bruijn sequences give a practical way (for reasonable values of k, w) to calculate exactly the expected density of a minimizer scheme, rather than estimating it using a very long random string. The meaning of the next lemma is that a de Bruijn sequence has the same distribution of contexts as an infinite random string.

Lemma 25.17. *The expected density of a minimizer scheme is equal to the specific density on any circular de Bruijn sequence of order o as long as $o \geq k + w$.*

Proof: In random sequence S with characters chosen i.i.d., the expected number of occurrences of any given context (a subsequence of length $k + w$) is $|S|/\sigma^{k+w}$. Let $C_\mathcal{O}$ be the set of charged contexts for order $\mathcal{O}$. As $|S|$ grows, the specific density of this random string is

$$\frac{|C_\mathcal{O}|}{|S| - k + 1} \frac{|S|}{\sigma^{k+w}} \xrightarrow[|S| \to \infty]{} \frac{|C_\mathcal{O}|}{\sigma^{k+w}}. \tag{25.14}$$

In any linear de Bruijn sequence of order o (hence of length σ^o), every context occurs exactly $\sigma^{o-(k+w)}$ times and the specific density is exactly

$$\frac{\sigma^{o-(k+w)}|C_{\mathcal{O}}|}{\sigma^o} = \frac{|C_{\mathcal{O}}|}{\sigma^{k+w}}. \tag{25.15}$$

□

25.8 Set of unavoidable words

A minimizer scheme on k-mers is an ordering of the nodes of the de Bruijn graph of order k. We can define a particular order using the concept of *unavoidable words*.

A set of unavoidable words is a set of words U such that any long enough string must contain a word from U. We are interested in words of constant length, k-mers, and we assume that $U \subset \Sigma^k$. These sets of unavoidable k-mers, which are called *universal hitting sets*, are closely linked to minimizers.

First, a minimizer method defines a universal hitting set:

Lemma 25.18. *Let $\mathcal{U}_{k,w,\mathcal{O}}$ be the set of all selected minimizer k-mers (using any ordering $\mathcal{O}$) on a de Bruijn sequence of order $k + w - 1$. This set is a universal hitting set.*

Proof: By construction, every window has a smallest k-mer that was selected.   □

Second, a universal hitting set defines a family of minimizer methods:

Definition 25.19 (Compatible order). Let $U \subset \Sigma^k$ be a universal hitting set that hits every sequence of length $L = k + w - 1$. A *compatible order* $\mathcal{O}_U$ is an order on the k-mers such that if $m_1 \in U$ and $m_2 \notin U$, then $m_1 <_{\mathcal{O}_U} m_2$. ■

That is, a compatible order puts all the universal hitting set k-mers before all those k-mers that are not in the universal hitting set. Such an ordering of course gives a minimizer scheme, and further the size of the universal hitting is related to the density of the minimizer scheme that uses it:

Lemma 25.20. *Given a universal hitting set U and a compatible order $\mathcal{O}_U$, the set $\mathcal{U}_{k,w,\mathcal{O}_U}$ of Lemma 25.18 is a subset of U. Furthermore, the size of U gives an upper-bound on the expected density of the minimizer method:*

$$d_{k,w,\mathcal{O}_U} \leq \frac{|U|}{\sigma^k}. \tag{25.16}$$

Proof: Every window is a sequence of length $L = k + w - 1$, hence it contains a k-mer from U. By construction, the k-mers of U compare less than other k-mers, so the minimizers belong to U.   □

This means that finding universal hitting sets of small size is one method to create minimizer methods with guaranteed density. Finding U of small size is not enough in practice as the actual order of the elements within U has a great influence on the actual density.

25.9 Summary and notes

Much more information about the design and use of minimizers can be found in Zheng et al. [2023]. There has been significant work on both practical and theoretical improvements to minimizer schemes [Orenstein et al., 2017; Marçais et al., 2017, 2018; DeBlasio et al., 2019; Ekim et al., 2020; Hoang et al., 2022; Zheng et al., 2020, 2021a,b; Ekim et al., 2020]. A related approach, called syncmers, for selecting k-mers has also been explored [Dutta et al., 2022; Edgar, 2021]. Minimizers have also been used to create sparse suffix arrays [Grabowski and Raniszewski, 2013]. The problem of finding de Bruijn sequences that don't share substrings longer than k was considered in Lin et al. [2011]. The problem of finding the lexicographically first de Bruijn sequence was solved in Faro and Lecroq [2013].

> Presentation Notes
> ────────────
> The first version of this chapter was written by Guillaume Marçais.

25.10 Exercises

25.1   Give a de Bruijn sequence on alphabet $\{0, 1\}$ for $k = 3$.

25.2   Prove that Definition 25.13 is equivalent to that of Definintion 25.14.

25.3   Let $\Sigma = \{0, 1\}$. Prove that if $0u$ and $1u$ are two k-mers in a universal hitting set U, then the set $U' = (U \setminus \{0u, 1u\}) \cup \{u0, u1\}$ is also a universal hitting set.

25.4   A *PCR cycle* of the de Bruijn graph is a cycle where all the edges are of the form $au \to ua$, where $a \in \Sigma$ and $u \in \Sigma^{k-1}$. Prove that a universal hitting set must contain at least 1 k-mer from every PCR cycle. (The fact that any minimum-size universal hitting set contains *exactly* 1 k-mer from each PCR was proven in Mykkeltveit [1972].)

25.5   Using the definition of PCR cycles from Exercise 25.4, show that the PCR cycles partition the nodes of the de Bruijn graph (i.e., every node is in exactly 1 PCR cycle).

25.6   Describe how to initialize Q to make Algorithm 25.1 correct.

Part V

Graphs and Generative Models

CHAPTER 26

Regular Grammars and the Chomsky Hierarchy

A *grammar* is a set of rules to generate sets of strings. Specifically, a grammar to generate a subset of the strings in Σ^* consists of substitution rules of the form:

$$\alpha \to \beta$$

where α and β are strings. Each of these $\to$ rules is called a *production*. The meaning of the rule is that a substring equal to α can be replaced by the string β. If we place various restrictions on the forms of α and β we obtain grammars of various expressibility and power.

To begin to define those types of grammars, we split our alphabet into two types of characters: Σ is the alphabet of our application (i.e., English letters or DNA bases), and $\mathcal{N}$ is a set of symbols (not contained in Σ) that will be used in our grammar but not appear in any final string that it produces. The symbols in $\mathcal{N}$ are called *non-terminals*, and the letters in Σ are called *terminals*. We also introduce a special character ϵ that represents the empty string. We will require that α and β are strings over the alphabet $\Sigma \cup \mathcal{N}$.

We also assume that there is at least one production of the form $S \to \beta$, where $S \in \mathcal{N}$ is a distinguished starting non-terminal. We start our sequence of replacements on the string that consists only of the non-terminal S.

The operation of a grammar is to repeatedly use some production to replace an instance of α with an instance of β until no more such replacements are possible. For example, the following grammar generates from S a set of strings that consist of an alternating sequence of as and bs, followed by a c, and then the reverse alternating sequence of as and bs.

$$S \to aXa$$

$$aXa \to abXba$$

$$bXb \to baXab$$

$$X \to c$$

Here, $\Sigma = \{a, b, c\}$ and $\mathcal{N} = \{S, X\}$. This can generate *ababcbaba* for example or *abacaba*, but not *abacbab*.

The set of strings over Σ (that is, strings with no non-terminals) that can be generated from a grammar is called the *language* of the grammar.

26.1   Regular grammars

Noam Chomsky, while studying the way humans learn language, developed a classification system of grammars of increasing power based on the restrictions we place on α and β in the productions [Chomsky, 1956]. The lowest-level of Chomsky's hierarchy consists of the *regular* grammars that only contain rules of one of the forms:

$$W \to aU$$

$$W \to a$$

$$W \to \epsilon$$

where W and U are in $\mathcal{N}$ and $a \in \Sigma$. The lefthand side always consists of a single non-terminal, and the righthand side always consists either of a terminal followed by a non-terminal, or a single terminal or the special character ϵ, representing the empty string.

Regular grammars generate the string from left-to-right and can't encode long-range interactions between different parts of the string. The reason for this is that after any application of a production (i.e., replacement) the only non-terminal in the string (if there is any) will be the last character. That last non-terminal can continue to be replaced, possibly extending the string, but there is no way to make coordinated choices in different parts of the string.

26.1.1   Finite state automata

Regular grammars can be generated or recognized by finite state automata (FSA). We've seen FSAs somewhat informally when looking at the Knuth-Morris-Pratt algorithm (Chapter 4). Now, we give the formal definition:

Definition 26.1 (Finite State Automaton). A *finite-state automaton* (FSA) is a tuple $F = (V, E, \ell, s, t)$, where

- V is a set of nodes called "states,"
- E is a set of directed edges between states,
- ℓ is a labeling of the edges with symbols from the alphabet or ϵ. That is, $\ell : E \mapsto \Sigma \cup \{\epsilon\}$ so that $\ell(e)$ is the label of edge e,
- $s \in V$ is the start state, and
- $T \subset V$ is the set of accepting states.   ∎

We say an FSA *accepts* a string S if there is a path from s to $t \in T$ that spells out S. The set of strings that an FSA accepts is called the *language* of the FSA, and we say that the FSA *recognizes* its language.

An FSA is *deterministic* if each edge leading out of every node is labeled with a different character (where ϵ in this context is taken as a wildcard that equals any character)—that is the sequence of characters exactly specifies your walk through the FSA graph. An FSA is *non-deterministic* if multiple edges with the same character leave any node.

To construct an FSA from a regular grammar, we set V equal to the set of non-terminals that appear in any production plus a special state t. s is set to the starting non-terminal S. Productions of the form $W \rightarrow aU$ create edges in the FSA from state W to state U. The edge is labeled with character a; that is $\ell((W, U)) = a$. Rules of the form $W \rightarrow a$ create edges labeled a in the FSA from W to the terminal state t. Rules $W \rightarrow \epsilon$ lead to edges (W, t) labeled with ϵ.

To generate a string from an FSA, you start at the special state s and successively follow edges. When you cross an edge labeled $a \in \Sigma$, you output a. If you cross an edge labeled ϵ, you output nothing. If you reach state t then you stop.

Example. A regular grammar for generating strings with an odd number of "a"s is as follows.

$$S \rightarrow aX$$
$$S \rightarrow bS$$
$$X \rightarrow aS$$
$$X \rightarrow bX$$
$$X \rightarrow \epsilon$$

The state S corresponds to the state "string has an even number of 'a's." When we add an "a" from state S we switch to state X which corresponds to "string has an odd number of 'a's." Adding a "b" in either state doesn't change the state.

This corresponds to the following FSA.

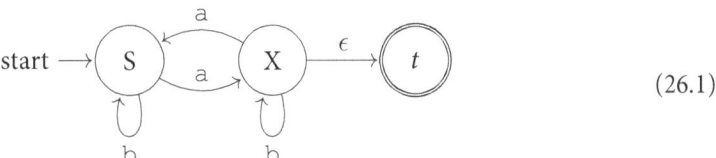

$$(26.1)$$

This correspondence goes in the other direction as well: any FSA can be converted into a regular grammar that accepts the same language:

Theorem 26.2 (Regular grammars and FSA are equivalent). *A grammar $\mathcal{G}$ is regular if and only if there is an FSA F that accepts the same language.*

Proof: The transformation in this example shows that any regular grammar can be converted into an equivalent FSA. For the other direction, we reverse the transformation. If F has multiple accepting states, create F' by adding a new accepting state q_f and connect the accepting states of F to q with ϵ edges and make q be the only accepting state. For each edge (q_1, q_2) labeled a in F' where $q_2 \neq q_f$, add a production to the grammar $q_1 \rightarrow aq_2$. For each edge (q_1, q_f) labeled a in F', add a production $q_1 \rightarrow a$. $\square$

Another natural problem encountered with grammars is to *parse* a given string T by providing a sequence of productions from the grammar that will turn the starting non-terminal S into T. To parse a string with an FSA, you proceed as if you were going to generate a string: start at state s, follow edges in the FSA, but now, you choose the edge corresponding to the next character in the string or an ϵ edge if one is available. If you have multiple choices about which edge to take (e.g., there is an edge labeled with the correct character and an ϵ edge or two edges labeled with the correct character), then you have to try all the acceptable choices. If there is a path from s to t that spells out the string then the string is *accepted* and the path gives the parsing of the string. If no such path exists, then the string is not part of the regular grammar recognized by this FSA.

26.1.2   Deterministic vs. non-deterministic FSAs

Surprisingly, deterministic and non-deterministic FSAs have exactly the same power: any non-deterministic FSA can be converted into an equivalent deterministic FSA. This transformation likely introduces many more states, but they will accept the same language.

For non-deterministic FSAs, the parsing algorithm is the same as the deterministic one, except that, since we have multiple choices, we may have to track multiple pointers to states: when we don't know which choice to make, we create pointers to all possible choices. At any point in time, we may have a pointer to any set of the states: these are all the possible places we might be in the parse. We move the pointers (and possibly increase the number of them) at each step. Hence, the operation of the non-deterministic FSA can be captured by recording the set of states we might be in at any moment. We can use this idea to eliminate the non-determinism.

Theorem 26.3 (Non-deterministic $\rightarrow$ DFA). *Any non-deterministic FSA F can be converted into a deterministic FSA (DFA) D that accepts the same language.*

Proof: Let Q be the set of states of F. We will have one state in D for every non-empty *subset* of states from Q. So, in total, we will have $2^{|Q|} - 1$ states in D. For clarity, we will call states of D "super states" since they correspond to sets of states of F, but "super states" are just regular states of the deterministic FSA D. If a super state contains an accepting state of F, then it is an accepting state of D.

The edges between the super states will correspond to simulating the multi-pointer, non-deterministic parsing algorithm. Suppose while parsing using F we are at the set of possible states S, and on character c in T, we would move the pointers to a new set of states S'. The states in S' are those we could reach by parsing c. We could also additionally reach any states from S' that we can get to by following ϵ-only paths (since those don't use up any more characters) followed by a c edge. So let $E_{S'}$ be the set of states reachable from some state in S' following only edges labeled ϵ. Let $S'' = S' \cup E_{S'}$. Note that S'' is a super state in D. We add an edge in D from S to S'' (both super states) labeled c. Add these edges for every possible S and c.

The start state of D is the set of states reachable from the start state of F by 0 or more ϵ transitions in F. These are the places the pointers could be without reading any characters of a string.

D is deterministic since there is exactly one edge labeled with any character c and there are no ϵ edges. D accepts the same languages as F since D simulates the parsing algorithm of F by explicitly recording the state of the pointers during the non-deterministic parse algorithm.   □

Clearly, the DFA generated by this approach could be quite a bit bigger than the non-deterministic version. To find an equivalent, minimal-sized version of any DFA, one can use Hopcroft's algorithm [Hopcroft, 1971] (which we will not cover).

26.1.3   Regular grammars and regular expressions

Regular grammars correspond to traditional *regular expressions* with which you may be familiar.

Definition 26.4 (Regular Expressions). The *regular expressions* are strings over the alphabet $\Sigma \cup \{|, (,), ^*\}$ that are defined recursively. If A and B are regular expressions, then the following are also regular expressions:

1. a for any $a \in \Sigma$
2. (A)
3. AB (this is the concatenation operator);
4. $A \mid B$ (This is the "or" operator);
5. A^* (This is the repeat operator, sometimes called the Kleene star.)   ∎

Regular expressions define sets of strings that they *match*, where whether a string S matches a regular expression R is defined recursively using the same recursive cases as Definition 26.4:

1. If $R = a$ for some $a \in \Sigma$, then S matches R if $S = a$.
2. If $R = (A)$, then R matches S if A matches S.
3. If $R = AB$, where A and B are regular expressions, then R matches S if $S = \alpha\beta$ where A matches α and B matches β.
4. If $R = A|B$, where A and B are regular expressions, then R matches S if A matches S or B matches S.
5. If $R = A^*$, where A is a regular expression, then R matches S if $S = \alpha_1, \alpha_2, \ldots, \alpha_k$ (for some $k \geq 0$) where A matches α_i for each i. Under this definition, A^* always matches the empty string.

The set of strings that match a regular expression R is called the *language* of R.

In practical settings, the notation for regular expressions is often extended to include things like A^+ as a shorthand for AA^* and . (period) as a shorthand for $e_1 \mid e_2 \mid e_3 \mid \cdots \mid e_k$ where $\Sigma = \{e_1, \ldots, e_k\}$, but these do not increase the set of languages that we can specify with regular expressions. Most regular expression parsers in common libraries and languages also include extensions that make them not regular grammars anymore.

The set of languages that can be specified by regular expressions is the same set of languages that can be specified by finite state automata, and hence the same set as regular grammars. Suppose a language L is recognized by a regular expression R. We will show how we can recursively construct an FSA that recognizes it, using a proof by induction.

By definition, R is one of a, (A), AB, $A \mid B$, A^* for some regular expressions A, B, where the length of regular expressions A and B are less than the length of R. If $R = a$ we can construct an FSA (with a single acceptance state) that matches R.

$$\text{start} \longrightarrow \boxed{S1} \xrightarrow{a} \boxed{S2} \qquad (26.2)$$

This is the base case. For the general case, assume we can find an FSA (with a single accepting state) that accepts the same language as any regular expression shorter than R. In particular, this is true for the A and B subexpressions that make up R.

- If $R = AB$, by induction, we have FSAs that correspond to A and B. We connect them in serial:

$$\boxed{A} \longrightarrow \boxed{B} \qquad (26.3)$$

- If $R = A \mid B$, then we connect the FSAs for A and B in parallel by introducing a new start state q_0 and new accepting state q_1.

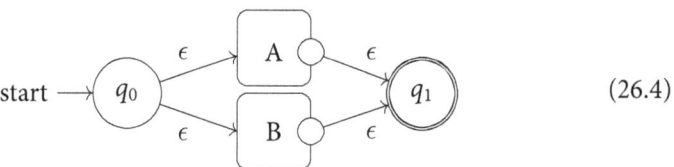

$$(26.4)$$

   We connect these new states up to the start and accepting states of A and B with ϵ edges.
- If $R = A^*$, we connect the accepting state of A to its start state with an ϵ edge. Since A^* also accepts 0 repetitions of A, we connect its start state directly to its accepting state with another ϵ edge.

$$\overset{\epsilon}{\boxed{A}} \qquad (26.5)$$

If $R = (A)$, then we need do nothing, since the FSA for A is already an FSA for R. Hence, using the (26.2–5) rules we can construct a (non-deterministic) FSA for any regular expression. Since the languages accepted by FSAs are exactly those accepted by regular grammars (Theorem 26.2), this shows that regular expressions are no more expressive than regular grammars:

Theorem 26.5. *Any regular expression R can be written as a regular grammar. It can also be expressed as an FSA.*

The converse is also true: any regular grammar can be written as a regular expression, although this is a little more involved to see. To prove it, we need to introduce a more expressive (but not more powerful) notation for FSAs called general FSAs:

Definition 26.6 (General FSA). A *general FSA* is an FSA where the edges are labeled with regular expressions instead of single characters. Further, we require that:

- there is a single start state with no incoming edges, and a single accepting state with no outgoing edges, and that the start and accepting states are different; and
- there is at most one edge between any pair of states.

A string S is accepted by the general FSA if S is matched by the concatenation of the regular expressions along some path from the start state to the accepting state. ■

It's easy to convert an FSA to an equivalent general FSA: we add a new start state, connecting it to the old start state with an ϵ edge (now there are no incoming edges to the start state); we add a new accepting state, and connect all the old accepting states to it via ϵ edges (now there is a single accepting state with no outgoing edges). In addition, if there are multiple edges in F between two states (q_1, q_2) labeled with $a_1, \dots, a_k$, we replace these edges with a single edge $a_1 \mid \cdots \mid a_k$. Given this transformation, we can assume any FSA is in general FSA form.

Theorem 26.7. *Any general FSA F can be converted to an equivalent regular expression.*

Proof: Let n be the number of states in F. We will construct a sequence of equivalent general FSAs $F_n = F, F_{n-1}, F_{n-2}, \dots, F_2$, each with 1 fewer state, until we reach a general FSA F_2 with only two states (necessarily the start and acceptance states). The edge between those two states will be a regular expression equivalent to F.

To obtain F_{i-1} from F_i we pick any state q in F_i that is not the start or accepting state. For every path in F_i of 3 nodes $q_i \to q \to q_j$ (including the case when $i = j$), we add an edge (q_i, q_j) that bypasses q. Let the regular expressions on the edges of the path be R_{iq}, R_{qj}, and let R_{qq} be the regular expression on the self loop of state q if such a loop exists. We label the new (q_i, q_j) edge with the regular expression $R_{iq}(R_{qq})^* R_{qj}$. (If q does not have a self-loop, we omit the $(R_{qq})^*$ part.) Now, we can accept the same substring as the $q_i \to q \to q_j$ path by taking the direct (q_i, q_j) edge and matching its regular expression. If we already had a (q_i, q_j) edge labeled with regular expression R_{ij}, we merge the old and new edges using the $\mid$ operator, so the final label of the new (q_i, q_j) edge is $R_{ij} \mid (R_{iq}(R_{qq})^* R_{qj})$. Now, state q is not needed, so we remove it, which yields F_{i-1} that accepts the same language as F_i.

After repeating this enough times, we will reach F_2. A regular expression that is equivalent to F will be on the only remaining edge. □

26.1.4 Summary of regular grammars

We've seen a number of ways of formulating the same concept, and shown that they are all equivalent:

$$DFA = NFA = \text{regular grammars} = \text{regular expressions}. \tag{26.6}$$

One of the most initially surprising results is that non-determinism doesn't really increase the power of FSA—in many other contexts non-determinism appears to significantly increase the class of languages that can be recognized. For regular languages, that is not the case, although a deterministic FSA might be much larger than the equivalent non-deterministic one.

26.2 Context-free grammars

The next level of grammars are context-free grammars (CFGs). Here, rules are all of the following form:

$$W \to \beta$$

where $W \in \mathcal{N}$ and β is any string in $(\mathcal{N} \cup \Sigma)^*$ (where ϵ will denote the empty string). Context-free grammars can encode long-range interactions. For example, here's a CFG for strings with balanced parentheses:

$$S \to ()$$
$$S \to (Q)$$
$$Q \to SQ$$
$$Q \to S$$

where Q expands into a sequence of some number of S, and each S expands either to $()$ or to a parenthesized list of some number of Ss.

To save space, we can re-write this CFG as

$$S \to () \mid (Q)$$
$$Q \to SQ \mid S$$

where the $\mid$ indicates options for the righthand side of the given production.

26.2.1 CFG parse trees

The generation of a string from a CFG corresponds to a parse tree where the root is the start symbol S of the CFG and the leaves of the tree correspond to the string. The internal nodes correspond to non-terminals and their children indicate the characters that were generated by those non-terminals.

Example. Suppose we are parsing the string *abbabbaabba* with the CFG.

$$S \rightarrow AA$$
$$A \rightarrow BB \mid aBa$$
$$B \rightarrow abb$$

Then we could have the following parse tree.

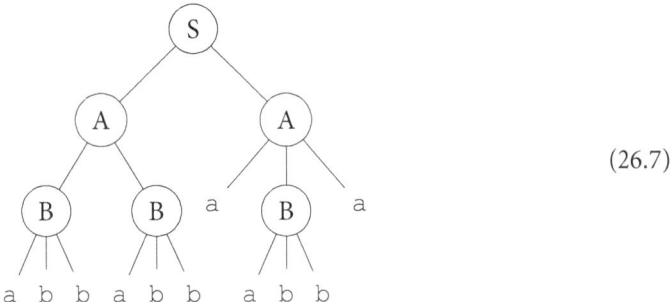

(26.7)

The string is spelled out by an in-order (left-to-right) traversal of the leaves of the tree. Parse trees are not necessarily unique for a given CFG and string.

26.2.2 Chomsky normal form

Any CFG can be rewritten so that it only uses productions of the following form:

$$W \rightarrow UV$$
$$W \rightarrow a$$

where $W, U, V \in \mathcal{N}$ and $a \in \Sigma$. This is called *Chomsky Normal Form* or CNF. The transformation of a general CFG to one in CNF is straightforward: nonterminals and productions are introduced to break up any production $W \rightarrow \beta$ where β is too long into a series of smaller productions. Additional rules $A \rightarrow a$ are introduced to fix productions that look like $W \rightarrow Ua$ and so on.

For a grammar in CNF, the parse tree will be a binary tree (each node has at most 2 children) and for, any node that has a single child, that child will be a leaf.

26.3 Context-sensitive and unrestricted grammars

The most powerful grammars in the hierarchy are the context-sensitive and the unrestricted grammars. A context-sensitive grammar has rules of the following form:

$$\alpha_1 W \alpha_2 \rightarrow \alpha_1 \beta \alpha_2$$

where $\alpha_1, \alpha_2, \beta \in (\mathcal{N} \cup \Sigma)^*$. The α_1 and α_2 strings (either of which may be empty) give the context where the rule to replace W can be applied—they are unchanged in the substitution. Whether a string is in a context-sensitive grammar is a decidable question. There is a parsing algorithm for context-sensitive grammars, but the only guarantee is that it will eventually terminate—it could take a very long time.

Unrestricted grammars allow any rules of the form:

$$\alpha_1 W \alpha_2 \to \beta$$

where $\alpha_1, \alpha_2, \beta \in (\mathcal{N} \cup \Sigma)^*$. The only requirement is that there is a non-terminal on the lefthand side. Unrestricted grammars require Turing machines to parse. It is in general an undecidable problem to determine whether a given string is in the language of a particular unrestricted grammar.

Context-sensitive and unrestricted grammars, because of the computational difficulty parsing them, are not often directly applicable in practical problems, though they are interesting from a theoretical viewpoint.

26.4 Summary and notes

The classes of grammars provide a connection between string algorithms and computational complexity. As the power of a class of grammars increases, the power of the corresponding computational model also increases. DFAs can parse regular grammars, but Turing machines are needed for unrestricted grammars (and some problems are undecidable). Hence, these grammars provide an alternative view into the power of various computational models.

Grammars have also been used to compress strings. A string that can be encoded in a small grammar can be transmitted by transmitting the grammar. This requires learning a small grammar that would generate the string, for which a number of algorithms have been developed [e.g., Larsson and Moffat, 2000; Nevill-Manning and Witten, 1997; Cherniavsky and Ladner, 2004].

They are also very useful in practice, with regular expressions often used ubiquitously for search and pattern matching and context-free grammars use for compression and simulation.

Presentation Notes

Durbin et al. [1998] covers the Chomsky hierarchy, as does nearly every elementary textbook on automata theory [e.g., Sipser, 2012] (which also tend to cover the equivalence between non-deterministic FSA and deterministic FSA and regular expressions).

26.5 Exercises

26.1   Show that productions of the form $W \to \epsilon$ where ϵ is the empty string are not necessary in regular grammars and any regular grammar can be written without them.

26.2   Give a context-free grammar to generate exactly the strings with balanced paren-thesis that have an even number of left parens. For example: () (( )) is not allowed but () () (( )) is.

26.3   Give a string S and a CFG that has more than one parse tree for S.

26.4   Describe how to transform a Lempel-Ziv parse of a string (Section 14.1) into a grammar.

CHAPTER 27

Hidden Markov Models

Regular languages (Chapter 26) could be generated by a finite state automaton (FSA). Hidden Markov Models (HMMs) extend this to become probabilistic generators of strings using finite state machines. They add an additional twist, which is that often when generating a string, it's not really the final string we care about, but rather the *states* that are used in the FSA to generate the string. HMM models give us a way to probabilistically generate strings and to infer how the model generated those strings.

27.1 Definition of hidden Markov models

Definition 27.1 (HMMs). A *Hidden Markov Model* (HMM) is a tuple (Σ, Q, A, E) where:

- Σ is an alphabet of symbols;
- Q is a set of states;
- A is a $|Q| \times |Q|$ matrix, where entry $A[k, \ell]$ is the probability of moving to state ℓ from state k; and
- E is a $|Q| \times |\Sigma|$ matrix, where entry $E[k, b]$ is the probability of emitting symbol b when entering state k. ∎

The matrix A can be thought of as a weighted adjacency matrix for a graph with nodes equal to the states. Every row of the matrix E provides a probability distribution on the characters of Σ.

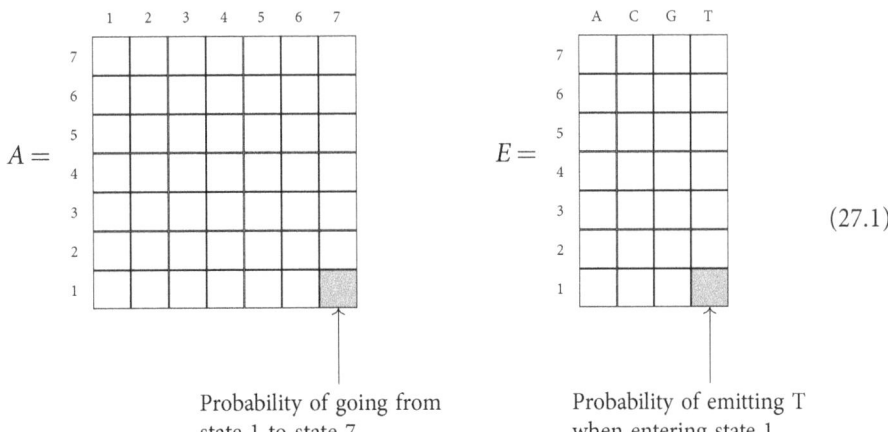

$$A = \qquad\qquad E = \qquad\qquad (27.1)$$

Probability of going from state 1 to state 7

Probability of emitting T when entering state 1

The sum of the entries in each row of A and E must be 1.

We can visualize an HMM as a graph with associated probabilities. Here's an example that models a sequence of coin flips where the hidden state is whether we are using a fair coin (probability of heads is the same as the probability of tails) or a coin that is biased toward heads.

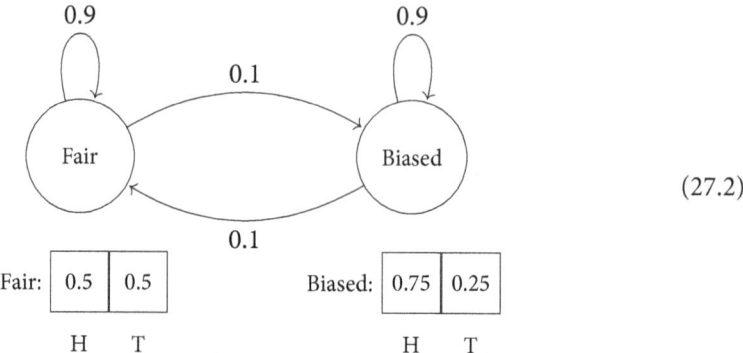

(27.2)

The HMM is a generative model where we walk from state to state. We assume there is a distinguished start state in which the HMM always starts. The next state that we visit is selected according to the probability distribution given by the appropriate row of A. This is why they are *Markov* models: the next state depends only on the current state and not the history of states we've visited. When entering each state (except the start state), a character is *emitted*, selected according to the probability distribution in the correct row of E.

The sequence of characters that are emitted is a string S. The sequence of states visited during this process is a path (ordered list) π of states. For example, a sequence of $\uparrow$ (heads) and $\downarrow$ (tails) could be generated by visiting a sequence of Fair and Biased states.

$$
\begin{array}{rcccccccccc}
S = & \downarrow & \uparrow & \downarrow & \uparrow & \uparrow & \uparrow & \downarrow & \uparrow & \uparrow & \uparrow \\
\pi = & F & F & F & B & B & B & B & F & F & F \\
\Pr[S_i \mid \pi_i] = & 0.5 & 0.5 & 0.5 & 0.75 & 0.75 & 0.75 & 0.25 & 0.5 & 0.5 & 0.5 \\
\Pr[\pi_i \to \pi_{i+1}] = & & 0.9 & 0.9 & 0.1 & 0.9 & 0.9 & 0.9 & 0.1 & 0.9 & 0.9
\end{array}
$$

(27.3)

Another example: to a first approximation, a Eukaryotic gene is a region of the genome that consists of interspersed regions called introns and exons. The introns are spliced out (between acceptor and donor sites), and the remaining string encodes the sequence of a protein using groups of 3 nucleotides called *codons*. The first codon is the start codon (usually ATG) and the last codon is a stop codon (e.g.,TAG).

(27.4)

One could create an HMM model with states that correspond to the various components of such a gene.

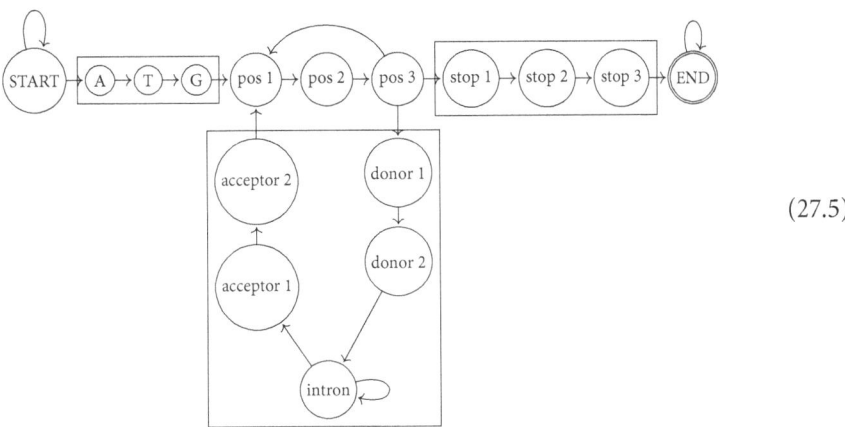

$$(27.5)$$

A straightforward HMM like (27.5) is likely to be very poor at recognizing real genes. The positions in each codon are treated independently: the probability of a base in a codon can't depend on what was emitted in the previous position. Only one strand of the DNA is considered at once, but real DNA is typically double stranded so genes could be read in either direction. In addition, the length distributions of the introns and exons are not modeled: the length of such a feature in this model will just be given by the probability of staying in the appropriate states. However, more sophisticated approaches building on the ideas of HMMs have very successfully modeled eukaryotic genes and simpler genes from bacteria [Salzberg et al., 1998; Delcher et al., 1999, 2007; Burge and Karlin, 1997].

27.2 HMM decoding problems

If we are given both the emitted string S and the path π, we can directly compute:

- The probability of generating string S given a particular path of states. This is $\Pr[S \mid \pi]$, which can be computed as the product of $\Pr[S_i \mid \pi_i] = E[\pi_i, S_i]$.
- The probability of following a given path π of states. This is $\Pr[\pi]$, which is the product of the state transitions $\Pr[\pi_i \to \pi_{i+1}] = A[\pi_i, \pi_{i+1}]$.
- The joint probability of both the observed S and the given path π being used by the operation of the HMM. This is $\Pr[S, \pi]$, which can be computed as the product of all the state transitions $\pi_i \to \pi_{i+1}$ and the probability of emitting the observed characters at the corresponding states $\Pr[S_i \mid \pi_i]$. This is:

$$\Pr[S, \pi] = \Pr[\pi_0 \to \pi_1] \prod_{i=1}^{n} \Pr[S_i \mid \pi_i] \Pr[\pi_i \to \pi_{i+1}] \qquad (27.6)$$

$$= A[\pi_0, \pi_1] \prod_{i=1}^{n} E[\pi_i, S_i] A[\pi_i, \pi_{i+1}]. \qquad (27.7)$$

However, these are *hidden* Markov models, meaning that the path taken is generally not observed or known. We observe only the emitted string S. In that case, we would like to infer the most likely set of states that were visited to generate S.

Problem 27.2 (HMM decoding problem). *Given a string $S = x_1, x_2, \ldots, x_n$ generated by an HMM (Σ, Q, A, E), find a path π through the HMM that maximizes $\Pr[S, \pi]$.* ◆

Note that

$$\arg\max_{\pi} \Pr[\pi \mid S] = \arg\max_{\pi} \Pr[S, \pi] / \Pr[S] \tag{27.8}$$

$$= \arg\max_{\pi} \Pr[S, \pi]. \tag{27.9}$$

Therefore, finding the π that maximizes $\Pr[S, \pi]$ is equivalent to finding the π that maximizes $\Pr[\pi \mid S]$, which is the more intuitive thing to maximize.

27.3 Viterbi algorithm

The Viterbi algorithm [Viterbi, 1967; G. D. Forney Jr., 1973] solves Problem 27.2. It is a dynamic programming algorithm to find the highest probability path given the observed emitted symbols. Let $S = x_1, \ldots, x_n$ be our observed string with the sequence of characters x_i. As with other times we've seen dynamic programming, the first step is to define the subproblems:

- $V[a, k]$ is the probability of the best path that for $x_1, \ldots, x_k$ that ends at state a.

With the answer to these subproblems, our probability of the best path can be computed as $\max_{a \in Q} V[a, n]$, where n is the length of the observed string S. We write $V[a, k]$ in terms of smaller subproblems. To emit the kth character from the ath state, we must have emitted the first $k - 1$ characters and ended up in some state b that will transition into a. We have:

- $V[a, k]$ is the probability of the best path for $x_1, \ldots, x_{k-1}$ that goes to some state b (i.e., $V[b, k - 1]$) times the probability of the transition from b to a (it might be that $a = b$), and then times the probability to output x_k from state a.

The recurrence for $V[a, k]$ is then:

$$V[a, k] = \max_{b \in Q} \left\{ \underbrace{V[b, k - 1]}_{\substack{\text{Best path for} \\ x_1, \ldots, x_{k-1} \\ \text{ending in state } b}} \times \underbrace{\Pr[b \to a]}_{\substack{\text{Prob of going} \\ \text{from state } b \text{ to } a}} \times \underbrace{\Pr[x_k \mid \pi_k = a]}_{\substack{\text{Prob of emitting } x_k \\ \text{given that the } k\text{th} \\ \text{state is } a}} \right\}. \tag{27.10}$$

For a two-state HMM, we end up with the following dynamic programming matrix.

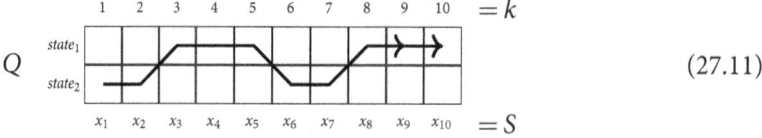

$$\tag{27.11}$$

Using the standard traceback approach let's recover the actual π that achieves the computed probability.

Running time. There are $O(|Q|n)$ cells in the array. Each subproblem takes $|Q|$ time to compute since we have to take the maximum over possible previous states. That leads to an overall runtime of $O(n|Q|^2)$ to find the best path. See Exercises 27.1 and 27.2 for refinements.

27.4 The Forward-Backward algorithm

A natural probability to be interested in is:

$$\Pr[\pi_i = a \mid S] = \frac{\Pr[\pi_i = a, S]}{\Pr[S]}. \tag{27.12}$$

This is the probability that the state that emitted the ith character is state a. This probability would let us answer a question such as, "what is the chance that a given nucleotide was emitted by an intron state?" To compute it, we need to compute both the numerator and denominator of (27.12).

The *Forward-Backward algorithm* computes the numerator by rewriting it into two parts:

$$\Pr[\pi_i = a, S] = \Pr[x_1, \ldots, x_i, \pi_i = a] \, \Pr[x_{i+1}, \ldots, x_n \mid \pi_i = a]. \tag{27.13}$$

This follows because the generation of $x_{i+1}, \ldots, x_n$ depends only on the state x_i and not on anything before by the Markov property of HMMs.

The Forward algorithm computes the first part: $\Pr[x_1, \ldots, x_i, \pi_i = a]$. The Backward algorithm computes the second part: $\Pr[x_{i+1}, \ldots, x_n \mid \pi_i = a]$. Again, both are dynamic programming algorithms.

27.4.1 Forward algorithm

Define:

- $F[a, k]$ is the probability of emitting characters $x_1, \ldots x_k$ using some path that ends at state a.

These subproblems are very similar to the Viterbi subproblems, except there we only cared about a "best" path, whereas now any path that ends at a is ok. Accordingly, the recurrence is very similar as well. The probability we are seeking is the total probability of ending up at some state b after emitting $k - 1$ characters and then transitioning to state a to emit the x_k character:

$$F[a, k] = \sum_{b \in Q} F[b, k-1] \times \Pr[b \to a] \times \Pr[x_k \mid \pi_k = a]. \tag{27.14}$$

This is exactly the same as the Viterbi recurrence, with the max replaced by a sum.

The first part of (27.13) is:

$$\Pr[x_1, \ldots, x_i, \pi_i = a] = F[a, i]. \tag{27.15}$$

This is computable in the same running time as Viterbi: $O(i|Q|^2)$. Visually, this is (27.16).

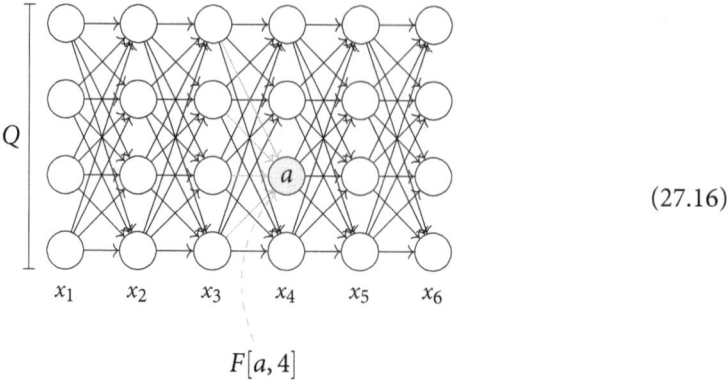

$$F[a, 4]$$

(27.16)

For example, $F[a, 4]$ computes the total probability of the paths ending at state a in step 4.

The Forward algorithm also lets us compute $\Pr[S]$ as a bonus. Since x_n must be emitted by some state, we have:

$$\Pr[S] = \sum_{b \in Q} F[b, |S|].$$

(27.17)

That takes care of the denominator of (27.12).

27.4.2 Backward algorithm

How about computing $\Pr[x_{i+1}, \ldots, x_n \mid \pi_i = a]$? Define the subproblems:

- $B[a, k]$ is the probability of generating string $x_{k+1}, \ldots, x_n$ starting from state a ($\pi_k = a$).

We can compute this by treating subproblems that use shorter suffixes of S as the smaller subproblems.

$$B[a, k] = \sum_{b \in Q} \left(\underbrace{\Pr[a \to b]}_{\substack{\text{Prob of going} \\ \text{from state } a \text{ to } b}} \times \underbrace{\Pr[x_{k+1} \mid \pi_{k+1} = b]}_{\substack{\text{Prob of emitting } x_{k+1} \\ \text{given that the } k+1\text{th state} \\ \text{is } b}} \times \underbrace{B[b, k+1]}_{\substack{\text{Prob of} \\ x_{k+1}, \ldots, x_n \\ \text{starting in state } b}} \right).$$

(27.18)

Now the second half of (27.13) is:

$$\Pr[x_{i+1}, \ldots, x_n \mid \pi_i = a] = B[a, i].$$

(27.19)

Putting the Forward and Backward algorithm together, we can compute

$$\Pr[\pi_i = a \mid S] = \frac{\Pr[\pi_i = a, S]}{\Pr[S]} = \frac{F[a, i] B[a, i]}{\sum_b F[b, n]}.$$

(27.20)

This allows us to compute the probability that state a was visited at character i:

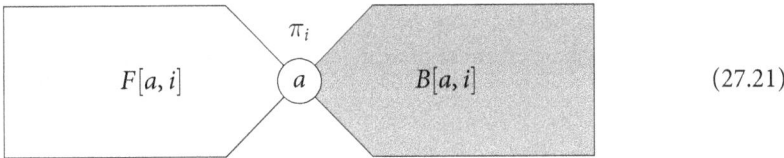

$$(27.21)$$

27.5 Estimating HMM parameters

The various probabilities that are part of an HMM need to be chosen somehow. In some situations they may be known and set by hand based on the application. But in most cases, learning the weights from data makes more sense. Next, we briefly introduce a few approaches for such data-driven training, although, as you can imagine, there are lots of approaches to tackle this problem.

27.5.1 Labeled training

When training examples are available that have both the emitted string and the path used to generate them, choosing good HMM parameters is relatively straightforward. If we have such a set:

$$(x^{(1)}, \pi^{(1)}) = \begin{pmatrix} x_1^{(1)} & x_2^{(1)} & x_3^{(1)} & x_4^{(1)} & x_5^{(1)} & \ldots & x_n^{(1)} \\ \pi_1^{(1)} & \pi_2^{(1)} & \pi_3^{(1)} & \pi_4^{(1)} & \pi_5^{(1)} & \ldots & \pi_n^{(1)} \end{pmatrix} \qquad (27.22)$$

$$\vdots$$

$$(x^{(K)}, \pi^{(K)}) = \begin{pmatrix} x_1^{(K)} & x_2^{(K)} & x_3^{(K)} & x_4^{(K)} & x_5^{(K)} & \ldots & x_n^{(K)} \\ \pi_1^{(K)} & \pi_2^{(K)} & \pi_3^{(K)} & \pi_4^{(K)} & \pi_5^{(K)} & \ldots & \pi_n^{(K)} \end{pmatrix} \qquad (27.23)$$

we can compute the probabilities by counting. Let $\#A_{ab}$ be the number of times the transition $a \to b$ is observed in the training paths, and let $\#E_{ax}$ be the number of times symbol x was observed to be emitted from state a. We can estimate the probabilities:

$$\Pr[a \to b] = \frac{\#A_{ab}}{\sum_{q \in Q} \#A_{aq}}, \qquad (27.24)$$

$$\Pr[x \mid a] = \frac{\#E_{ax}}{\sum_{x \in \Sigma} \#E_{qx}}. \qquad (27.25)$$

The more typical case, however, is that we have observed strings, but we do not have the corresponding paths for those strings. In that case, we need to do something more sophisticated to estimate good HMM parameters.

27.5.2 Viterbi training

The ability to compute the best paths using the Viterbi algorithm suggests an iterative refinement approach to estimate parameters:

Algorithm 27.1. Viterbi training.

1. Choose a random set of parameters.
2. Repeat:
   (a) Find the best paths using the Viterbi algorithm for each training string.
   (b) Use those paths to estimate new parameters.

This is a local search-type algorithm. One would stop once, for example, the change in the parameters is small for some number of iterations.

27.5.3 Baum-Welch algorithm

The issue with Viterbi training is that it commits to a single best path for each training string. The Baum-Welch algorithm [Baum, 1972] instead uses all the possible paths for each training string to estimate. To do this, it computes the expectation of the counts over all paths. Specifically,

$$\Pr[\pi_i = a \text{ and } \pi_{i+1} = b \mid S] = \frac{F[a, i]A[a, b]E[b, x_{i+1}]B[b, i+1]}{\Pr[S]}. \tag{27.26}$$

The righthand side uses all the various subproblems and HMM parameters we've defined so far. It is the chance of getting to state a at position i, then transitioning to state b, emitting the required symbol from state i, and then going from state b to the end of the string.

That allows us to estimate:

$$\mathbb{E}A_{ab} = \text{expected \# of times transition } a \to b \text{ is observed} \tag{27.27}$$

$$= \sum_{S_j} \sum_i \Pr[\pi_i = a \text{ and } \pi_{i+1} = b \mid S_j]. \tag{27.28}$$

That sums over all the training strings S_j and all the positions i in those training strings and obtains the probability that an $a \to b$ transition happened between positions i and $i + 1$ using (27.26). This gives the expected number of times the transition happened.

Similarly, we can compute:

$$\mathbb{E}E_{ax} = \text{expected \# of times } x \text{ is emitted when in state } a \tag{27.29}$$

$$= \sum_{S_j} \frac{1}{\Pr[S_j]} \sum_{i : S_j[i]=x} F[a, i]B[a, i]. \tag{27.30}$$

Here, $F[a, i]B[a, i]/\Pr[S]$ is $\Pr[\pi_i = a \mid S]$ from (27.20). The sum is only over positions that emitted character x, so this is the expected number of emissions of x from state a.

Using these expectations, we can now outline the Baum-Welch algorithm.

Algorithm 27.2. Baum-Welch.

Input: Training strings $S_1, \ldots S_m$, and HMM states Q and alphabet Σ.
Output: Model probabilities A and E.

1. Set the parameters to a random initial starting point (or uniform probabilities).
2. Repeat:
    (a) Compute $\mathbb{E}A_{ab}$ and $\mathbb{E}E_{xa}$ using (27.27) and (27.29).
    (b) Using the labeled training approach of Section 27.5.1, treating the computed expected values as the observed counts, compute new model parameter matrices A and E.
3. Stop when the probability of generating the training strings converges.

27.6 Summary and notes

Hidden Markov Models (HMM) model the generation of strings. They are governed by a string alphabet (Σ), a set of states (Q), a set of transition probabilities A, and a set of emission probabilities for each state (E). They are called "hidden" because the sequence of states is not known—we must either integrate over the choice of states (using sums over the possibilities as in the Forward algorithm) or pick a single likely path (as in the Viterbi algorithm). More history on the development and use of the Viterbi algorithm can be found in Viterbi [2006].

Given a string and an HMM, we can compute, among other things: (1) the most probable path the HMM took to generate the string (Viterbi); (2) the probability that the HMM was in a particular state at a given step (Forward-Backward algorithm). All of these algorithms are based on dynamic programming.

Finding good parameters when constructing the model is a harder problem. The Baum-Welch algorithm is an oft-used heuristic for that, but there are many other ways to estimate these parameters.

Presentation Notes

Our presentation of HMMs was informed by a discussion in Pevzner [2000] (from where we borrow some notation) and from Durbin et al. [1998]. Durbin et al. [1998] also contains a much broader discussion of probabilistic string models. Both use the fair coin flip example as a motivating problem for HMMs. More details about Markov models (and hidden Markov models) can be found in Schwartz [2008].

27.7 Exercises

27.1   Suppose the maximum in-degree of a node in an HMM is d (that is, no state u has more than d transitions $v \to u$ with non-zero probability). Describe an implementation of the Viterbi algorithm that achieves a runtime of $O(|Q|nd)$.

27.2   The naïve implementation of the Viterbi algorithm uses space $O(n|Q|)$, where Q is the set of states of the HMM and n is the length of the input string, if you want to backtrack to find the nodes in the optimal path. Show how to reduce this to $O(\sqrt{n}|Q|)$ space while still being able to output the nodes on the optimal path in the same amount of time as normal. *Hint: subsample every $\sqrt{n}$ position in the input string.*

27.3   Suppose you are designing an HMM to output the strings of the form: $ab^k c$ where k is drawn from a discrete distribution D given by the user. That is, the probability of generating a string of k "b"s in the middle of the string should be equal to $D[k]$. Suppose $\Pr[k=0]=0$ and $\Pr[k>6]=0$. Describe how to construct an HMM for any such D that the user provides.

27.4   Many of the algorithms for HMMs require multiplying many numbers that are between 0 and 1. Once a problem gets big enough, this is likely to lead to very small numbers, which can underflow on a computer. A solution to this is to work in log-space: transform all the probabilities by taking their log. Rewrite the Forward-Backward recurrences to work with the log probabilities, and explain why this helps with underflow.

CHAPTER 28
Stochastic Context-Free Grammars

A grammar, as defined in Chapter 26, specifies a set of strings: those that can be produced by the grammar. We can extend this to assign a probability to every string produced by the grammar by giving each production a probability. When applied to CFG, this produces *Stochastic* Context-Free Grammars (SCFGs). This mirrors the addition of probabilities to automata to produce HMMs. Such grammars have been used to model and predict RNA structures, where high-probability parses correspond to low-energy (good) RNA folding [Eddy and Durbin, 1994; Sakakibara et al., 1994].

Specifically, we assign a probability $p_{W,\beta}$ to each rule $W \to \beta$ with the constraint that

$$\sum_{\beta} p_{W,\beta} = 1 \tag{28.1}$$

for each non-terminal W. This ensures that we end up with a probability distribution. The interpretation of these probabilities is the chance that a particular production will be applied to expand a particular non-terminal. In other words:

$$p_{W,\beta} = \Pr[\beta \mid W]. \tag{28.2}$$

In a parse tree (Section 26.2.1), this looks like (28.3).

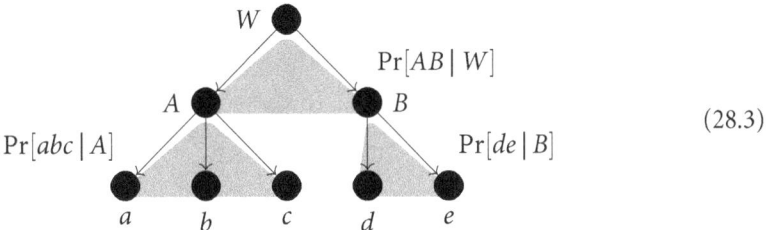

$$\tag{28.3}$$

The probability of a particular tree $\mathcal{T}$ is

$$\Pr[\text{root is } S] \prod_{u \in \mathcal{T}} \Pr[\text{children} \mid u], \tag{28.4}$$

where u ranges over the non-leaf nodes of u, and where S is the distinguished starting non-terminal. Taking $\Pr[\text{root is } S] = 1.0$, the probability of a particular parse tree is the product of the probabilities of the transitions that have been used.

SCFGs provide an expressive framework for specifying distributions of strings and for generating strings from that distribution. This is useful, for example, to generate testing data for programs (where the SCFG models a reasonable distribution of inputs) or for serving as null hypotheses to empirically assess statistical significance of some event (if an observed event is rare among the SCFG modeling the background distribution, perhaps the event corresponds to something real).

28.1   The Inside algorithm

The first computational task that arises with SCFGs is:

Problem 28.1 (String probability). *Let G be a stochastic context-free grammar. What's the probability that G will generate a given string T?* ◆

Problem 28.1 is solved by the Inside algorithm. We assume that G is in Chomsky normal form (CNF; Section 26.2.2).

We will apply dynamic programming treating each substring of T as a subproblem. We add a parameter to our DP recurrence to track which non-terminal gave rise to the substring. Specifically, let

$$I(i, j, W) := \text{"probability of generating} \tag{28.5}$$
$$\text{substring } T[i \ldots j] \text{ starting}$$
$$\text{from non-terminal } W.\text{"}$$

When $i = j$, we have the base case which is easy to compute:

$$I(i, i, W) = \Pr[W \to T[i]] = p_{W, T[i]}. \tag{28.6}$$

This is given directly by the probabilities on the productions.

For the general case, the first production at the top of the parse subtree for $T[i, j]$ must be a rule of the form $W \to UV$ since the substring is of length greater than 1 and our grammar is in CNF. Schematically, we have:

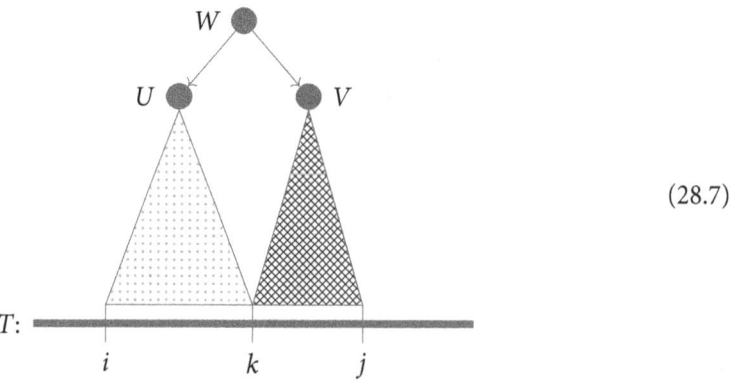

$$\tag{28.7}$$

where k is the point that divides the substring into the part generated by U and the part generated by V. Of course, we don't know k or which rule would be applied in the parse tree, so we have to sum up over everything to get the general recurrence:

$$I(i, j, W) = \sum_{U \in \mathcal{N}} \sum_{V \in \mathcal{N}} \sum_{k=i}^{j-1} I(i, k, U) \times I(k+1, j, V) \times \Pr[W \to UV]. \tag{28.8}$$

The sums iterate over all the different productions that could apply to generate the substring. The products inside the sums recursively compute the probability of the substrings of the substring. We take $\Pr[W \to UV]$ to be 0 if that production doesn't exist in the grammar. Otherwise, it is $p_{W,UV}$. Taking the sums is correct since each of the events inside the sums is independent.

At the end, $I(1, |T|, S)$ will contain the probability $\Pr[T \mid \mathcal{G}]$ that the grammar will generate the given string T.

28.2 Cocke-Younger-Kasami (CYK) algorithm

We also would like to be able to find the *most probable* parse tree $\mathcal{P}^*$, leading to the problem:

Problem 28.2 (Best parse tree). *What is the most probable parse tree for T using a given stochastic context-free grammar G?* ◆

This problem is solved by the CYK algorithm. We modify the Inside algorithm to replace the sums with "max" operations. Instead of integrating over all the possible parses, the algorithm keeps only the highest probability parse at each step.

The base case is again:

$$CYK(i, i, W) = p_{W,T[i]}. \tag{28.9}$$

The general case is:

$$CYK(i, j, W) =$$

$$\max_{U \in \mathcal{N}} \max_{V \in \mathcal{N}} \max_{k=i}^{j-1} CYK(i, k, U) \times CYK(k+1, j, V) \times p_{W,UV}. \tag{28.10}$$

Then $CYK(1, |T|, S)$ is $\Pr[T, \mathcal{P}^* \mid \mathcal{G}]$, which is the probability of generating T using the most probable parse tree $\mathcal{P}^*$.

28.3 Estimating the production probabilities

Another natural question is how to learn the probabilities for an SCFG given a set of training strings in the language [Lari and Young, 1990]. If we are given a set of strings that have been generated by a given grammar $\mathcal{G}$, we want to estimate the probability of each production. When we have some initial estimates for each $p_{W,\beta}$, we can try to improve on them using these training strings.

Let $\#(W)$ be the expected number of times that non-terminal W will be used in a parse tree randomly generated from $\mathcal{G}$. Similarly, let $\#(W \to \beta)$ be the expected number of times that production $W \to \beta$ should be used. If we have these values, then we can estimate the probabilities via:

$$p_{W,\beta} = \frac{\#(W \to \beta)}{\#(W)}$$

$$= \Pr[\beta \text{ is generated} \mid \text{we are currently expanding } W]. \tag{28.11}$$

Of course, we don't normally have $\#(W)$ and $\#(W \to \beta)$ laying around. But we could compute $\#(W)$ *given the current probabilities* by seeing how many times W appears in parses of the training strings we've been given.

Again, we will deploy dynamic programming, using the following definition of a subproblem:

- $Q(i,j,W) :=$ "probability of generating the string T except for the substring $T[i \ldots j]$, which we assume is generated by W."

That is $Q(i,j,W)$ is

$$\Pr[\text{Completing the rest of the parse tree} \mid \\ \text{substring } T[i \ldots j] \text{ is generated from } W]. \tag{28.12}$$

Pictorially, this looks like (28.13).

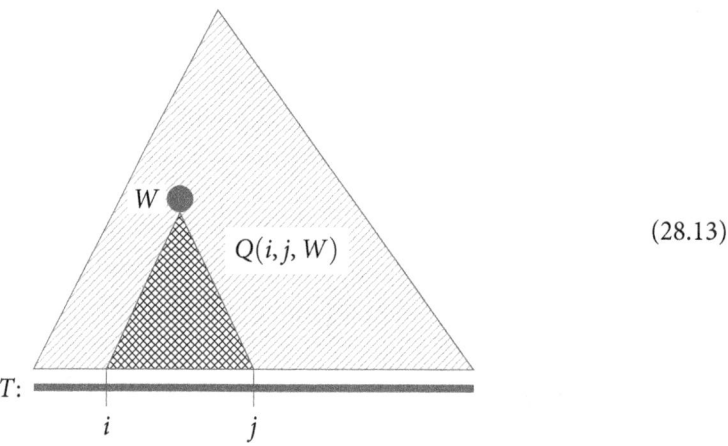

$$(28.13)$$

We can use these values to estimate the probability of generating the entire string T assuming $T[i \ldots j]$ is generated from W by combining the Q values with an application of the Inside algorithm (Section 28.1) to estimate the probability of generating $T[i \ldots j]$ from W. Let $W \implies T[i \ldots j]$ denote the event that W is used to generate $T[i \ldots j]$. Then,

$$I(i,j,W) \times Q(i,j,W) = \Pr[T \text{ is generated and } W \implies T[i \ldots j] \mid \mathcal{G}]. \tag{28.14a}$$

Rewriting this using the definition of conditional probability:

$$= \Pr[T \mid \mathcal{G}] \Pr[W \implies T[i \ldots j] \mid \mathcal{G}]. \tag{28.14b}$$

Dividing by $\Pr[T \mid \mathcal{G}]$, we have:

$$\Pr[W \implies T[i \ldots j] \mid \mathcal{G}] = \frac{1}{\Pr[T \mid \mathcal{G}]} I(i, j, W) \times Q(i, j, W). \tag{28.15}$$

Equation (28.15) gives us the probability that W is used for a particular substring.

Since we're interested in the expected number of uses of W across the entire parse tree, we need to consider every substring. Let $B_{W,i,j}$ be a random variable that is 1 if the event that W was used to generate $T[i \ldots j]$ occurs and is 0 otherwise. By (28.15),

$$\mathbb{E}[B_{W,i,j}] = \frac{1}{\Pr[T \mid \mathcal{G}]} I(i, j, W) \times Q(i, j, W). \tag{28.16}$$

Therefore, the expected number of uses of W is

$$\#(W) = \sum_{i=1}^{|T|} \sum_{j=i}^{|T|} \mathbb{E}[B_{W,i,j}] \tag{28.17}$$

$$= \frac{1}{\Pr[T \mid \mathcal{G}]} \sum_{i=1}^{|T|} \sum_{j=i}^{|T|} I(i, j, W) \times Q(i, j, W). \tag{28.18}$$

Similarly, we can estimate the number of times a particular transition $W \to UV$ is used:

$$\#(W \to UV) = \frac{1}{\Pr[T \mid \mathcal{G}]} \sum_{i=1}^{|T|-1} \sum_{j=i+1}^{|T|} \sum_{k=i}^{j-1}$$

$$Q(i, j, W) \times I(i, k, U) \times I(k+1, j, V) \times p_{W,UV}. \tag{28.19}$$

Formally, this follows from a similar argument to (28.18). See Exercise 28.3.

These values allow us to apply an expectation-maximization-style algorithm to optimize the probabilities:

Algorithm 28.1. Iterative improvement of SCFG probabilities.

- Repeat until the probabilities converge:
  1. Using the current probabilities (or random starting probabilities), estimate $\#(W)$ and $\#(W \to \beta)$ for each transition and non-terminal using equations (28.18) and (28.19).
  2. Estimate updated production probabilities using equation (28.11).

This gives us updated production probabilities. What remains is to determine how to compute Q.

28.3.1 The Outside algorithm

To compute Q, we again use a dynamic programming approach. The base cases are the situation where the "black hole" is the entire string:

$$Q(1, |T|, S) = 1, \tag{28.20}$$

$$Q(1, |T|, U) = 0 \qquad \text{for } U \neq S, \tag{28.21}$$

since we require the entire string to be generated from S. Now consider $Q(i, j, V)$ in general. V must have a parent (since it doesn't generate the whole string) and 1 sibling since our grammar is in CNF. V's sibling could be to the left of V or to the right of V in the rule that generated this instance of V, depending on whether that rule was $W \to \bullet V$ or $W \to V \bullet$.

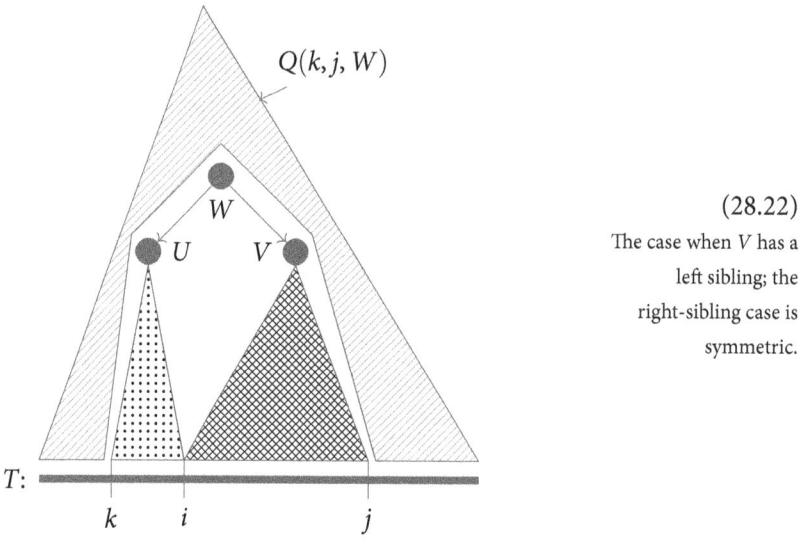

(28.22)

The case when V has a left sibling; the right-sibling case is symmetric.

The general equation has two parts corresponding to these two cases:

$$Q(i, j, V) = \sum_{U} \sum_{W} \sum_{k=1}^{i-1} \left(I(k, i-1, U) Q(k, j, W) p_{W, VU} \right) +$$

$$\sum_{U} \sum_{W} \sum_{k=j+1}^{|T|} \left(I(j+1, k, U) Q(i, k, W) p_{W, UV} \right). \tag{28.23}$$

The terms inside the parentheses correspond to each of the parts of the parse tree that we need to consider: applying the particular rule that generated the V, generating the subtree rooted at V's sibling, and computing the rest of the parse tree. We don't need to have a term for the subtree rooted at V, since the Q values take as given that that substring was generated by V.

28.4 Summary and notes

Stochastic context-free grammars provide a nice framework for modeling distributions of strings. The grammars are expressive, we can modify the distribution by setting rule probabilities, and there are polynomial-time algorithms for natural problems over these grammars. The Inside-Outside algorithm is from Lari and Young [1990]. A drawback is that the algorithms' running times have quite a large exponent, so they may be impractical for large grammars.

The CYK algorithm was discovered independently by several people [Cocke and Schwartz, 1970; Kasami, 1966; Younger, 1967; Sakai, 1962].

Presentation Notes

Our presentation of these algorithms was informed by Durbin et al. [1998], which contains much more detail on probabilistic approaches for string analysis.

28.5 Exercises

28.1   What is the running time of the CYK algorithm to find a parse tree for a context-free grammar that has m productions, n nonterminals, and an alphabet size of σ? Give the best running time you can.

28.2   Describe how to retrieve the optimal parse tree in the CYK algorithm.

28.3   Write out the derivation of (28.19).

CHAPTER 29

Genome Graphs

Pan-genomics is the use and study of the genomes of many individuals of a species simultaneously. While the genome of every human is very similar, each contains variants and unique sequences that help to give a person their specific characteristics. Rather than use a single string to represent "the" human genome, it would be more informative to use a collection of sequences from many people. This is the human pan-genome.

Since each of the sequences is very similar, we must find a way to store and use them that is more efficient than merely recording them all individually. One approach to this is to use *genome graphs*, which encode a set of strings as paths in a labeled graph [Novak et al., 2017; Paten et al., 2017; Sherman et al., 2019; Sherman and Salzberg, 2020; Computational Pan-Genomics Consortium, 2018; Pandey et al., 2021; Qiu and Kingsford, 2021]. This approach has begun to gain prominence within genomics, and a number of tools have been developed to create and manipulate genome graphs [Garrison et al., 2018; Li et al., 2020].

29.1   Some types of genome graphs

There are a number of types of genome graphs and software for creating them [e.g., Paten et al., 2011; Mäkinen et al., 2020; Li et al., 2020; Garrison et al., 2018; Myers, 2005; Gagie et al., 2017; Equi et al., 2023; Lee et al., 2002]. While they each have the goal of representing a collection of strings, their benefits and applications vary.

29.1.1   De Bruijn graphs

We have already seen one very fundamental type of genome graph: de Bruijn graphs (Section 25.7). The *classical de Bruijn* graph of order k over an alphabet Σ is defined by:

$$V = \Sigma^k, \tag{29.1}$$

$$E = \{(u, v) : u[2 \ldots k] = v[1 \ldots k - 1]\} \tag{29.2}$$

where substrings are indexed starting at 1. In other words, the graph contains a node for every length-k string (k-mer) over the alphabet, and we have an edge from u to v if those substrings overlap in their length $k - 1$ suffixes and prefixes.

$$u: \underline{} \;\; \overline{} : v \tag{29.3}$$

For example, the de Bruijn graph of order 3 over $\Sigma = \{0, 1\}$ is illustrated here.

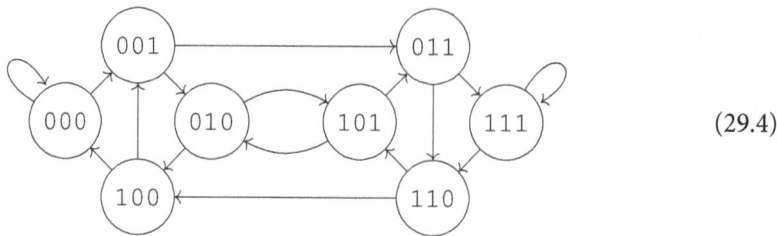

(29.4)

This classical de Bruijn graph has been of theoretical interest for some time, originally to enumerate de Bruijn sequences, which are the sequences that contain every k-mer exactly once.

More recently [Pevzner et al., 2001], de Bruijn graphs were recognized as an important element in genome assembly. In the assembly problem, you are given a large set of short fragments $R = \{r_1, \ldots, r_m\}$ that are random substrings of the true, unknown genome string S. Usually, each $|r_i|$ is much smaller than $|S|$. The assembly task is to infer S from R.

To do this, one can construct a subgraph of the de Bruijn graph that is based on R, assuming the strings in R are each $\geq k + 1$ long:

$$V' = \{u \in \Sigma^k : u \in r_i \text{ for some } i\}, \tag{29.5}$$

$$E' = \{(u, v) : u \circ v[k] \in r_i \text{ for some } i\}, \tag{29.6}$$

where $\circ$ indicates concatenation and $\in$ indicates presence as a substring. In other words, we have a k-mer node if and only if the k-mer occurs in some fragment in R, and we have an edge only if k-mer v follows k-mer u in some fragment. This is a subset of the classical de Bruijn graph because $V' \subseteq V$ and $E' \subseteq E$.

For assembly, we then extend this graph to a multi-graph by duplicating edges so that there are c_{uv} edges between nodes u and v if the $(k + 1)$-mer $u \circ v[k]$ occurs i times in S. We have to estimate c_{uv} from the frequency of the $(k + 1)$-mer in R.

A path in a graph G is *Eulerian* if it walks across edges of a graph, using each edge exactly once. If the strings of R cover the entirety of S and if they are sampled uniformly enough that we can determine how many times each $(k + 1)$-mer appears in S from them, then S will be exactly spelled out by some Eulerian path in this subgraph. Any Eulerian path of the subgraph represents a string that is consistent with the evidence provided by R and, hence, is a candidate for the assembly of the fragments.

Many practicalities must be dealt with that add complications to this idealized formulation (such as errors in the r_i strings, imperfect random sampling, selecting from multiple Eulerian paths, etc.), so simply finding Eulerian paths is not the way assembly is done in practice. Instead, algorithms are developed to output long paths through the de Bruijn graph that we can be confident in—these chunks are called *unitigs*.

The de Bruijn graphs represent a set R of strings but we really don't care about the individual r_i except in so far as they tell us about some subsequence of a single string S.

29.1.2   Colored de Bruijn graphs

In the case where we are given multiple (long) strings $S_1, \ldots, S_n$ that we want to represent, we can extend the de Bruijn graph idea further by *coloring* each edge with the string from which it came. This "color" will be a number from 1 to n. This is a *colored de Bruijn graph* [Iqbal et al., 2012]. To reconstruct S_i, one seeks an Eulerian tour in the graph restricted to only edges colored i.

These two ideas can be combined: if you have a set of fragments R_i sampled from each unknown string S_i, you can construct the edges of colored de Bruijn graph from the $(k+1)$-mers within $R = \bigcup_i R_i$, coloring them based on which R_i they appear in. Hence, colored de Bruijn graphs can be viewed as sets of sets of $(k+1)$-mers.

One might question whether there is any space savings achieved by storing strings this way. First, in some cases, space savings is not the point: if your main goal is to reconstruct the unknown string or strings, the point of the de Bruijn graph is to turn this reconstruction problem into one of finding a path in the graph.

Second, consider the set of colors C_{uv} that appear on edges between two nodes u and v. If your set of strings $\{S_i\}$ consists of a few highly similar subsets of strings, then most C_{uv} will be one of a small number of possible sets. For example, $(k+1)$-mers that appear in every S_i will have $C_{uv} = \{1, \ldots, n\}$. Hence, rather than storing these colored edges explicitly, we need only store a code that tells us which set occurs between each pair of nodes. In the simplest form, imagine an array C, with $C[i]$ storing a subset of colors. Each $C[i]$ is a *color class*. If $C[j] = C_{uv}$, we could label the edge (u, v) with j and that would tell us which colored edges appear between u and v. If our sets are similar enough, the array C will be short, and we can therefore store our colored de Bruijn graph by recording (the small) C and a (small) integer on an edge in the de Bruijn subgraph. A line of work has pursued this idea for encoding color classes compactly [Holley and Melsted, 2020; Muggli et al., 2019; Almodaresi et al., 2020, 2017] and otherwise encoding de Bruijn graphs compactly [Muggli et al., 2019, 2017].

29.1.3   Variation graphs

Variation graphs are another type of genome graph. Such a graph has a set of nodes V, each associated with a string label $\ell : V \mapsto \Sigma^*$. These graphs are typically used to represent variants relative to a reference genome. Hence, there is usually a "backbone" path between two distinguished nodes s and t that spells out that reference. Variants (often derived from genomic variant calling software) are then added that represent edits to this reference. Insertions are additional longer paths between two backbone nodes, while deletions are paths that skip over backbone nodes.

Often, variation graphs are required to be acyclic although this is not universally true. An example of a variant graph is (29.7).

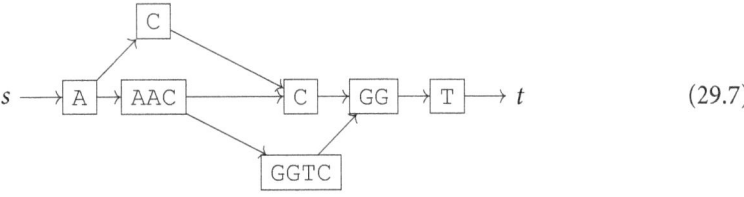

$$\tag{29.7}$$

Edges in variation graphs can be colored as in colored de Bruijn graphs to indicate the string to which they belong.

29.2 Constructing genome graphs via compression

One approach to creating a genome graph is to exploit the correspondence between compression and genome graphs: both seek to identify common substrings. We will present a method from Qiu and Kingsford [2021] that uses a particular class of compression algorithms from Storer and Szymanski [1982]:

Definition 29.1 (External pointer macro compression). Given an alphabet Σ, a reference string R, and a string T, an *external pointer macro* (EPM) compression of T is a string of the form $C = R\#p_1p_2p_3 \ldots p_k$, where

- R is a string over Σ^*,
- $\#$ is a character that is not in Σ, and
- each p_i is a pair (pos_i, len_i) where pos_i is a position in R and len_i is a length of substring,

such that when each p_i is replaced with the substring it points to in R, the string T is spelled out.

Each of the p_i is called a *pointer*. ∎

In pictures, a compressed string in this framework looks like (29.8).

$$(29.8)$$

This should remind you of the Lempel-Ziv compression scheme (Section 14.1), with the difference that rather than using the string itself as the "dictionary," we use a known reference string R as the target of the pointers. The problem of finding a minimum-sized compressed form under the EPM scheme is NP-hard, as was shown in Storer and Szymanski [1982].

Relative Lempel-Ziv (RLZ) [Ferrada et al., 2014; Kuruppu et al., 2011; Gagie et al., 2016] is an example of an implementation of such an EPM scheme: in RLZ, we do Lempel-Ziv, but instead of using the prefix of the string we have already encoded as the target for the pointers, we use a given reference string.

Using RLZ, we can construct a graph from a string T using the following three-pass algorithm. To construct a graph that contains multiple strings we can take $T = T_1\$T_2\$ \ldots$ using some separation character $.

Algorithm 29.1. RLZ genome graph creation.

1. Run relative Lempel-Ziv using reference R to encode T, produce an encoding of pointers $p_1, p_2, \ldots, p_t$.
2. Break R at the start and end of every pointer, creating a sequence of substrings $s_1, \ldots, s_m$, such that $s_1 \circ s_2 \circ \cdots \circ s_m = R$.
3. Create a node for each s_i.
4. For each pointer p_i, if it covers $s_i, \ldots, s_{i+j}$, add edges from $s_\ell \to s_{\ell+1}$ for $\ell \in \{i, i + j - 1\}$.
5. Add an edge (s_i, s_j) if s_i is the end of some pointer p_k and s_j is the start of pointer p_{k+1}.
6. Remove any node that has no edges adjacent to it.

The RLZ encoding of reference $R = ATCGATAGA$ and input string $T = TCGAGATGA$ is:

$$ATCGATAGA\#(1,4)(3,3)(7,2). \tag{29.9}$$

The graph that is created is shown in (29.10). In this case, the graph stores both R and T: the horizontal path encodes R, the bold path encodes T:

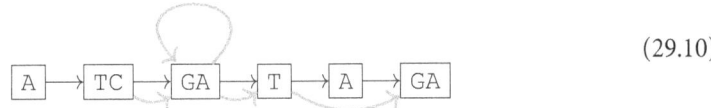

$$\tag{29.10}$$

By construction, the graph resulting from algorithm (29.10) will contain a path that spells out the input string T. If the compressed form C is small, then the size of the genome graph created in this way is also small. Let t' be the number of unique pointers in C, and let $\alpha = \min\{2t', |R|\}$. The constructed graph then has size that is bounded by:

$$size(G) \leq |R| \log |\Sigma| \tag{29.11}$$

$$+ 2\alpha \log |R| \tag{29.12}$$

$$+ 2((\alpha + 1)t) \log \alpha. \tag{29.13}$$

Each unique pointer introduces at most 2 new cuts in R. The number of nodes is also $\leq |R|$ since the nodes partition R. So, the number of nodes created is $\leq \alpha$.

A node is uniquely specified by the location of its cuts (or equivalently a (*position, length*) pair), each of which can be specified in $\log |R|$ bits.

Therefore, the space to store all the nodes is $\leq 2\alpha \log |R|$ bits. Since the node strings partition R, the total space for all the strings in the nodes is $|R| \log |\Sigma|$ bits.

Finally, there are $t - 1$ edges between adjacent pointers in the encoding (step 5), where t is the number of pointers in the encoding. Each pointer spans at most α edges connecting adjacent substrings of R (step 4). So at most $\alpha t + t - 1 \leq (\alpha + 1)t$ edges are created. An edge is specified (in a naïve way) via two ($\log \alpha$)-bit integers giving the node identifiers for the endpoints of an edge.

What this means is that if we start with a small EPM-based compressed version of the string, we can create a genome graph that has bounded size. The running time to execute Algorithm 29.1 is linear in $|R| + |T|$, so this graph can be created efficiently.

29.3 Alignment of strings to genome graphs

A natural task is to find the location in a genome graph of a query string q. There are a number of different formulations of this problem. We describe one:

Problem 29.2 (Sequence-to-graph alignment). *Let $G = (V, E)$ be a genome graph, with node labels given by a function $\ell : V \mapsto \Sigma$. Let q be a non-empty string over Σ. For a path p in G, let $\ell(p)$ be the string represented by the path, i.e., the concatenation of the labels of the nodes along p.*

The sequence-to-graph alignment *problem is to find the path p in G that minimizes $edit'(q, \ell(p))$, where $edit'(x, y)$ is the traditional edit distance between x and y but restricted to only allow insertions, deletions, and substitutions in x.* ◆

Two things to note about this problem: first, most crucially, in the $edit'$ measure, edits are allowed only in the query string q—we are not allowed to introduce edit operations on the graph. When edit operations are allowed on the graph, the problem becomes NP-hard, even for $|\Sigma| = 2$ [Jain et al., 2020]. Second, we've defined the problem so that the label of each node is a single character. This is not a big restriction because we can break longer nodes into a series of nodes of length 1.

29.3.1 Alignment graph

To solve Problem 29.2, we use a sequence-to-graph alignment graph (as defined in Jain et al. [2020]).

Let $G = (V, E)$ be a directed graph with single-character node labels, and let q be a string of length m. The *sequence-to-graph alignment graph* $\mathcal{A}(q, G)$ has vertices

$$V = \{1, \ldots, m\} \times (V \cup \{\delta\}) \cup \{s, t\}, \tag{29.14}$$

where δ is a new dummy vertex, and s and t are new source and sink vertices.

We divide the nodes into $m + 2$ layers: layer 0 contains only s, layer $m + 1$ contains only t, and layer i contains nodes (i, u) for $u \in V \cup \{\delta\}$.

The edges of $\mathcal{A}(q, G)$ are defined as follows. In layer i, we have "insertion" edges that mirror the edges of G:

$$E_{\text{ins}} = \{(i, u) \to (i, v) : (u, v) \in E\}. \tag{29.15}$$

The weight of these edges is the cost of an insertion. These let us move across an edge in G without using a character of q.

Between layers $i-1$ and i (for $i = 2, \ldots, m$), we have "deletion" edges between corresponding nodes:

$$E_{\text{del}} = \{(i-1, u) \to (i, u) : u \in V \cup \{\delta\}\}, \tag{29.16}$$

with weight equal to the cost of a deletion. These let us "use up" a character of q without moving in G.

Between adjacent layers, we also have "match" edges:

$$E_{\text{mat}} = \{(i-1, u) \to (i, v) : (u, v) \in E\}. \tag{29.17}$$

We define $match(i, v) = 0$ if $q[i] = \ell(v)$ and the cost of a substitution otherwise. The weight of edge $(i-1, u) \to (i, v)$ is $match(i, v)$. These edges allow us to both use up a character of q and move across an edge in G.

We connect s to the first layer and the mth layer to t:

$$E_s = \{s \to (1, u) : u \in V \cup \{\delta\}\} \tag{29.18}$$

$$E_t = \{(m, u) \to t : u \in V \cup \{\delta\}\}. \tag{29.19}$$

The cost of edges $s \to (1, \delta)$ is the cost of deletion. The cost of edges $s \to (1, u)$ for $u \in V$ is $match(1, u)$. The cost of the edges entering t is 0.

Finally, from each dummy node (i, δ), we add edges to each $u \in V$ in the next layer of cost $match(i, u)$:

$$E_\delta = \{(i, \delta) \to (i+1, u) : u \in V\}. \tag{29.20}$$

Suppose $q = ACGA$ and G is

$$\tag{29.21}$$

then $\mathcal{A}(q, G)$ is

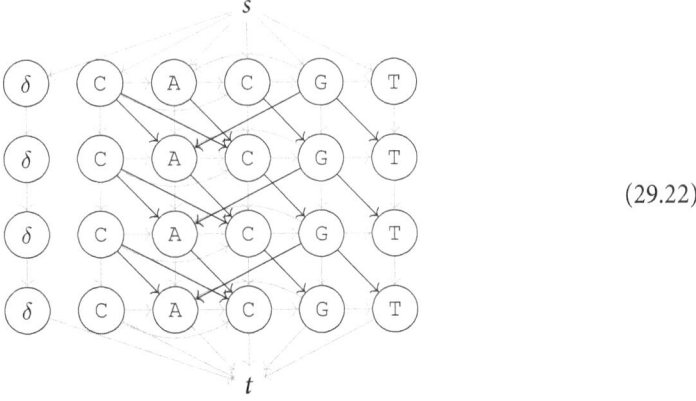

$$\tag{29.22}$$

For clarity, in the (29.22) picture, we have not drawn the edges E_δ from the δ node in layer i to all of the non-δ nodes in the layer $i + 1$. They look like (29.23).

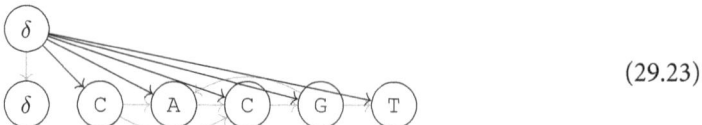

(29.23)

Moving into layer i corresponds to "aligning" $q[i]$ with the node in layer i. The "dummy node" column lets us jump into any layer, paying the deletion cost for skipping (deleting) a prefix of q. Any $s - t$ path corresponds to an alignment between q and G. If we find the shortest (lowest cost) path from s to t this will give us the optimal alignment.

We can use Dijkstra's shortest path algorithm, which in this case runs in time $O(m|V| \log(m|V|) + m|E|)$ since we have $O(m|V|)$ nodes and $O(m|E|)$ edges. In fact, this can be sped up to $O(|V| + m|E|)$ by exploiting the special structure of $\mathcal{A}(q, G)$. A different, more complex alignment graph can be used to support affine gap penalties. For details on both of these improvements, see Jain et al. [2020].

29.4 Comparing genome graphs

An interesting open problem is how to compare two genome graphs effectively. Suppose you are given graphs G_1 and G_2 derived from fragments from two different individuals. How can you assess how similar the genomes of the two individual are?

You could assemble each genome (using techniques for genome assembly starting from G_1 and G_2) and then use, e.g., edit distance, between the estimated strings. This has several problems: first, it's computationally slow to assemble a genome; second, it's computationally slow to compute the edit distance for large strings; and third, the assembly process is error-prone and is likely to introduce errors, so you won't be certain about the two strings you inferred to compare.

Or, on the other hand, you could extract a set of k-mers from the graph and apply a sketching method (Section 23.2) to estimate the distance. This could be efficient, but it doesn't use the information encoded in the graph about the relationships between the k-mers.

It would be nice to have an approach that compares the graph directly and that would give a reasonable estimate of the distance that exploits adjacencies encoded in the graphs.

One measure was introduced by Ebrahimpour Boroojeny et al. [2020]:

Definition 29.3 (Graph traversal edit distance (GTED)). Let G_1 and G_2 be two Eulerian graphs, with edges labeled with characters from Σ. Then

$$\mathrm{GTED}(G_1, G_2) = \min_{\substack{T_1 \in \mathcal{T}(G_1) \\ T_2 \in \mathcal{T}(G_2)}} \mathrm{edit}(T_1, T_2), \qquad (29.24)$$

where $\mathcal{T}(G)$ is the set of strings that are spelled by some Eulerian tour of G, and edit is the traditional edit distance (Chapter 7). ∎

This is an elegant approach, since it does not require committing to a single potential string that either graph represents. Because any Eulerian tour could be the true string, absent external information, we don't really have a reason to prefer one particular tour, and so the measure picks the two tours (strings) that are closest in terms of edit distance.

To solve GTED, we can formulate the problem based on the *graph alignment graph*:

Definition 29.4 (Graph alignment graph). Let $G_1 = (V_1, E_1)$ and $G_2 = (V_2, G_2)$ be two genome graphs with edges labeled with single characters from Σ. The *graph alignment graph* $\mathcal{A}(G_1, G_2) = (V, E)$ has $V = V_1 \times V_2$ and edges defined by the union of:

$$(u_1, u_2) \rightarrow (v_1, u_2) \text{ for } (u_1, v_1) \in E_1 \text{ and } u_2 \in V_2 \qquad (29.25)$$

$$(u_1, u_2) \rightarrow (u_1, v_2) \text{ for } u_1 \in V_1 \text{ and } (u_2, v_2) \in E_2 \qquad (29.26)$$

$$(u_1, u_2) \rightarrow (v_1, v_2) \text{ for } (u_1, v_1) \in E_1 \text{ and } (u_2, v_2) \in E_2. \qquad (29.27)$$

∎

The graph alignment graph is reminiscent of the alignment matrix for standard edit distance and the alignment graph of Section 29.3.1. Edges (29.25) are the "horizontal edges" that represent using an edge of G_1 without using an edge in G_2; edges (29.26) are "vertical edges" that represent using an edge of G_2 without using an edge in G_1; and edges (29.27) are "diagonal edges" that represent using an edge of both graphs. It is also the natural extension of the sequence-to-graph alignment graph we saw in (29.22), extended to compare two graphs. We can assign arbitrary costs $\delta(e)$ to each of these edges e (e.g., 0 to diagonal edges that represent a match and 1 to all other edges).

An example of the alignment graph between two graphs is shown in (29.28).

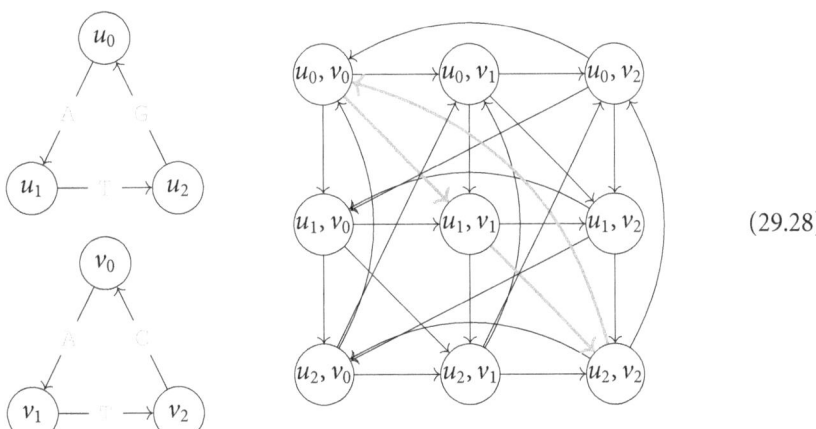

$$(29.28)$$

Each edge in $\mathcal{A}(G_1, G_2)$ corresponds to an edge in G_1, an edge in G_2, or both. We define the corresponding edge in G_i of an edge e in $\mathcal{A}(G_1, G_2)$ as $\Pi_i(e)$. This is the *projection* of edge e to one of the input graphs.

We can now formulate GTED as the problem of finding a particular path in the alignment graph [Ebrahimpour Boroojeny et al., 2020]:

Lemma 29.5. *Let G_1 and G_2 be directed, Eulerian graphs, with edges labeled from Σ. Let $\delta(p)$ be sum of cost of edges in a trail in $\mathcal{A}(G_1, G_2)$. Then:*

$$GTED(G_1, G_2) = minimize_p \, \delta(p) \tag{29.29}$$

$$\text{subject to: } p \text{ is a path in } \mathcal{A}(G_1, G_2) \tag{29.30}$$

$$\Pi_i(p) \text{ is an Eulerian trail in } G_i \text{ for } i = 1, 2, \tag{29.31}$$

where $\Pi_i(p)$ is the projection of edges in $\mathcal{A}(G_1, G_2)$ to G_i.

In other words, the goal is to find a path in $\mathcal{A}(G_1, G_2)$ that, when projected to each of the graphs G_1 and G_2 forms an Eulerian path. GTED is the cost of that path in $\mathcal{A}(G_1, G_2)$. The intuition here is that such a trail in $\mathcal{A}(G_1, G_2)$ induces Eulerian tours in each of the input graphs (since its projection in both is an Eulerian trail by definition). Also, it accounts for the edit distance between those trails because whenever it uses a horizontal or vertical edge in $\mathcal{A}(G_1, G_2)$ it pays the cost for an indel, and whenever it uses a diagonal edge, it pays for the match or mismatch between the edges in G_1 and G_2.

Optimizing this measure is NP-hard [Qiu et al., 2024]. While integer linear programs exist to solve this problem [Qiu et al., 2024], they are either exponential in size or difficult to solve in practice. Hence, it is an open problem to find efficient heuristics or other approximation approaches to these problems.

29.5 Summary and notes

The strategy of using genome graphs is a recent topic, and we are able to only cover a sampling of the ideas and problems encountered building and using them. There are a number of open problems, both theoretical and practical, making them an interesting area for research. Faster algorithms for GTED, approximation algorithms, and heuristics for NP-complete versions of sequence-to-graph algorithms [e.g., Ivanov et al., 2020], and even more practical construction approaches are all important lines of investigation. Development of the unifying principles behind the various types of genomes graphs is also important [Cicherski and Dojer, 2023].

> ### Presentation Notes
>
> No single reference covers the emerging field of genome graphs. Mäkinen et al. [2015] has a long discussion of various questions on string graphs.

29.6 Exercises

29.1   Draw the classical de Bruijn graph of order 4 (meaning $k = 4$) over the alphabet $\Sigma = \{0, 1\}$.

29.2   Let $q = $ "ACT" and G be the genome graph (29.32).

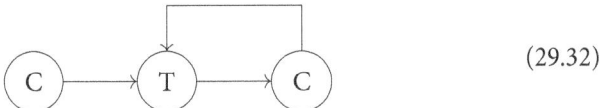

$$(29.32)$$

Draw the string-to-graph alignment graph $\mathcal{A}(q, G)$.

29.3   Give an example string and graph where you would get the wrong (suboptimal) alignment if you deleted the δ nodes from $\mathcal{A}(q, G)$ and explain why.

CHAPTER 30

BWT on Labeled Trees

Often we have a collection of related strings that we want to search. For example, we may have the genomes of many individuals of a species or several versions of a program's source code. Rather than search these strings individually, we could get some computational improvement by putting multiple strings into a single data structure. This is the opposite case from what we saw with Aho-Corasick (Chapter 6), where we had multiple patterns, and similar to what we saw with generalized suffix trees (Section 13.4.3).

In this chapter, we are going to look at string trees, where the strings are encoded on a rooted tree. We'll see how to create a BWT-like data structure for these trees that allows for answering the query, "how many times does substring q appear in one of the encoded strings?" The ideas for the tree-based data structure can be extended to support DAGs so that they are applicable to some types of genome graphs. They also give us insight into how the ideas of the BWT can be extended to graph structures.

30.1 String trees

A string tree is a rooted, labeled tree $T = (V, E, r, \ell)$ where $r \in V$ is the root and $\ell : V \to \Sigma$ is a function that labels each node u with a character $\ell(u)$. Here's an example:

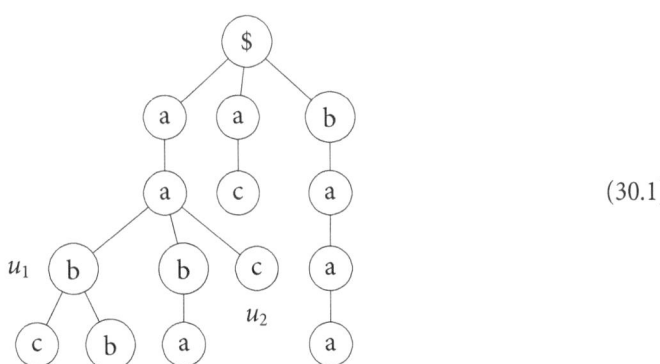

$$(30.1)$$

where the character in each node gives $\ell(u)$. We further assume two things about string trees: (1) the root is labeled with a special character $ that is not used elsewhere, and (2) the children of each node are ordered left-to-right, and a depth-first search tree traversal will visit those edges in that defined order. Property (2) means that there is a total ordering of the nodes and of the edges: the order they would be visited in such a DFS.

The interpretation of a string tree is that it contains a collection of strings, one for each node u. The string $L(u)$ associated with node u is the string that is spelled out by the characters in the nodes on the path from u to the root. Note that the direction you read off each string differs between string trees and suffix trees.

30.2 String tree BWT

We will construct an array, Tree BWT, that has BWT-like properties that allow us to use nearly the exact same backward-search "LF"-like algorithm to search the tree that we used to search a BWT string. This extension was first proposed in Ferragina et al. [2009b]. Let's start by connecting some of the ideas of the BWT to string trees.

First, the items in the BWT are the characters in a string. Analogously, for string trees, the items in our Tree BWT will be the nodes (that contain the characters). In the BWT, we sorted the items by their right context: the string that follows the character in the string. We will do something similar in Tree BWT, where the *right context* of a node u is the string spelled out from the parent of u to the root.

More formally, for each node u, define its *data* to be

$$D(u) = \langle L(parent(u)), \ell(u), last(u) \rangle, \tag{30.2}$$

where $last(u)$ is 1 if u is the rightmost child of its parent, $parent(u)$, and 0 otherwise. In (30.1), $D(u_1) = \langle \text{``aa\$''}, \text{``b''}, 0 \rangle$ and $D(u_2) = \langle \text{``aa\$''}, \text{``c''}, 1 \rangle$.

We now construct an array B that is our Tree BWT by sorting the set $\{D(u)\}$ for every u by $L(parent(u))$, breaking ties based on the total order of nodes given by the DFS. The array B stores only the pairs $(\ell(u), last(u))$, throwing away the primary sort keys $L(parent(u))$ (just as we threw away the sort key for the string BWT). For the example in (30.1), we have as follows.

$$
\begin{array}{|c|c|l|}
\hline
0 & a & \$ \\
\hline
0 & a & \$ \\
\hline
1 & b & \$ \\
\hline
1 & a & a\$ \\
\hline
1 & c & a\$ \\
\hline
0 & b & aa\$ \leftarrow u_1 \\
\hline
0 & b & aa\$ \\
\hline
1 & c & aa\$ \\
\hline
1 & a & aab\$ \\
\hline
1 & a & ab\$ \\
\hline
1 & a & b\$ \\
\hline
0 & c & baa\$ \\
\hline
1 & b & baa\$ \\
\hline
1 & a & baa\$ \\
\hline
\end{array}
\tag{30.3}
$$

String intervals. There is an interval $\mathcal{I}_{\mathrm{str}}(S)$ corresponding to every substring S starting at position >1 of a string. For example $\mathcal{I}_{\mathrm{str}}(\text{"aa"})$ is represented by rows $6\dots9$ in the (30.3) example. If a substring doesn't exist anyplace, then the interval will be empty.

There's a bit of an annoying asymmetry between internal nodes and leaf nodes, since leaf nodes v could have an empty $\mathcal{I}_{\mathrm{str}}(L(v))$. This is because we sort by *parent* strings not the whole string. We solve this by adding dummy leaf nodes, labeled with a special character # hanging off of each leaf node before constructing B. Now every former leaf node becomes an internal node, and thus all the real nodes we care about are internal nodes. For simplicity, we don't show the # nodes in the array shown in (30.3).

After adding these new # leaf nodes, there is a single, non-empty contiguous interval $\mathcal{I}_{\mathrm{str}}(S)$ in the array for every substring S in the tree, corresponding to the rows for which $L(u)$ has that string as a prefix. This is because we sort by $L(parent(u))$, and all the nodes that share a prefix of $L(parent(u))$ will be grouped together.

Node intervals. The children of a node u are all within the interval $\mathcal{I}_{\mathrm{str}}(L(u))$ since all the children of u share the same sort key $L(u)$. Furthermore, since we broke ties using the DFS order, the children of any node u are all consecutive in the list.

Every internal node u is "represented" in two ways: (1) it's in the list of nodes, sorted by $L(parent(u))$, and (2) it is represented by the interval $\mathcal{I}_{\mathrm{node}}(u)$ that corresponds to its children (the bars at the left of B in figure (30.3)). $\mathcal{I}_{\mathrm{node}}(u)$ is the interval containing the labels of the children of u. This interval $\mathcal{I}_{\mathrm{node}}(u)$ for node u is delimited by the 1-bits of the *last* array. The rows r where $last[r]=1$ are the last rows in the continuous sets of rows corresponding to each node.

Schematically, we now have the following situation:

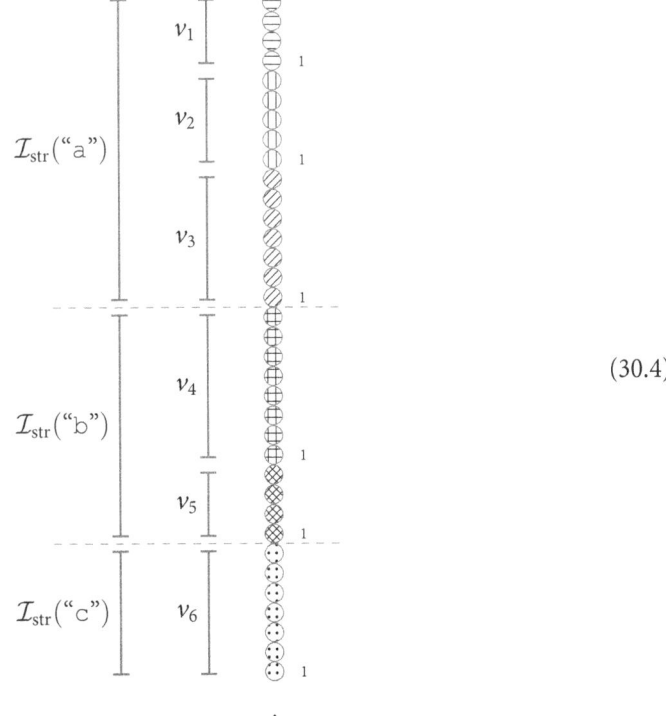

$$(30.4)$$

where every internal node v both appears in the list and corresponds to an interval $\mathcal{I}_{\text{node}}(v)$.

30.2.1 The interval corresponding to the ith node

We index nodes by their intervals. Node 1 in the Tree BWT corresponds to the interval from position 1 to the position of the first 1 bit in *last*.

If we know that u is the ith node in the Tree BWT, then we can obtain this interval via:

$$\mathcal{I}_{\text{node}}(u) := IthNodeInterval(i) := [\text{select}_1(last, i-1) + 1, \text{select}_1(last, i)]. \quad (30.5)$$

This finds the starting index of the interval as the location just after the $i - 1$st 1 bit in *last* and the ending index of the interval as the ith 1 bit in *last*. (We assume that the intervals are numbered starting at 1.) If we store *last* in a data structure that supports select_1 (such as an RRR bitvector from Chapter 19), we can compute function (30.5) in $O(1)$ time.

30.2.2 A child interval

Consider the problem of "walking down the tree" when given only the Tree BWT. In other words, if we have some information about node u, how can we find the range in the Tree BWT that corresponds to a particular child of u? We can formalize this problem as follows:

Problem 30.1 (Find child interval). *Given:*

- *a node u,*
- *its interval $\mathcal{I}_{node}(u)$,*
- *a character c, and*
- *an integer i,*

find the interval of the ith child of u that is labeled with character c. That is, u may have many children labeled c, and we want to find the interval corresponding to the ith one of them:

- *Child(u, c, i) is the interval of the ith child v of u with $\ell(v) = c$.*   ◆

In other words, we have $\mathcal{I}_{\text{node}}(u)$ and are looking for

$$\mathcal{I}_{\text{node}}(i\text{th "}c\text{"-labeled child of } u). \quad (30.6)$$

Problem 30.1 can be used iteratively to walk down the tree by repeatedly moving to the interval of one of the current interval's children. We now see how to solve Problem 30.1.

All of u's children are in interval $\mathcal{I}_{\text{node}}(u)$, and one of them v is the ith child labeled with c. Its location j will be the ith entry within $\mathcal{I}_{\text{node}}(u)$ such that $B[j] = c$. The i is v's "local" index relative to the other children of u. Here's an example with $i = 2$.

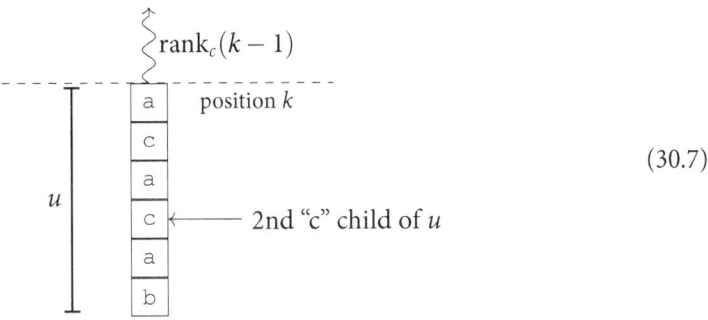

(30.7)

Our first step is to convert this local index i to a "global" index that gives which c this is in B using rank_c:

$$cIndex(\mathcal{I}_{\text{node}}(u), c, i) := \text{rank}_c(B, \mathcal{I}_{\text{node}}(u).\text{start} - 1) + i. \qquad (30.8)$$

Here, $\mathcal{I}_{\text{node}}(u).\text{start}$ is the start of u's interval. This counts the number of cs before the start of u's interval and then adds i because we're looking for the ith c child. This changes our problem into finding the interval for the node that represents the $cIndex(\mathcal{I}_{\text{node}}(u), c, i)$th c in B.

What is the point of this? The interval corresponding to

$$Child(u, c, i) = \text{the interval for } v \qquad (30.9)$$

must be a subset of the interval for c since each of v's children are sorted by $L(v)$.

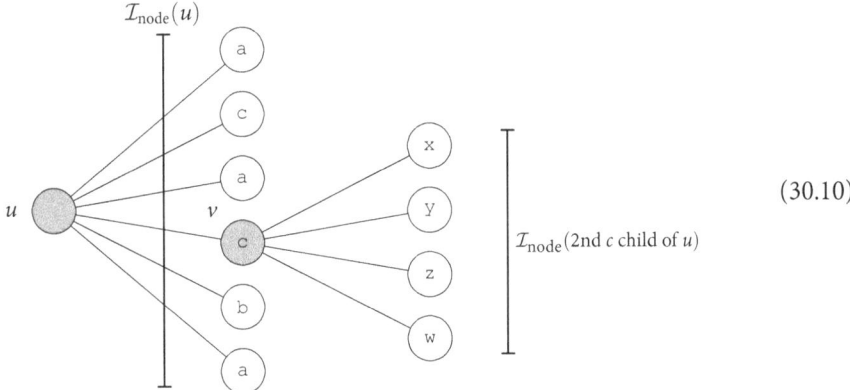

(30.10)

The interval we seek is within $\mathcal{I}_{\text{str}}(\text{"c"})$. How do we know which of the node intervals in $\mathcal{I}_{\text{str}}(\text{"c"})$ correspond to v? Luckily, we have a property similar to the LF property (Theorem 18.2) of the standard BWT:

Theorem 30.2 (Tree order). *Let v_1 and v_2 be nodes in our tree with $\ell(v_1) = \ell(v_2) = c$, and let $j(v)$ be the index of node v in B. Then:*

$$j(v_1) < j(v_2) \implies \mathcal{I}_{node}(v_1) < \mathcal{I}_{node}(v_2), \qquad (30.11)$$

where "$\mathcal{I}_{node}(v_1) < \mathcal{I}_{node}(v_2)$" means the first interval entirely precedes the second interval.

Proof: For a given v_1 and v_2 the situation is as follows.

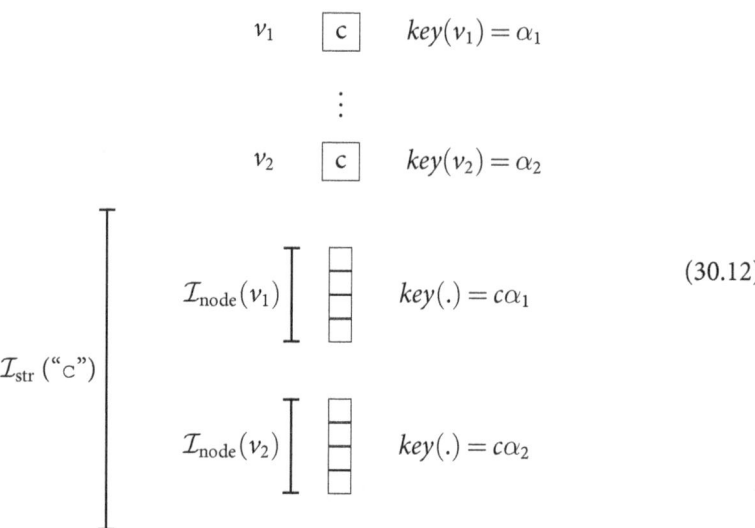

$$(30.12)$$

The sort keys for v_1 and v_2 are α_1 and α_2. Since $j(v_1) < j(v_2)$ by assumption, we have $\alpha_1 \le \alpha_2$. Consider the nodes in $\mathcal{I}_{\mathrm{node}}(v_1)$: their sort keys are $c\alpha_1$. Similarly, the sort keys for nodes in $\mathcal{I}(v_2)$ are $c\alpha_2$. Clearly, $c\alpha_1 \le c\alpha_2$. This means that the position of each node $w_1 \in \mathcal{I}_{\mathrm{node}}(v_1)$ is $\le$ the position of each node $w_2 \in \mathcal{I}_{\mathrm{node}}(v_2)$. By the DFS property, $\mathcal{I}_{\mathrm{node}}(v_1)$ and $\mathcal{I}_{\mathrm{node}}(v_2)$ are each contiguous and disjoint. Therefore, $\mathcal{I}_{\mathrm{node}}(v_1) < \mathcal{I}(v_2)$. $\square$

What this is saying is that if node u comes before node v in our sorted Tree BWT, then the *interval* for node u comes before node v. Analogously to the LF-mapping property of BWT, this lets us jump from a node in our Tree BWT to its interval.

To find the interval for $Child(u, c, i)$ corresponding to node v: if the global index of v is j, then v must be the jth node interval contained within $\mathcal{I}_{\mathrm{str}}(\text{"}c\text{"})$. Similar to standard BWT, we keep an array $C[x]$ that gives the number of nodes w with $\ell(w) < x$ for each $x \in \Sigma$. Thus, there are $C[c]$ nodes whose intervals precede $\mathcal{I}_{\mathrm{str}}(\text{"}c\text{"})$.

Therefore, the solution to Problem 30.1 is:

$$Child(u, c, i) = \mathcal{I}_{\mathrm{node}}(v) = IthNodeInterval\left(C[c] + cIndex(\mathcal{I}_{\mathrm{node}}(u), c, i)\right). \qquad (30.13)$$

$C[c]$ tells us how many non-c node intervals to skip, $cIndex(B, c, i)$ tells us how many c-node intervals to skip, and *IthNodeInterval* finds the corresponding node interval in the array.

30.3 Extending to substring intervals

Recall that every string S corresponds to a (possibly empty) interval $\mathcal{I}_{\mathrm{str}}(S)$ containing the nodes w that have S as a prefix of their key $L(parent(w))$.

Problem 30.3 (Left extension). *Find $\mathcal{I}_{str}(cS)$ for a given $c \in \Sigma$, given that we know $\mathcal{I}_{str}(S)$ for the string S and that $\mathcal{I}_{str}(S)$ is non-empty.* ◆

We want to find the interval corresponding to the string "cS".

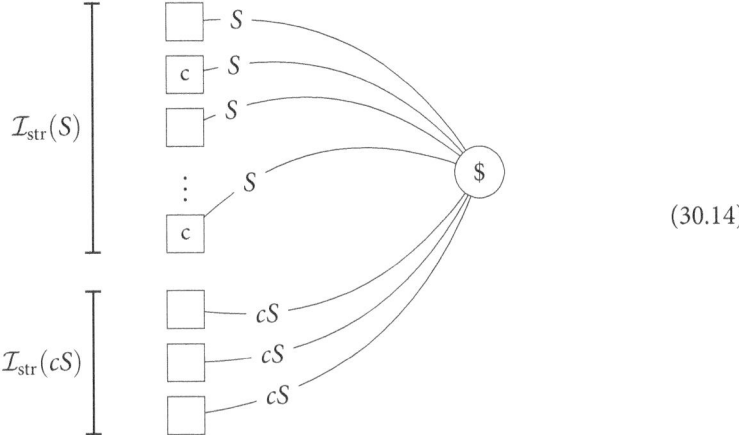

$$(30.14)$$

Remember: the interval $\mathcal{I}_{str}(S)$ always deals with an interval of the sort key of a node, which is $L(parent(u))$; that means that the characters in B (our sorted tree BWT array) within $\mathcal{I}_{str}(xS)$ need not be x since the label of a node itself is not used in its sort key.

To go from $\mathcal{I}_{str}(S) \rightarrow \mathcal{I}_{str}(cS)$, we follow the same idea as in the previous section: we ask, within $\mathcal{I}_{str}(S)$, which c characters are there, measured by their global indices? The global index of the last c that appears before $\mathcal{I}_{str}(S)$ is:

$$cIndex(\mathcal{I}_{str}(S), c, 0) = \text{rank}_c(B, \mathcal{I}_{str}(S).\text{start} - 1). \qquad (30.15)$$

The global index of the last c that appears in the range is:

$$\text{rank}_c(B, \mathcal{I}_{str}(S).\text{end}). \qquad (30.16)$$

We can then find the interval within $\mathcal{I}_{str}(\text{"c"})$ corresponding to cS via:

$$extendLeft(c, \mathcal{I}_{str}(S)) := \mathcal{I}_{str}(cS) =$$
$$[\text{select}_1 \, (last, C[c] + cIndex(\mathcal{I}_{str}(S), c, 0)) + 1,$$
$$\text{select}_1 \, (last, C[c] + \text{rank}_c(B, \mathcal{I}_{str}(S).\text{end}))]. \quad (30.17)$$

The $+1$ occurs because the 1 bits mark the *end* of the previous node's interval; $+1$ takes us to the start of the relevant node's interval. The addition of $C[c]$ translates from the global c count into indexes of node ranges in the "c" interval.

30.4 Backwards search

Now that we have all the machinery built, we can solve the search problem using an algorithm very similar to the one we used with the string BWT:

Problem 30.4 (Tree BWT search). *Determine if a query string $q = q_1 q_2 \ldots q_n$ occurs as a substring of any of the strings $L(u)$ for any node u in a string tree.* ◆

We start with the interval for the last character of q, which can be found using our $C[x]$ array. Then we iterate:

Algorithm 30.1. Find query in string tree.

function EXISTSQ(T, q)
    $W \leftarrow \mathcal{I}_{\text{str}}(q[|q|])$
    for $i \leftarrow |q| - 1$ **downto** 1 **do**
        $W \leftarrow$ EXTENDLEFT($q[i], W$)
    return $W \neq$ empty

This is nearly identical to the standard BWT search, but using the tree-based *extendLeft*: we search for q from the end to the start of q, iteratively finding the interval for the node corresponding to the next left extension. If the interval isn't empty at the end, then q appears in the tree.

30.5 Summary and notes

String trees can be used to efficiently store strings that share many common suffixes. For strings that share other substrings that are not necessarily suffixes, we really need string DAGs (as in Chapter 29). Fortunately, the ideas for trees [Ferragina et al., 2009b] can be extended to support BWT-like searches over DAGs [Sirén et al., 2011, 2014]. Unfortunately, we don't have space to cover this.

Presentation Notes

Our presentation of the Tree BWT index was informed by Chapter 9 of Mäkinen et al. [2015], which contains many more details about the BWT, its extension to trees, and graphs, and other extensions.

30.6 Exercises

30.1   Give the Tree BWT for the following String Tree.

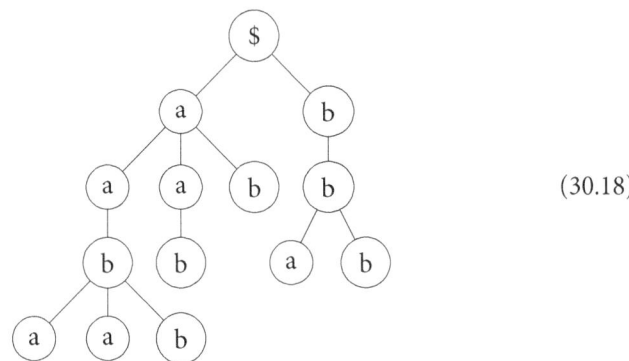

(30.18)

30.2   For a Tree BWT containing n nodes over an alphabet of size $O(1)$, what is the running time of the *extendLeft* operation in (30.17)?

30.3   Given a Tree BWT of a string tree $\mathcal{T}$, show how to compute:

- $child(\mathcal{I}_{\mathrm{node}}(u), k) =$ the interval corresponding to the kth child of u.

That is the children of the kth child of u are in positions

$$child(\mathcal{I}_{\mathrm{node}}(u), k).\text{start}, \ldots, child(\mathcal{I}_{\mathrm{node}}(u), k).\text{end}$$

in the Tree BWT. We saw a similar function that focused on a child with a given label c. This function is different in that doesn't take into account what label the child has.

30.4   Suppose you are given a Tree BWT of a string tree $\mathcal{T} = (V, E)$. Let W be an interval of the Tree BWT that corresponds to positions with keys that begin with w. Give an expression (way to compute) for

- the number of nodes $u \in V$ such that $L(u) = cw\alpha$ for $c \in \Sigma, \alpha \in \Sigma^*$

where $L(u)$ is the string represented by node u.

30.5   Describe how to report a node u for which q is a prefix of $L(u)$ using Algorithm (30.1). (In other words, report a location in the tree, not just whether q exists or not.)

CHAPTER 31

Transformers

Transformers [Vaswani et al., 2017] are a neural network architecture that has found wide success in many sequence-based tasks, from natural language translation (e.g., English to German) to modeling DNA and protein sequences. As with other artificial intelligence (AI) approaches, they have many weights (parameters) that are trained using training examples of the correct output. Once these weights are learned, the transformer architecture produces predictions for the output given new inputs.

As you might expect from the name, transformers "transform" one sequence of *tokens* $x_1, \ldots, x_n$ into another sequence of tokens $y_1, \ldots, y_n$. For natural language applications, the tokens are usually words or parts of words. The training examples are pairs of sequences $(\vec{x}, \vec{y})$, and through the training process, the transformer learns to map $\vec{x}$ to $\vec{y}$. The hope is—and this is true empirically—that this training process leads to a transformer that can output a reasonable $\vec{y}$ for new, previously unseen inputs $\vec{x}$.

In this chapter, we will build up the ideas behind the transformer architecture. Although there are many technical choices in its design, the main idea is that of *attention* (Section 31.4) which allows the transformer's neural network to learn relationships between tokens in each of the sequences.

In contrast to many other topics we have covered, our focus isn't on provable bounds (which are challenging to obtain in these kinds of contexts), but rather to understand the "algorithm" that transformers implement that works so well in practice.

31.1   Embeddings

First, let's discuss the general concept of *embeddings* as used in AI. Suppose $\vec{v}$ is a d-dimensional vector that represents a data item. For example, $\vec{v}$ might be a 1-dimensional vector (i.e., a scalar) that is a number that corresponds to a particular word, or it might be a $k \times k$-dimensional vector that represents the pixel intensity for a picture that is $k \times k$. The vector $\vec{v}$ is an embedding (in dimension d) of whatever data item $\vec{v}$ represents.

An *embedding function* is a function f that maps items like $\vec{v}$ to some other space with possibly different dimension d':

$$f(\vec{v}) : \mathbb{R}^d \to \mathbb{R}^{d'}, \tag{31.1}$$

where $\mathbb{R}$ denotes the real numbers. d' might be smaller than d, in which case we say that the embedding function is doing *dimensionality reduction*—finding a smaller representation for input $\vec{v}$. d' might equal d, in which case f is converting one embedding (the input representation) into a different representation of the same dimension. d' can also be bigger than d, in which case we are expanding the dimension, putting our input into a larger space that gives us more "room" to play with.

Why would we want to expand the dimension? Suppose $d = 1$ and $\vec{v}$ represents a number that encodes a word. Maybe $\vec{v} = \langle 32 \rangle$, and 32 means the word "cat". Perhaps 33 is the number for the word "car". The difference between these two input vectors is small, but the concepts of "car" and "cat" are pretty different. By expanding the dimension, we have room to find an embedding function that puts "car" and "cat" far apart. Using only a single dimension, *some* unrelated words are likely to be near each other. A larger dimension can avoid that.

In the case that $d = 1$ and $d' = 2$, this looks like (31.2).

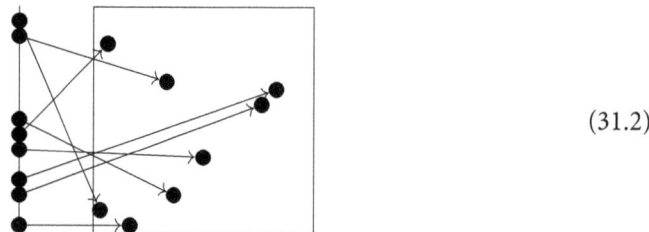

$$(31.2)$$

What makes a good or useful embedding depends on what it is going to be used for. One of the simplest embedding functions is a *linear map*. We pick a matrix W of dimension $d \times d'$ and multiply $\vec{v}$ (as a row vector) against it:

$$f(\vec{v}) = \vec{v}W. \tag{31.3}$$

This gives us a d' dimensional vector, where each entry is the dot product of $\vec{v}$ with the corresponding column of W.

$$(31.4)$$

In this case, we can think of the embedding function f as a function of two different things: the input $\vec{v}$ and the weight matrix $W: f(W; \vec{v})$. When we want to learn (train) the embedding function, $\vec{v}$ and its corresponding desired output are given, and W is what we are trying to learn. When using the embedding, W becomes fixed, and f is treated as a function only of the input $\vec{v}$.

A large part of any neural network architecture (including transformers) is composing various learned embedding functions (sometimes called *layers* in deep-learning speak). Some of the embedding functions used by transformers are a little more complex (we will see) than the simple linear map of (31.3), but they all build on this general concept.

31.2 Encoder-decoder architectures

Transformers use an *encoder-decoder* architecture. Encoder-decoders conceptually break apart a sequence of embeddings into two parts: the encoder that embeds the input into some intermediate representation in $\mathbb{R}^{d'}$ and a decoder that converts items in that intermediate space into the desired final representation.

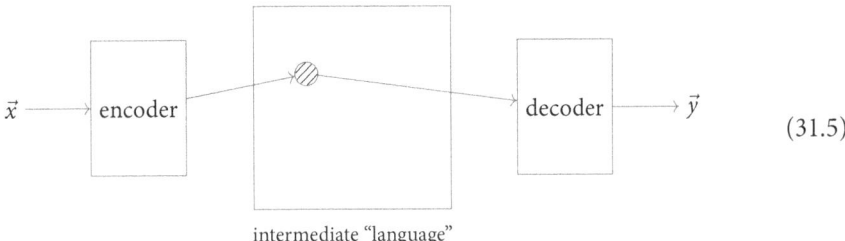

intermediate "language"

$$(31.5)$$

We can think of the encoder as translating our natural input representation ($\vec{x}$) into some intermediate "language" (representation) that the decoder speaks. The decoder takes this intermediate "language" and translates it into our desired output. Each of the encoder and decoder themselves are typically built from the composition of some embedding functions (and perhaps other operations). The encoder and decoder are trained together so that they learn to "agree" on the meaning of the intermediate language.

Transformers use a slightly more complex encoder-decoder architecture because they solve the following problem: given all of $\vec{x}$ and a prefix of the corresponding $\vec{y}$ sequence $y_1, y_2, \ldots, y_{i-1}$, predict y_i. This requires that the decoder know about the prefix $y_1, y_2, \ldots, y_{i-1}$. Therefore, they use the following high-level architecture.

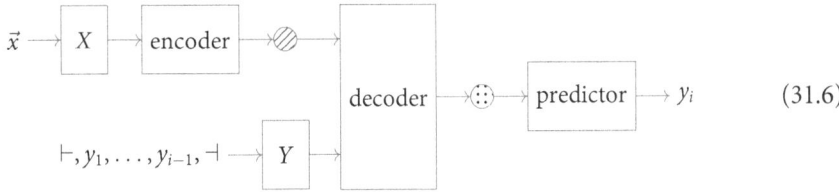

$$(31.6)$$

The boxes X and Y are components that prepare the inputs for entering into the encoder and decoder, respectively. The ⬰ is the embedding of $\vec{x}$ into the intermediate representation created by the encoder. The ⬭ is the decoded representation created by the decoder. The predictor box predicts from this decoded representation the next token y_i.

To predict the entire sequence $\vec{y}$, we prepend a special "start of sequence" symbol ⊢ to $\vec{y}$. We first run the transformer with input $(\vec{x}, \langle \vdash \rangle)$, which will produce a prediction $\tilde{y}_1$ for y_1. We then iterate: next using the input $(\vec{x}, \langle \vdash, \tilde{y}_1 \rangle)$, which will produce a prediction $\tilde{y}_2$ for y_2, and so on. How do we know when to stop? We add a special stop symbol ⊣ to the possible outputs of the predictor, and we stop the iteration when ⊣ is predicted.

Another technical bit we must consider: the encoder and decoder take a fixed input size. That is they require that the inputs $\vec{x}$ and $\vec{y}$ are some known length. But, of course, we want to be able to (say) translate sentences of different lengths, and we don't know a priori the length of the predicted output sequence. To overcome this, we pick some maximum length n, implement everything to assume all the inputs have length n, and add a new symbol $\square$ to our token alphabet that serves as padding at the end for any sequence shorter than n. $x_1, \ldots, x_k$ becomes the length-n sequence $x_1, \ldots, x_k, \dashv, \square, \ldots, \square$, and $y_1, \ldots, y_\ell$ becomes $\vdash, y_1, \ldots, y_\ell, \dashv, \square, \ldots, \square$. This way all the input sequences are the same length, and the predictor's goal is to guess what to replace $\dashv$ with in the current $\vec{y}$ input.

Next, we will look at the two main types of embedding function layers that transformers use. These are feedforward neural networks and an attention mechanism.

31.3 Feedforward neural networks

A neural network is a collection of nodes and weighted edges between them. The nodes are called *neurons*, an analogy with biological neurons. The weights on the edges may be positive or negative (or 0). Neurons with no incoming edges are called *input neurons*, and those with no outgoing edges are called *output neurons*. Each neuron has a value that is computed from the values of the neurons pointing to it, weighted by the edge that a value "travels along."

Transformers use fully connected, feedforward neural networks, which look like (31.7) (though of course with many more nodes).

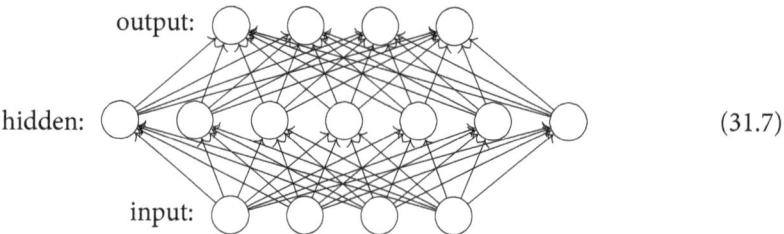

$$\text{(31.7)}$$

Such an arrangement is called feedforward because all the edges go in one direction: from the input towards the output; there are no loops or back edges. It's called fully connected because each neuron is connected to all the neurons in the next layer (with weighted edges).

To apply such a network to a vector $\vec{u}$, we need the input (bottom) layer to have a number of neurons equal to the number of entries in $\vec{u}$. We put the value u_i on input neuron i. Each of the neurons in the next layer then grab the values from the previous layer, multiply them by the weight on the edge the value must travel along, and sum them up. More formally, if $\vec{L}_{i-1}$ are the values on the neurons in layer $i-1$, the value going into neuron a in layer i is:

$$\sum_j L_{i-1}[j] \times W_{j,a}, \tag{31.8}$$

where $W_{j,a}$ is the weight on the edge going from neuron j in layer $i-1$ to neuron a in layer i. This is the dot product between a's incoming weights and the values on the previous layer.

Equation (31.8) is the input to neuron a, but it's not yet a's value. That is because we don't just want a weighted sum of the previous layer, we want to transform it. We want there to be some suppressing of low sums (if the input signal is low, we want to lower it) and some amplification of large sums (if the input signal is high enough, we want to keep it high or make it higher). This is again a broad analogy with biological neural networks: they don't fire at all until there is sufficient signal coming from their neighbors, and once there is, they fire at a high value. To achieve this, the value of neuron a in layer i is computed as:

$$L_i[a] = activation \left(\sum_j \vec{L}_{i-1}[j] \times W_{j,a} \right), \tag{31.9}$$

where *activation* is a chosen activation function that determines a neuron's value given the weighted sum of its input signals.

Many activation functions have been designed. One of the most popular because of its computational efficiency and empirically good performance is the *rectified linear unit* or *ReLU* for short, which is simply:

$$activation(x) = ReLU_b(x) = \max(0, x + b), \tag{31.10}$$

where b is a "bias" term (that may be different for each neuron). If the value of $x < -b$, return 0; otherwise return $x + b$. This function has the property we want: low signals (defined as $\leq -b$) are turned into a 0-value signal, and high signals (defined as $> -b$) are passed through amplified (shifted) by b.

Using (31.9), we can compute the value for every node in the network given input u. Input nodes are assigned the input values from u. We apply (31.9) to each neuron, layer-by-layer, to get the value of each internal neuron (often called *hidden neurons* because they are neither the input nor the output). For the output neurons, we do the same thing, but skip applying the max in the activation function—this is because for the output we want all the information content computed by the network, and don't want to throw away values $< -b$.

As you can see, the workings of such a network are pretty simple, and built on simple operations. It looks even simpler if we translate it into matrix notation. Consider (31.8). We have to apply this dot product to each neuron in layer i, and because the network is fully connected, the only difference between what we do for neuron a and neuron a' is the vector of weights $W_{\cdot,a}$ vs. $W_{\cdot,a'}$.

We can gather all these different weight vectors as columns in a matrix W and get the dot products for every node in layer i all at once by multiplying W by the values of the previous layer. (See again (31.4).) Hence, for the 3-layer (input, hidden, output) network used by transformers, we can compute the output vector as:

$$ffn(\vec{u}) = ReLU_{\vec{b}_1}(\vec{u}W_1)W_2 + \vec{b}_2 \tag{31.11}$$

$$= \max(0, \vec{u}W_1 + \vec{b}_1)W_2 + \vec{b}_2, \tag{31.12}$$

where W_1 is a matrix of the weights between the input and hidden layer, W_2 gives the weights between the hidden layer and the output, and $\vec{b}_i$ ($i = 1, 2$) are the vectors of the bias terms for the neurons in the hidden and output layers.

31.4 Attention

Transformers use an attention component to model how important each token is to the other tokens in the input. For example, in the two sentences:

The fox was quiet. Its fur was brown.

The word "fox" is quite relevant to the word "Its" while, despite being closer, the word "quiet" isn't too related to "Its." We want to estimate, for every token, how "relevant" the other tokens are. Transformers do this with a dot-product-based similarity matrix.

Let's look at the use of attention in the encoder (the use in the decoder is similar with one important difference we'll see later). In the overview of (31.6), the box X translates (via a 1d to d-dimensional embedding) each of our tokens x_i into a d-dimensional vector $rep(x_i)$, so we will end up with a matrix Q that looks like (31.13).

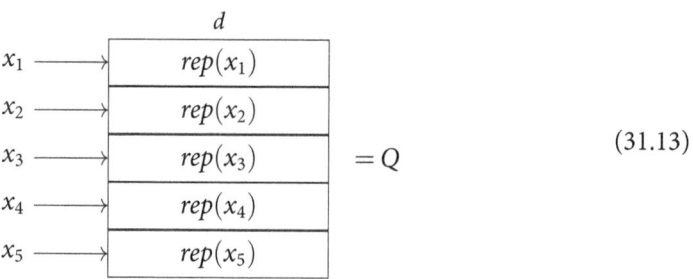

$$(31.13)$$

That is, every input token is now represented by a row of Q. The expression $S = QQ^T$ produces a matrix S where $S_{i,j}$ is the dot product between $rep(x_i)$ and $rep(x_j)$. The dot product is a reasonable measure of how similar two vectors are, so S is a good candidate for a similarity matrix between our input tokens.

Transformers apply two normalization steps to S. First, they divide the entires of S by $\sqrt{d}$ (where d is the length of each of the vectors representing a token). Second, they turn each row into a probability distribution using the *softmax* function:

$$softmax(\vec{v}) = \frac{\exp(\vec{v})}{\sum_{i=1}^{|v|} \exp(v_i)}. \qquad (31.14)$$

Here, the denominator is a scalar, and the $\exp(\vec{v})$ in the numerator means apply the exp function element-wise to each entry in $\vec{v}$ to obtain another vector. The elements in $softmax(v)$ will clearly sum to 1 because the denominator normalizes by the sum of the numerator's elements. Also true is that $\exp(x) \geq 0$ for any x, so no matter whether the items in $\vec{v}$ were positive or negative, the elements in $softmax(\vec{v})$ will always be between 0 and 1. These two facts are the definition of a probability distribution.

After these normalization steps, we have a similarity matrix:

$$S' = softmax\left(\frac{QQ^T}{\sqrt{d}}\right). \qquad (31.15)$$

What does the encoder do with S'? It multiplies it by Q again! The rows of S' are probability distributions, so after $S'Q$ we end up with a matrix Q' where the rows correspond to tokens (just as with Q) but the jth entry of a row i is the weighted sum of the jth entries of all the tokens, where the weights are given by the ith row of S'. In this way, each row i in Q' contains an average of the original tokens' representations, weighted by their similarity to the original representation of token i. Formally,

$$EncAttention(Q) = softmax\left(\frac{QQ^T}{\sqrt{d}}\right)Q. \tag{31.16}$$

Finally, we take the output of the attention mechanism to be $Q + Q'$, so that Q' is assumed to tell us how much to perturb the original representation of our tokens based on the similarity of the token's representations. (There is an additional normalization step that we won't describe.) If token i has a similar representation as token j, then token i will be "shifted" to be more similar to token j. (In the impossible extreme, if $S'[i,j] = 1$, then token j would simply be added to token i.) In other words, the attention mechanism takes our input representation Q, and modifies it to incorporate information contained in other, relevant tokens.

In preparation for seeing how attention is used in the decoder, we can generalize (31.16) a bit. Even though Q is used three times in that equation, it's really playing several different roles. The first Q contains the vectors to compare with the vectors in the second Q, but there is no reason these have to be the same set of vectors. We could compare one Q with another Q-like matrix, which is what the decoder will do. The third Q plays a third role: it is the vectors we want to average based on the similarities—this too could be a different matrix, and transformers exploit this with their "multi-head attention scheme," but in the interest of simplification, we won't describe this. So, we can define the general attention function as:

$$attention(R, Q) = softmax\left(\frac{RQ^T}{\sqrt{d}}\right)Q. \tag{31.17}$$

We'll see in Section 31.5.2 where this generalization is useful.

31.5 Transformers

31.5.1 The transformer encoder

We now have most of the pieces to look at the encoder portion of transformers. The encoder has the following general structure.

$$Q \longrightarrow \boxed{\begin{array}{c} \text{Attention} \\ attention(Q,Q) \end{array}} \longrightarrow \oplus \xrightarrow{Q'} \boxed{\begin{array}{c} \text{Feedforward} \\ \textit{ffn}(Q'_{1,\cdot}),\ldots,\textit{ffn}(Q'_{n_x,\cdot}) \end{array}} \longrightarrow \oplus \longrightarrow Q'' \tag{31.18}$$

The encoder takes as input a matrix of the type shown in (31.13), where each row represents the encoding for a token in $\vec{x}$.

This Q matrix is processed according to $Q + attention(Q, Q)$, leading to a new Q' matrix, which is fed into a feedforward network of the form of (31.12) (with trainable weights W_1, W_2). This feedforward network has input size d and is applied to each token representation in Q' independently.

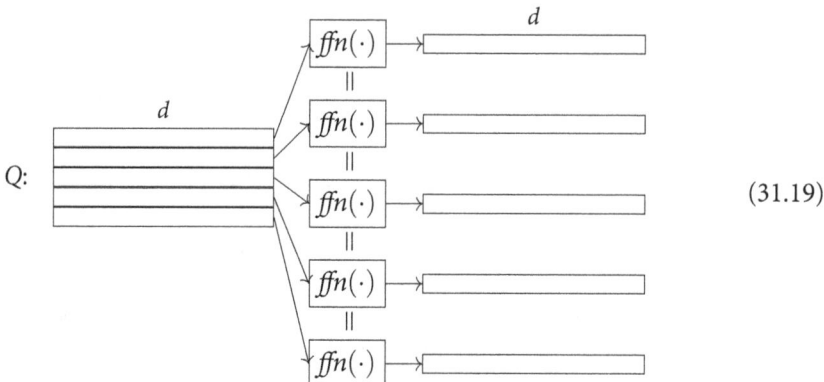

(31.19)

Each copy of the feedforward network is identical: they all share the same weights.

This yields a new $n \times d$ matrix Q'', which has the same dimensions and interpretation as the input to (31.18), so we do this 5 more times, each time taking the output of one instantiation of (31.18) as the input to the next copy. In each copy, we have a new feedforward network that has weights that are independent of the weights in other copies. This ultimately leads to the output matrix Q_{enc} that contains rows encoding each input token.

Because of the attention mechanism, the final encoding of each token x_i has been informed by the encodings of every token in the input. But, on the other hand, because we apply the feedforward network independently to each row in Q, each row in the output is still ultimately interpretable as corresponding to one of the input tokens.

31.5.2 The transformer decoder

The decoder is similar to the encoder. It has the architecture shown in (31.20).

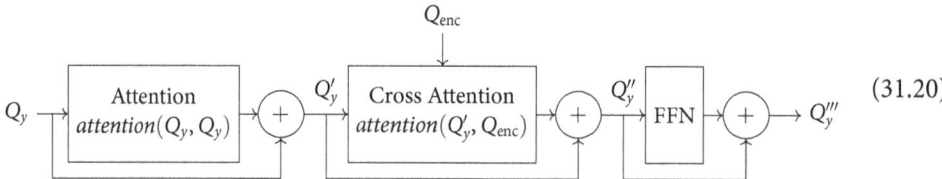

(31.20)

We again apply an embedding function (this time to $\bar{y}$) to create a matrix Q_y similar to the Q matrix of (31.13) that is one of the inputs to the decoder. The main part of the decoder, as with the encoder, starts with the standard attention mechanism: $Q'_y = Q_y + attention(Q_y, Q_y)$. But the difference is that, after this first attention adjustment, we apply an additional attention adjustment that is informed by the encoder:

$$Q''_y = Q'_y + attention(Q'_y, Q_{enc}).$$

(31.21)

That's where we use the generalization in (31.17). This is called *cross attention* because it uses information from both the encoder output and the decoder. The similarity matrix it computes is the similarity between the representations Q'_y computed for $\vec{y}$ and *the representations (Q_{enc}) computed by the encoder for $\vec{x}$.*

Because of this, the decoder representations Q'_y are modified by a weighted average of the *encoder* representations. The weights are based on the similarity between the encoder and decoder representations. Every time this cross-attention component is encountered, the decoder's representations are affected by the encoder's final encoding of $\vec{x}$.

After this, we again apply the same type of position-wise independent feedforward network (FFN) that we used for the encoder to end up with a Q_y-shaped matrix Q'''_y. As with the encoder, we repeat this 5 more times, using the output of one instantiation as the input of the next instantiation. This ultimately leads to a Q-shaped matrix as the output of the decoder (the ⊕ of (31.6)).

31.6   Training and using transformers

To make a prediction for the next token, the output of the decoder Q_{dec} is used as the input to a standard machine learning task: predict the next token y_i given Q_{dec}. Suppose we have d_Σ possible choices for y_i. The original implementation of transformers used a linear classifier and an application of *softmax* to transform Q_{dec} into a d_Σ-dimensional vector of probabilities. The highest probability entry was used as the prediction for y_i.

31.7   Summary and notes

We have glossed over many details for the sake of getting at the big picture. First, transformers actually use a little more complex approach to attention than is outlined in Section 31.4. They employ what is called *multi-head attention*, where instead of a single application of *attention*$(\cdot, \cdot)$ they use h applications of a slightly more general function on smaller vectors created from Q. Second, to avoid leaking information about y_j values for $j \geq i$ when predicting y_i, transformers apply a mask to the similarity matrix computed by the attention mechanism in the first layer of the decoder in order to zero out the positions corresponding to "future" positions in $\vec{y}$. Third, there are several other details about how the various embeddings are constructed that we've glossed over. Finally, when constructing the initial Q and Q_y matrices in boxes X and Y of (31.6), a vector that depends only on the position of a token in the input ($\vec{x}$ and $\vec{y}$) is added to each row. This means that the initial input embedding contains information about both the token and its location within the sequence.

We haven't talked at all about how to learn the various weights for the embeddings functions (such as the W matrices for each of the feedforward networks). Since the entire network uses simple, differentiable operations, it can be trained via standard back propagation methods to optimize the weights to achieve the best accuracy on the training samples. Some other training tricks are used as well.

The originally surprising thing about transformers is how powerful the attention mechanisms are. Work prior to transformers used convolution networks or recurrent networks (where there are back loops, unlike feedforward networks) to capture relationships between input tokens. The paper [Vaswani et al., 2017] that introduced transformers showed that that complication was not needed: in the words of the title of that paper, "Attention is all you need."

One might reasonably ask, why this particular architecture? The best answer is that this one worked. Here, "worked" means both excellent empirical results and computational efficiency. A nice thing about this architecture is that it uses computationally efficient, parallelizable operations. But the design space is huge, and there are other architectures that can also perform well, as evidenced by the growing literature on large language models (LLMs). There is no claim that this is the *best* architecture, just that this is a feasible and successful one.

Driven in part by the success of transformers (and other deep-learning architectures), LLM techniques have begun to be applied to the "language" of the genome. Genomic large language models (such as DNABERT [Ji et al., 2021], codonBERT [Li et al., 2023], scGPT [Cui et al., 2024], Geneformer [Theodoris et al., 2023], and many others), rather than being trained on human language, are trained on biological sequences such as the genome, transcriptome, or protein sequences (see [Consens et al., 2023; Liu et al., 2024] for reviews). These models are then used for tasks such as predicting biological function of genes, comparing individuals' or species' genomes, predicting stability and potency of RNA vaccines, and annotating regions of the genome in interpretable ways and many other tasks.

> **Presentation Notes**
>
> We largely follow the paper that introduced transformers [Vaswani et al., 2017], but with many simplifications and reordering of the material.

31.8 Exercises

31.1   Estimate the number of parameters (weights) that need to be trained in a transformer encoder (as presented in this chapter) for $d = 512$. Estimate the same thing for the decoder (as presented).

31.2   Give the value of $softmax(\langle 3, -10, 100, -60 \rangle)$.

31.3   Give the value of $softmax(\langle 1, 1, 1, -\infty \rangle)$.

31.4   Is the *ReLU* function differentiable everywhere? Explain.

Part VI

Miscellaneous

CHAPTER 32

Huffman Codes

In this chapter, we look at a particular compression scheme called Huffman codes that achieves optimal compression under some settings. Huffman encoding will use a variable number of bits for each character in our alphabet, with characters that are used more rarely in a text encoded with more bits than characters that are used more frequently. Huffman encoding is a character-by-character compression method: it defines an (invertible) function $h : \Sigma \to \{0, 1\}^*$ that maps each uncompressed character into its variable-length code word.

Huffman coding [Huffman, 1952] is widely used because of its efficiency and simplicity (although one can do better if the character-by-character requirement is dropped). We saw a use of Huffman coding when we discussed the FM-index in Chapter 21.

32.1   Prefix-free codes

A prefix-free code is a mapping $h : \Sigma \to \{0, 1\}^*$ such that for any two symbols $x \neq y$ in Σ, $h(x)$ is not a prefix of $h(y)$. The advantage of prefix-free codes is that if you are reading a stream of bits, you know when you have read an entire code word of a symbol. As long as there is ambiguity about which symbol is encoded, you know you have to read more bits. If you have read the bit string b, and b is a prefix of two possible symbols x and y, you must not have read all of the codeword yet.

One can imagine a prefix-free code as a binary tree, where the leaves correspond to the symbols in the alphabet.

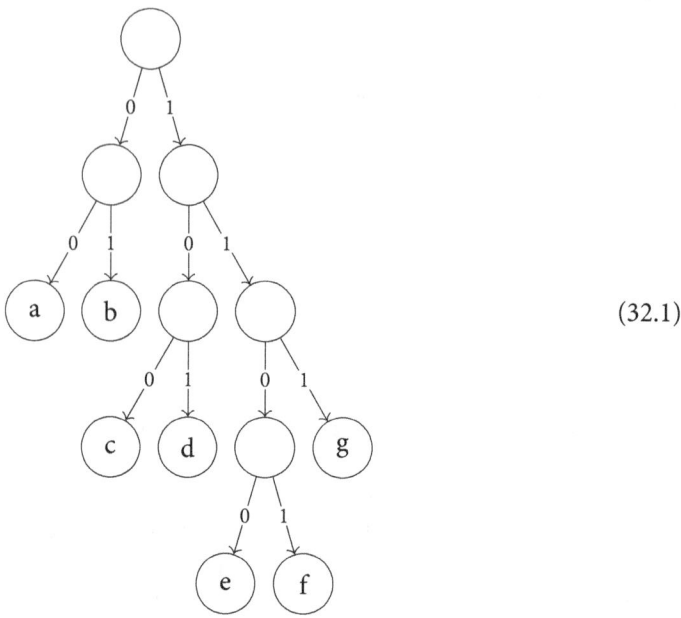

(32.1)

In such a tree, label the left edge leaving every node with a 0 and the right edge with a 1. There is a unique path from the root to any leaf v, and the sequence of 0s and 1s encountered on the edges along this path gives an encoding for v. This code is a prefix-free code because any string that is a prefix of two code words corresponds to an internal node, which doesn't (completely) encode for any symbol in the alphabet.

Decoding such a code is a simple matter of simultaneously walking down a received bit string and the tree, following the edge with the label of the next symbol in the bit string. Whenever a leaf is reached, that character is decoded and the traversal starts back at the root of the tree.

32.1.1   UTF-8 encodings

Prefix-free codes have very practical uses. Unicode is a system mapping bit strings to "characters," where the characters are logical ideas representing symbols used in human communication. For example the character 😀 is encoded as the number $0x1F600$ (where the notation $0x$ indicates the number is written in hexadecimal). Unicode allows numbers up to $1{,}114{,}112$, which equals $0x10FFFF$ in hexadecimal (although not all these numbers are valid characters). It encodes 20-bit numbers up to $0xFFFFF$ plus an additional 2^{16} numbers encoded as $0x10abcd$ where $abcd$ is a 4-digit hexadecimal number. This allows a huge number of characters to be represented, but has the problem that a naïve encoding would use 3 bytes (24 bits) per character, which is a lot if most of the characters being encoded are from a smaller set of commonly used characters. This is the problem that UTF-8 encoding [The Unicode Consortium, 2011] is designed to solve: it specifies a way to encode any of the $0x10FFFF$ character numbers (called "code points" in Unicode's terminology) using a variable number of bits (between 1 and 4 bytes), where more common characters use fewer bits.

Specifically, if a Unicode character is ≤ 127 (requiring at most 7 bits), the character is encoded as a single (8-bit) byte with a leading zero:

$$0abcdefg \tag{32.2}$$

where *abcdefg* are the 7-bits encoding the Unicode character. Since these first 128 characters equal the ASCII encoding standard (that includes the Latin characters), the currently most-used characters can be represented like this. In fact, any file that is simply a string of bytes encoding ASCII characters of value ≤ 127 is already UTF-8 encoded.

For larger unicode numbers (> 127), a multi-byte representation is used. The first byte of this representation starts with a number of 1 bits that indicate how many bytes are going to be used to represent the character, followed by a 0. For example, the character ﬚ is represented by $0x05E9$ and will need two bytes, so its UTF-8 representation will start with `110`. The remaining 5 bits of the first byte will be used to encode the first 5 bits of the character's Unicode code. The second byte starts with a "continuation" signal consisting of the two bits `10` followed by the next 6 bits of the character's Unicode code. Characters requiring more bits use additional continuation bytes.

Range (Hex)	Byte 1	Byte 2	Byte 3	Byte 4
00 – 7F	0*abcdefg*			
80 – 7FF	110*abcde*	10*fghijk*		
800 – FFFF	1110*abcd*	10*efghij*	10*klmnop*	
10000 – 10FFFF	11110*abc*	10*defghi*	10*jklmno*	10*pqrstu*

This table mirrors Table 3-6 of the Unicode 6.0 specification [The Unicode Consortium, 2011].

The 4-byte code can encode numbers that require up to 21 bits, so the full range of Unicode characters can be encoded. Unicode characters with lower numbers are more common (by design of Unicode), and these use fewer bits, so we save space. Finally, this encoding is prefix free: no representation is a prefix of any other due to the length-signaling bits in the first byte.

One might ask why we need the `10` bits at the start of the continuation bytes if the prefix of the first byte tells us how many bytes follow. With these bits, if we start at any byte in stream, we can find the start of the next character by skipping to the first byte that does not start with `10`. In addition, if a byte is dropped, that will only affect one character.

32.2 Huffman coding algorithm

The Huffman coding algorithm creates a prefix-free code for any set of symbols. The input to the algorithm is the alphabet Σ and also probabilities $p(x)$ for each $x \in \Sigma$ that estimate how likely x will be used in messages sent with the encoding—a higher probability would ideally translate into a shorter codeword. ,

The algorithm builds a prefix-free code by building a tree such as in (32.1) by merging nodes from the bottom up. It starts with a forest of single-node trees, each corresponding to a symbol. It repeatedly merges the trees with the smallest probabilities on their root nodes, until a single tree remains. Since low-probability trees are merged first, their nodes will be farther from the eventual final root node of the tree, and thus have longer codewords.

Algorithm 32.1. Build Huffman code.

1. Create a node x for each $x \in \Sigma$.
2. Repeat the following until all nodes are connected into a single tree:
    (a) Let u and v be the two nodes with the lowest probabilities that have no incoming edges.
    (b) Create a new node uv.
    (c) Make u and v children of this node (now u and v have incoming edges).
    (d) Assign $p(uv) \leftarrow p(u) + p(v)$.
3. Arbitrarily label the two edges leaving each non-leaf node with 0 and 1.

Here's an example with $\Sigma = \{a, b, c, d, e, f\}$. Nodes higher on the page were created later in time by the algorithm.

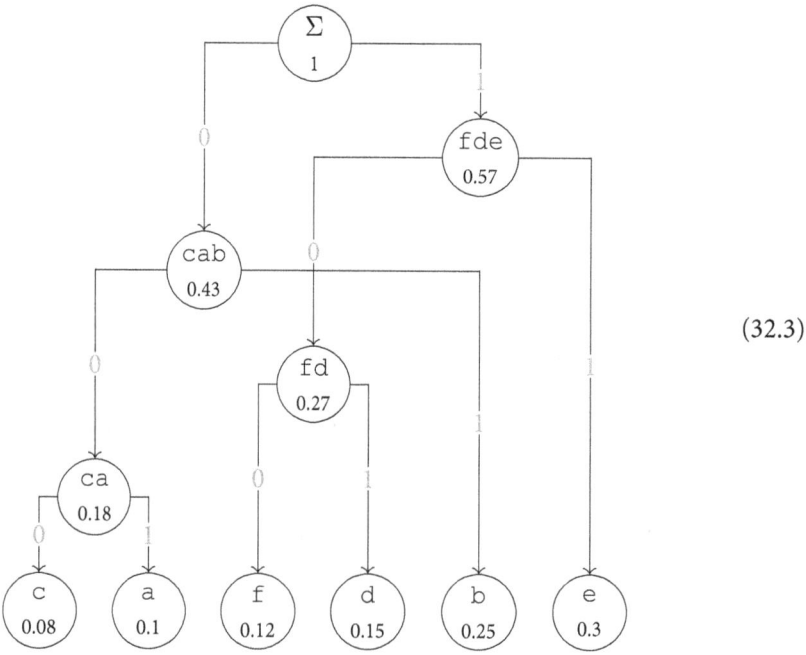

(32.3)

The Huffman code construction algorithm can be implemented efficiently with a priority queue. In step 1, we enqueue the nodes corresponding to the symbols, using their probabilities as their priority. In step 2, we dequeue the two lowest weight items in the priority queue. In step 2, we enqueue the newly created node. In this way, the priority queue always contains the nodes that are available for merging, and the lowest priority items correspond to the lowest probability nodes. Each of these operations on a priority queue can be implemented in time $O(\log n)$, where n is the number of items in the queue. In this case,

$n = O(|\Sigma|)$. We iterate at most $O(|\Sigma|)$ times, leading to an $O(|\Sigma| \log |\Sigma|)$ total running time to construct the code.

32.2.1 Proof of optimality

Fix a code, and let $L(a_i)$ be the length of the codeword (in bits) for symbol a_i in this code. The expected length of a message $x_1, \ldots, x_m$ in this code is:

$$\sum_{i=1}^{m} p(x_i) L(x_i). \tag{32.4}$$

This is the quantity we want to minimize to obtain an optimal code.

Following Huffman's [Huffman, 1952] original proof, we will derive some properties that any optimal, prefix-free code must have, otherwise we could find a code with a smaller value of (32.4). We will show first that the code constructed by the algorithm (32.1) has those properties, and then argue that any code that has those properties must be a variation of the code constructed by the algorithm. The properties are also instructive, since they immediately motivate the type of greedy, bottom-up merging that the algorithm does.

Assume $\Sigma = \{a_1, \ldots, a_k\}$ and that

$$p(a_1) > p(a_2) > \cdots > p(a_{k-1}) > p(a_k), \tag{32.5}$$

meaning both that (a) without loss of generality, we've reordered the alphabet into decreasing probability order, and (b) *with* loss of generality (for now) we assume all the probabilities are distinct.

Lemma 32.1 (Property 1). *In an optimal code, it must be the case that*

$$L(a_1) \leq L(a_2) \leq \cdots \leq L(a_{k-1}) \leq L(a_k). \tag{32.6}$$

Proof: Otherwise, one could swap the codewords for two symbols a_i, a_j with $i < j$ but $L(i) > L(j)$ and obtain a better encoding. $\qquad\square$

Lemma 32.2 (Property 2). *In an optimal code, the lengths of the codewords for a_{k-1} and a_k are the same. That is, $L(a_{k-1}) = L(a_k)$.*

Proof: Suppose not, and $L(a_k) > L(a_{k-1})$. Let $\gamma\alpha$ be the codeword for $L(a_k)$, where γ is the prefix of length $L(a_{k-1})$ bits and α is the rest. Since the code is prefix-free, the prefix γ doesn't encode for any other symbol (including, in particular, a_{k-1} which has a codeword of length $|\gamma|$ by definition). Nor can γ be the prefix of any other symbol, since the codeword lengths are monotonically non-decreasing (by Lemma 32.1), so all the codewords for a_i, with $i < k$ are of length $\leq |\gamma|$. Hence, the bits in the string α in the assumed $\gamma\alpha$ codeword for a_k are pointless: γ itself is a unique, unambiguous encoding for a_k. This contradicts the optimality of the code, since we could replace the code for a_k with γ to obtain a better code.

In a picture, we have (32.7).

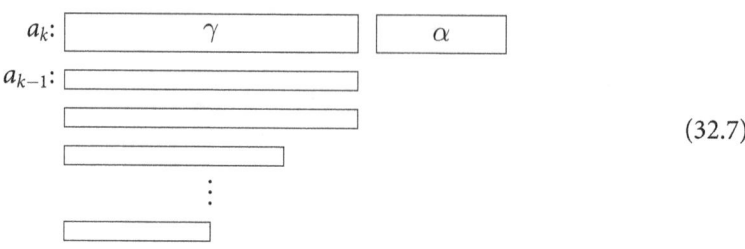

$$(32.7)$$

□

We can already see the algorithm start to emerge from these properties: if the two symbols with the smallest probabilities have the same length codewords (which are also the two longest codewords), perhaps merging them into a single unit that differs in the last bit makes sense. Indeed, in an optimal code, two codewords must have the first $L(a_k) - 1$ bits in common, as the next lemma argues.

Lemma 32.3 (Property 3).   (a)   *At least two codewords of length $L(a_k)$ share the same $L(a_k) - 1$ prefix.*

(b)   *Conversely, no more than two codewords of length $L(a_k)$ share any particular $L(a_k) - 1$ prefix.*

(c)   *Finally, without loss of generality, we can assume the codewords for a_k and a_{k-1} share the same length-$L(a_k) - 1$ prefix.*

Proof: Consider the codewords of length $L(a_k)$:

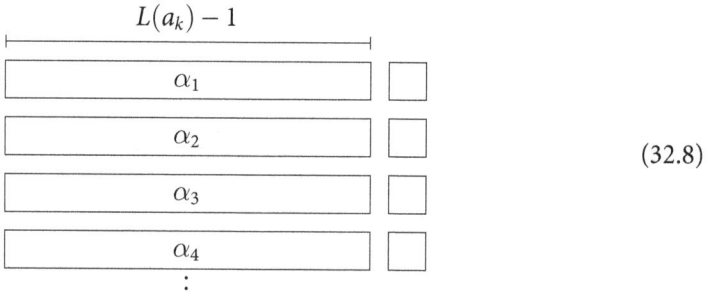

$$(32.8)$$

and their length $L(a_k) - 1$ prefixes $\alpha_1, \alpha_2, \ldots$. For a, if all of these α_i prefixes were unique, then the extra bits would not be needed to distinguish between these symbols. Since the code is prefix-free, no other symbol is encoded by any of the α_i, so we could just remove these bits, obtaining a better code, contradicting that the code was optimal.

On the other hand, for b, there can't be more than 2 codes of length $L(a_k)$ that share an $L(a_k) - 1$ prefix: we have only 1 bit to distinguish them (more generally, there can't be more than 2 codewords of any length ℓ that share a $\ell - 1$ prefix for the same reason).

Finally, for c, since the lengths are non-decreasing, all the codewords of length $L(a_k)$ are used for the last i symbols for some i. Reassigning codewords among these last i symbols doesn't change (32.4), so we may assume the last two codewords are two codewords that share a common $L(a_k) - 1$ prefix.   □

This is enough to both motivate the design of the algorithm and to prove that it is optimal.

Theorem 32.4 (Huffman optimality). *Huffman's algorithm produces an optimal code.*

Proof: Lemma 32.3 enables the use of a recursive algorithm for creation of the code. The codes for a_k and a_{k-1} are the same length (by Lemma 32.2) and can be assumed to differ only in their last bit (by Lemma 32.3). We can therefore consider their shared $L(a_k) - 1$ prefix as a code for the combined symbol $a_k a_{k-1}$. This prefix cannot be the prefix of any other codes by Lemma 32.3b. Now, we're looking for an optimal code with a new alphabet with 1 less symbol: $\Sigma \setminus \{a_k, a_{k-1}\} \cup \{a_k a_{k-1}\}$, and with the probability of the new symbol being $p(a_k) + p(a_{k-1})$. Once we have this "smaller" optimal code, we can obtain the codewords for a_k and a_{k-1} by extending the codeword for $a_k a_{k-1}$ by 1 bit. To find this smaller code, we're in the same position as we started: the same properties apply, and we can repeatedly merge the lowest two probability nodes. This is exactly what Huffman's algorithm does.

In the case when the probabilities are not distinct, we may have several choices for the two lowest probability symbols. But in that case since the probabilities are the same, they are indistinguishable in (32.4) and any choice of the two symbols are equivalent.   □

32.2.2 Linear-time code construction

Using a priority queue, the running time for constructing the code is $O(|\Sigma| \log |\Sigma|)$. This is the same time it would take to initially sort the symbols by their probabilities, so it's optimal in that sense. But if your symbols are already sorted, there is a neat linear-time algorithm to construct the tree. It relies on the fact that the weights of the constructed nodes are monotonically non-decreasing: the node created at time t will always have probability greater than or equal to any node created at time $< t$. Hence, only the earlier created nodes are candidates for having the lowest probability.

This leads to the algorithm (from Van Leeuwen [1976]):

Algorithm 32.2. Linear-time Huffman construction algorithm.

1. Add the nodes corresponding to symbols into a queue Q_1 in increasing order of their probability. Now the head of the queue has the lowest probability symbol.
2. Create an empty queue Q_2.
3. Run Huffman's algorithm. At step 1, to find the lowest probability nodes, look at the two nodes at the head of Q_1 and the two nodes at the head of Q_2. At step 2, enqueue the newly created node onto Q_2.

Each iteration of the algorithm takes $O(1)$ time. We still always find the lowest two probability nodes because Q_1 contains at its head the lowest probability leaf nodes, and Q_2 contains at its head the lowest probability non-leaf nodes. The two lowest probability nodes must be some combination of those.

32.3 Summary and notes

Huffman coding produces an optimal prefix-free code when each symbol must be encoded by a single, unique codeword. Relaxing these restrictions can produce better compression. For example, the codeword could change based on context or sets of letters may be encoded together.

We've explained Huffman coding and prefix-free codes in terms of binary alphabets, but that need not be the case. Even in Huffman's original paper, encoding into larger alphabets was considered, and the algorithm works similarly (rather than merging the two smallest trees, we merge the d smallest trees if we're encoding into a d-ary alphabet). See Exercise 32.2.

Finally, we never really made any use of the fact that the probabilities were in fact probabilities. We just used the fact that we could compare and add them. Huffman's algorithm works in any such situation, so the $p(x)$ could be character counts or other addable and comparable properties of the symbols.

> Presentation Notes
>
> Our presentation of Huffman's algorithm follows his original paper introducing it.

32.4 Exercises

32.1   Prove that in an optimal code, every bit string of length $\leq L(a_k) - 1$ must either code for a symbol or be used as the prefix of a code for a symbol.

32.2   Describe in detail how to extend Huffman's algorithm to support encoding messages into a non-binary alphabet with D symbols. That is, the codewords should be strings over the alphabet $\{1, \ldots, D\}$.

32.3   Give a code created by Huffman's algorithm for the following alphabet and probabilities: a $= 0.17$, b $= 0.02$, c $= 0.07$, d $= 0.09$, e $= 0.28$, f $= 0.05$, g $= 0.04$, h $= 0.13$, i $= 0.15$.

CHAPTER 33

Shortest Superstring

In this chapter, we discuss the following problem:

Problem 33.1 (Shortest Superstring). *Let $\mathcal{S} = \{s_1, \ldots, s_k\}$ be a set of k strings. A superstring of $\mathcal{S}$ is a string S such that each s_i is a substring of S. The shortest superstring problem seeks S^*, a superstring of $\mathcal{S}$ of the shortest length.* ◆

Problem 33.1 is useful in data compression contexts, since each string s_i can be encoded by a position in S^* and a length. If S^* can be made small, the strings $\mathcal{S}$ can be transmitted efficiently.

This problem is NP-hard (see Chapter 34), meaning that it is likely hard to find the smallest string S^* in polynomial time. As a consequence, we resort to approximation algorithms, where we guarantee that the length of the returned string S' will be within some multiple of the optimal length $|S^*|$.

We can assume that no string $s_i \in \mathcal{S}$ is a substring of any other string $s_j \in \mathcal{S}$. If that were the case, then s_i could be removed from $\mathcal{S}$ since any solution that contains s_j as a substring would necessarily contain s_i. We will use the fact that no input strings are substrings of any other in several of the proofs.

The proof we give of an approximation algorithm for this problem is due to Blum et al. [1994].

33.1 Relationship to the traveling salesman problem

The traveling salesman problem is a famous combinatorial optimization problem that arises in several contexts. Shortest superstring has a close relationship to it.

Problem 33.2 (Traveling salesman (TSP)). *Let V be a set of cities,, and let $d(i, j)$ be a positive distance between each ordered pair of cities $i, j \in V$. The optimal traveling salesman tour is an ordering π of the cities such that $d(\pi(|V|), \pi(1)) + \sum_{k=1}^{|V|-1} d(\pi(k), \pi(k+1))$ is minimized.* ◆

In other words, the optimal tour is one that visits every city exactly once, returns to the starting city, and uses the shortest total distance. This general problem doesn't require that the distances between cities are symmetric or that they obey the triangle inequality.

Let's turn the shortest superstring problem into a TSP problem. We define a weighted, directed graph $G = (V, E)$ with a vertex (aka city) in V for each string $s_i \in S$. We have an edge in E for every pair (s_i, s_j) and the length of each such edge is

$$d(i, j) := |s_i| - |\text{suffixprefix}(s_i, s_j)|, \qquad (33.1)$$

where $\text{suffixprefix}(x, y)$ is the longest suffix of x that is also a prefix of y (see Section 13.4.4). In other words, $d(i, j)$ is the number of characters one must "spend" in order to travel from s_i to s_j since they will not be included in s_j.

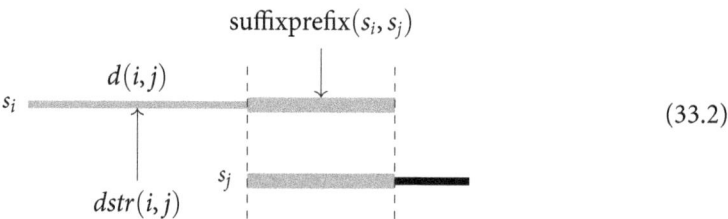

$$(33.2)$$

Let $dstr(i, j)$ denote the $d(i, j)$-long prefix of s_i; it is the part of s_i that is not covered by the maximum suffix-prefix overlap between s_i and s_j.

We call G the *prefix graph* of S. Of course, it may be that $d(i, j) \neq d(j, i)$, and this distance $d(i, j)$ doesn't necessarily satisfy the triangle inequality. Nevertheless, this graph provides an instance of a traveling salesman problem, where the goal is to visit every vertex of G exactly once (that is using every string in S exactly once), and using edges of minimum total length (where the total length will be related to the length of the superstring).

We've seen similar ideas in Chapter 29 where we talked about genome graphs: paths in those graphs spelled out strings that we were interested in. Here, paths in the prefix graph will be used to construct a superstring.

Theorem 33.3 (TSP lower bound). *Let TSP* be the length of a shortest traveling salesman tour on prefix graph G. Then:*

$$TSP^* \leq SS^*, \qquad (33.3)$$

where SS* is the length of the shortest superstring.

Proof: Let S^* be the shortest superstring, and let $s_1, \ldots, s_k$ be the strings of S ordered by their starting positions in S^*. Then S^* can be written as

$$dstr(s_1, s_2) \circ dstr(s_2, s_3) \circ \cdots$$

$$\circ\, dstr(s_{k-1}, s_k) \circ dstr(s_k, s_1) \circ \text{suffixprefix}(s_k, s_1), \quad (33.4)$$

where $\circ$ denotes concatenation. A picture makes this clear.

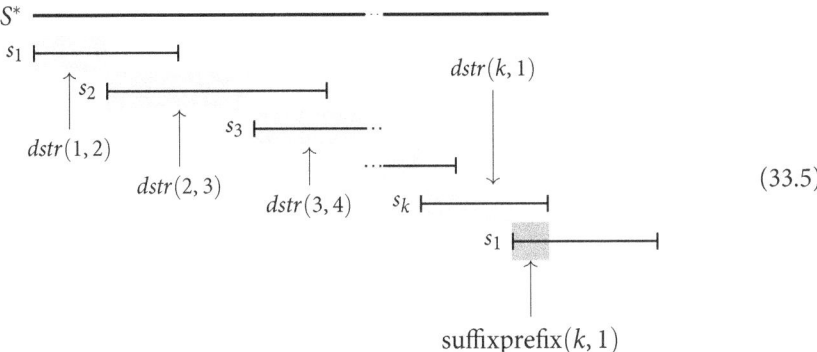

$$(33.5)$$

suffixprefix$(k, 1)$

It has to be the case that there are no gaps between s_i and s_{i+1}, otherwise we could make S^* shorter by removing those characters. It *can* be the case that $|\text{suffixprefix}(s_i, s_{i+1})| = 0$ and therefore that two consecutive strings abut end-to-start, but that doesn't change the (33.4) statement. The following arrangement, where (say) s_{i+2} overlaps both s_{i+1} and s_i, is possible, and also doesn't change the statement.

$$(33.6)$$

We also know that each string starts at a different location; otherwise, if two strings s_i and s_j started at the same point, the shorter one would be a substring of the longer one.

Based on the ordering of strings in (33.4), consider the cycle $\mathcal{C} = s_1 \to s_2 \to \cdots \to s_k \to s_1$ in G. This cycle has length $length(\mathcal{C}) = d(1,2) + d(2,3) + \cdots + d(k,1)$, which is $\leq$ the length of the string in (33.4) (by $|\text{suffixprefix}(s_k, s_1)|$ characters). Therefore,

$$TSP^* \leq length(\mathcal{C}) \leq |S^*| = SS^*. \tag{33.7}$$

$\square$

33.2 The Cycle Cover algorithm

Given this tight connection between TSP and shortest superstring, it's natural to use techniques for approximating TSP to find a reasonable (but not necessarily optimal) superstring. We do that using the Cycle Cover algorithm.

A *cycle cover* of a graph G is a collection of cycles $\{\mathcal{C}_1, \ldots, \mathcal{C}_m\}$ in G such that each vertex is in exactly one of the cycles $\mathcal{C}_i$. A Hamiltonian path, for example, is a cycle cover that consists of a single cycle. A traveling salesman tour is similarly a cycle cover with only one cycle that has minimum total cost. Finding a minimum weight cycle cover is easy (it turns out), but finding one that is required to have only one cycle seems to be hard. The Cycle Cover algorithm for the shortest superstring problem uses the polynomial-time algorithm for finding a minimum weight cycle cover:

Algorithm 33.1. Cycle cover for SHORTEST SUPERSTRING.

1. Find a minimum-weight cycle cover $\{C_1, \ldots, C_m\}$ of the prefix graph G.
2. Suppose C_i contains the nodes $s_i, \ldots, s_{p_i}$, then define

$$\alpha(C_i) := dstr(s_i, s_{i+1}) \circ dstr(s_{i+1}, s_{i+2}) \circ \ldots$$
$$\circ dstr(s_{p_i-1}, s_{p_i}) \circ dstr(s_{p_i}, s_i), \quad (33.8)$$

and

$$\beta(C_i) := \alpha(C_i) \circ s_i. \quad (33.9)$$

3. Output $\beta(C_1) \circ \cdots \circ \beta(C_m)$.

In step 2, we've picked an arbitrary "start" node for each cycle. In the algorithm, and throughout the rest of the chapter, $\alpha(C)$ denotes the string spelled out by the edges of cycle C starting at this arbitrary (but deterministically) chosen starting node.

Finding the minimum weight cycle cover is equivalent to finding the minimum weight perfect matching, which can be solved in polynomial time using, for example, network flow. We won't go into details here, but a description of how to do this can be found in most algorithms textbooks. The important point is that finding a minimum weight cycle cover can be done in polynomial time.

Let a^∞ mean a string that consists of infinitely many repetitions of string a. As a warmup, we can show that:

Lemma 33.4. *Let C_i be some cycle in the prefix graph and $\alpha(C_i)$ is the string spelt by traversing the cycle. If the node representing s_j is $\in C_i$, then s_j is a substring of $\alpha(C_i)^\infty$.*

Proof: Let s_j and s_{j+1} be successive pairs of strings along C_i. We must have:

$$|\text{suffixprefix}(s_j, s_{j+1})| < |s_j|, \quad (33.10)$$

otherwise s_j would be a substring of s_{j+1}, which we disallow.

At least 1 character must be
added by each edge.

(33.11)

Hence, $d(j, j+1)$ as defined in (33.1) must be ≥ 1, which means that at least one additional letter of s_j is spelled out on each traversal of an edge of the cycle, and eventually s_j must be spelled out. $\qquad\square$

Theorem 33.5 (Correctness of the Cycle Cover algorithm). *If the node representing $s \in C_i$, then s is a substring of $\beta(C_i)$, where β is defined in Algorithm 33.1.*

Because the algorithm finds a cycle cover, every node is in some cycle C_i. Hence, Theorem (33.5) implies that every string s is a substring of the output, and therefore the output is a true superstring.

Proof of Theorem 33.5: Let $s_1, \ldots, s_p$ be the nodes in the order they are visited in $\beta(C_i)$. The string

$$dstr(s_1, s_2) \circ dstr(s_2, s_3) \circ \cdots \circ dstr(s_{p-1}, s_p) \circ s_p \tag{33.12}$$

is a superstring of $s_1, \ldots, s_p$.

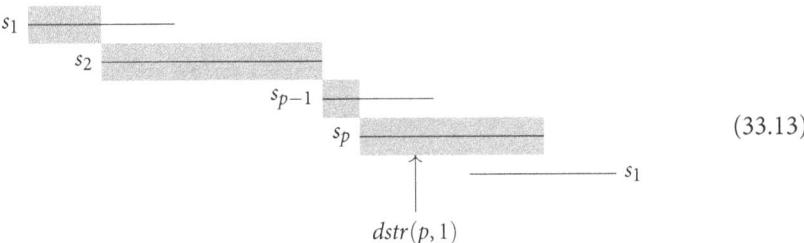

$$(33.13)$$

$dstr(p, 1)$

But s_p is a prefix of $dstr(s_p, s_1)s_1$ as shown in the shaded boxes. The right end of the second occurrence of s_1 (light box) must be to the right of the right end of s_p, otherwise, s_1 would be a substring of s_p. Therefore, the strings in C_i are also substrings of $\beta(C_i)$. □

If CC^* is the length of the minimum cycle cover of the prefix graph and A is the length of the string output by the cycle cover algorithm, then we have shown the first three inequalities:

$$CC^* \leq TSP^* \leq SS^* \leq A \leq 4 \cdot SS^*, \tag{33.14}$$

where the first $\leq$ comes from the fact that a TSP is a cycle cover; the second $\leq$ is due to Theorem 33.3, the third $\leq$ follows from Theorem 33.5, since what the cycle cover algorithm outputs can't be better than the optimal superstring. The $\leq 4 \cdot SS^*$ part is what remains to be proven.

To prove that part, we will take the following road: Algorithm 33.1 concatenates the $\beta(C_i)$ strings for each cycle. We want to ultimately show that strings from different cycles in an optimal cycle cover can't overlap by too much (since we didn't use those overlaps at all). Since any cycle is like a string repeated infinite times (going around the cycle as many times as we want), we first characterize repeated strings.

33.3 Some lemmas about repeated strings

Lemma 33.6. *Let a and b be two strings, and let $m = |a| + |b|$. Suppose*

$$\bar{a} \text{ is a prefix of } a^{\infty} \tag{33.15}$$

$$\bar{b} \text{ is a prefix of } b^{\infty}. \tag{33.16}$$

If $\bar{a}[1 \ldots m] = \bar{b}[1 \ldots m]$, then $a \circ b = b \circ a$ (in other words, the strings commute under the concatenation operation).

Proof: If $|a| = |b|$ then, because $m > |a|$ and $m > |b|$, it must be that $a = b$. So, without loss of generality, assume $|b| < |a|$. The situation we have is (33.17).

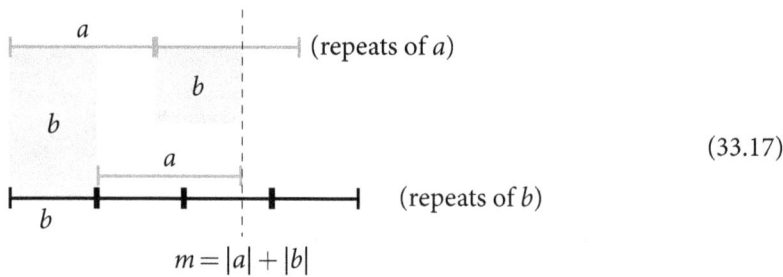

$$m = |a| + |b| \tag{33.17}$$

Because $\bar{a}[1 \ldots m] = \bar{b}[1 \ldots m]$, a prefix of a is b, and hence, the second copy of a in $\bar{a}$ also starts with b, and $a \circ b$ is a prefix of $\bar{a}$. This means that $a \circ b$ is also a prefix of $\bar{b}$ because $\bar{a}[1 \ldots m] = \bar{b}[1 \ldots m]$.

Also, a is a prefix of b^k, for some k big enough to make b^k longer than a. There is a copy of b^k following the first b in $\bar{b}$. So $b \circ a$ is a prefix of $\bar{b}$. Hence, $a \circ b$ and $b \circ a$ are both equal-length prefixes of $\bar{b}$. So, $a \circ b = b \circ a$. □

We can use the following lemma to build longer equivalent strings from strings that commute.

Lemma 33.7. *Let a and b be two strings. If $a \circ b = b \circ a$, then*

$$a^k \circ b^k = b^k \circ a^k \tag{33.18}$$

for any $k > 0$.

Proof: Consider $a^k \circ b^k$:

$$\underbrace{aaaaa \ldots aaaaa}_{k} \circ \underbrace{bbbbb \ldots bbbbb}_{k}. \tag{33.19}$$

Since a and b commute, this is equal to the following by swapping the innermost (a, b) pair:

$$\underbrace{aaaaa \ldots aaaaa}_{k-1} \, b \circ a \, \underbrace{bbbbb \ldots bbbbb}_{k-1}. \tag{33.20}$$

By repeatedly swapping adjacent copes of a and b, the innermost a and b pair can be driven to the start and end of the string:

$$baaaaa \ldots aaaaa \circ bbbbb \ldots bbbbba. \tag{33.21}$$

This can then be repeated for each subsequent innermost pair until arriving at $bbbbb \ldots bbbbb \circ aaaaa \ldots aaaaa$ which equals $b^k \circ a^k$. □

Lemma 33.8. *Let a and b be two nonempty strings such that $a \circ b = b \circ a$. Then*

$$a^\infty = b^\infty, \tag{33.22}$$

meaning that for any length i, the prefix $a^\infty[1 \ldots i] = b^\infty[1 \ldots i]$.

Proof: By Lemma 33.7, we have

$$a^k \circ b^k = b^k \circ a^k \tag{33.23}$$

for any $k > 0$. Let $i > 0$ be given, and choose k such that

$$\min\{|a^k|, |b^k|\} > i. \tag{33.24}$$

Then, by (33.23), $a^k[1 \ldots i] = b^k[1 \ldots i]$ since i falls into the repetitions of a in the lefthand side and into the repetitions of b on the righthand side. Because i was chosen arbitrarily, $a^\infty = b^\infty$. □

Lemma 33.9. *Let a be a string, and let S be a set of strings (none of which is a substring of any other) such that each $s \in S$ is a substring of a^∞. Let G be a prefix graph that contains nodes for the strings S. Then there is a cycle C in G of length $\leq |a|$ such that $\alpha(C)$ contains each $s \in S$ as a substring.*

Proof: Order the strings in S by the order in which they first appear in a^∞. This provides a well-defined ordering of these strings since no two can start at the same position. Let C be the cycle formed by traversing the nodes of G corresponding to the strings of S in this order. This looks like (33.25).

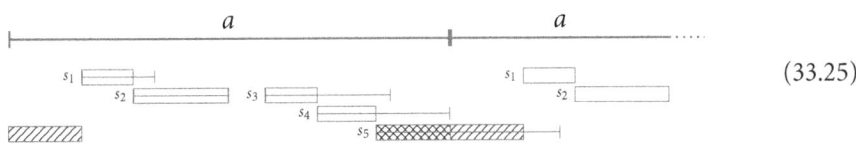

$$\tag{33.25}$$

The ☐ boxes denote $dstr(i, i+1)$. Each of them in the first instance of a corresponds to one edge in the cycle C, and the length of each box equals the length of the edge in the cycle. What about the last edge which is represented by ▨▨▨▨? The length of that edge can't be more than the ▨▨▨ box (which extends to the end of the first copy of a) plus the length of the ▨▨▨ box. The length of the ▨▨▨ box can't be more than the distance from the start of the string to the start of location for s_1. Hence, the sum of the lengths of all the edges is $\leq |a|$.   ☐

33.4 The Cycle Cover algorithm gives a 4-approximation

Our mission is to show that if the overlap between the strings of any two cycles in the minimum cycle cover is too big, then it can't have been a minimum cycle cover:

Theorem 33.10. *Let C_1 and C_2 be two cycles in the minimum cycle cover used by the Cycle Cover algorithm, and let $x \in C_1$ and $y \in C_2$ be two strings represented by nodes in these cycles. Then*

$$|suffixprefix(x, y)| < d(C_1) + d(C_2), \tag{33.26}$$

where $d(C)$ is the sum of the lengths of the edges in cycle C.

In other words, no two strings in separate cycles from the minimum-weight cycle cover can have too large of an overlap.

Proof: Suppose not, and the overlap between x and y is $m \geq d(C_1) + d(C_2)$.

Since the overlap is a substring of a string that is representable by C_1 and is bigger than $d(C_1)$, it must be made up of repeating instances of some string a of length $d(C_1)$. That is, it is a prefix of a^∞. Similarly, since the overlap is bigger than $d(C_2)$, it must be the prefix of b^∞ for some string b of length $d(C_2)$.

Lemma 33.6 applies since the first m characters of these two strings are equal, and we have that $a \circ b = b \circ a$. Applying Lemma 33.8, we have that $a^\infty = b^\infty$.

By Lemma 33.4, each $s_i \in C_1$ is a substring of a^∞ and each $s_j \in C_2$ is a substring of b^∞. But by the previous paragraph, $a^\infty = b^\infty$, meaning that every $s \in C_1 \cup C_2$ is a substring of a^∞.

By Lemma 33.9, there is then a cycle C of length $|a| = d(C_1)$ that contains the strings of both C_1 and C_2; replacing C_1 and C_2 with C gives a smaller-length cycle cover, contradicting that C_1 and C_2 came from the minimum cycle cover.   ☐

We now have the main theorem:

Theorem 33.11. *The length of the string output by the Cycle Cover algorithm is $\leq 4 \cdot SS^*$, where SS^* is the length of the optimal superstring.*

Proof: Let $C_1, \ldots, C_m$ be the cycles found in the algorithm. Let $s_1', \ldots s_m'$ be the "first" strings in each cycle (chosen in step 2 of Algorithm 33.1).

The string the algorithm outputs has length

$$\sum_{i=1}^{m} |\beta(\mathcal{C}_i)| = \left(\sum_{i=1}^{m} |\alpha(\mathcal{C}_i)|\right) + \left(\sum_{i=1}^{m} |s_i'|\right) \tag{33.27}$$

$$= CC^* + \sum_{i=1}^{m} |s_i'| \tag{33.28}$$

$$\leq SS^* + \sum_{i=1}^{m} |s_i'|. \tag{33.29}$$

We need then to bound $\sum_{i=1}^{m} |s_i'|$ by $3 \cdot SS^*$, where these s_i' are the arbitrarily chosen "first" strings of each cycle.

To do that, reorder s_i' into the order in which they appear in the optimal shortest superstring. Then

$$SS^* \geq \sum_{i=1}^{m} |s_i'| - \sum_{i=1}^{m-1} |\text{suffixprefix}(s_i', s_{i+1}')|. \tag{33.30}$$

This is because

$$\text{suffixprefix}(s_i', s_{i+1}') \tag{33.31}$$

is the maximum possible savings that can be achieved by following s_i' with s_{i+1}'. So $\sum_{i=1}^{m-1} |\text{suffixprefix}(s_i', s_{i+1}')|$ is an overestimate of the benefit, and this gives a lower bound on the length of the optimal, since the optimal must contain at least these strings.

We now use Theorem 33.10 to rewrite (33.30):

$$SS^* \geq \sum_{i=1}^{m} |s_i'| - 2CC^* \tag{33.32}$$

$$\geq \sum_{i=1}^{m} |s_i'| - 2SS^*, \tag{33.33}$$

since $2CC^*$ over-counts the weight of cycles appearing in applications of Theorem 33.10, and we already have seen that $CC^* \leq SS^*$.

Finally, rewriting (33.33), we have:

$$\sum_{i=1}^{m} |s_i'| \leq 3SS^*, \tag{33.34}$$

and plugging this into (33.29) gives the length of the string output by the cycle cover algorithm to be $\leq 4 \cdot SS^*$. $\square$

33.5 Summary and notes

Other work on the shortest superstring problem includes Alanko and Norri [2017]; Turner [1989]; Guo et al. [2014].

Algorithm 33.1 is related to the greedy algorithm: "repeatedly merge the two strings that have the largest overlap." That is because the cycle cover on the class of prefix graphs can be found using a greedy approach, and so it is possible to show that the pure greedy approach also is a 4-approximation algorithm. In fact, it's conjectured (but not proven) that this greedy algorithm is really a 2-approximation algorithm.

It is possible to modify the cycle cover algorithm to return a solution that is no more than 3 times the optimum. The way to do this is to apply the greedy merging heuristic to the strings $\beta(C_i)$ created by the cycle cover algorithm. Nevertheless, there is a gap between the conjectured existence of a 2-approximation algorithm and the known 3-approximation algorithm.

> Presentation Notes
> _____
>
> Our presentation follows Blum et al. [1994].

33.6 Exercises

33.1   Give an example where the cycle cover algorithm returns a string that is as far off from optimal as possible. In other words, if the optimal of an instance S has length $OPT(S)$ and the cycle cover algorithm returns a string of length $CC(S)$, then the distance from optimal is $\gamma(S) = OPT(S)/CC(S)$. Find an example set S that makes $\gamma(S)$ as small as possible.

33.2   The greedy approximation algorithm for shortest superstring starts with the input set of strings $S_0 = \{s_1, \ldots, s_k\}$. During its ith iteration, the greedy algorithm picks two strings s_{j1} and s_{j2} in S_{i-1} that have the longest suffix-prefix overlap, merges them into a new string s_{j1j2} using their suffix-prefix overlap, and forms S_i by removing s_{j1} and s_{j2} and adding s_{j1j2}. It stops when S_i contains a single string. Give an example set of input strings where this algorithm does as bad as possible, meaning that the length of the final string it creates divided by the length of the optimal shortest superstring is as large as you can make it.

CHAPTER 34

Limits on What's Possible

In this chapter, we'll see some hardness results about how fast certain problems can be solved. Since so much is not known for certain about computational complexity, many of these results will be conditional on (widely believed) conjectures. At the very least, the results presented here show that finding fast algorithms for these problems is likely to be very hard (and impossible if the conjectures are true).

34.1   A brief introduction to NP-completeness

A computational *decision problem* Π is a set of strings. While this definition seems quite broad, it is actually fundamental to computational thinking. Let's look at an example.

The problem SHORTEST SUPERSTRING (from Chapter 33) can be put into this framework by setting Π equal to the set of strings:

$$\{\text{ENCODING}((\{S_i\}, m))\}, \tag{34.1}$$

where S_i are strings, m is an integer, and $\{S_i\}$ has a superstring of length $\leq m$.

In (34.1), ENCODING is some function that encodes the pair into a string; for example, it could return the string of bits used to represent the pair in the computer's memory. From now on, we will omit mentioning the ENCODING step—which is assumed to be done—since it is simpler to talk directly about the underlying thing being encoded (in this case, the pairs $(\{S_i\}, m)$).

An *instance* of this problem is some pair $I = (S, m)$. An instance can be either in Π or not.

The definition encompasses only *decision* problems because the answer is either "yes" or "no" (is it in the set or not?). An instance I is a *yes-instance* if and only if $I \in \Pi$. In our example, that means that S has a superstring of length $\leq m$. An instance is a *no-instance* otherwise.

Typically in this book we've dealt with the broader class of *optimization* problems, where we are trying to minimize or maximize some objective (i.e., finding the shortest superstring). This restriction to decision simplifies many conclusions, but it's not really consequential when trying to prove a problem is hard: if you can't even solve the decision problem, you clearly cannot solve the corresponding optimization problem. Conversely, if you can solve the decision version, you can usually do a binary search over some parameter (like m in (34.1)) to find its optimal value (i.e., is there a shortest superstring of length $\leq m = (\sum_i |S_i|)/2$? If so, halve m, if not increase m, and ask again).

A decision problem is in the class P if there is an algorithm that runs in polynomial time to decide whether a problem is a yes or no instance. Here, polynomial time means the runtime is polynomial in the length of the encoding of the input instance. Throughout this book, we've mostly focused on describing algorithms that are in P (and showing that they are not just polynomial-time, but practically fast). In this chapter, we tackle the opposite question: do we have evidence that some problem *isn't* in P?

The set NP consists of the decision problems that have polynomial-time algorithms to *check* the answer if they are given a hint called a *certificate*. Formally, a problem Π is in NP if there an algorithm V (called a *verifier*) such that:

- If $I \in \Pi$, then there exists a string c_I, of polynomial length in the length of the encoding of I, such that $V(I, c_I)$ outputs "yes." In other words, if you give V the "right" hint c_I, it will correctly say yes for yes-instances.
- If $I \notin \Pi$, then for any polynomial-length string c, $V(I, c)$ outputs "no." In other words, we can't fool V into accepting a no-instance no matter what hint string we give it.

The definition of NP doesn't say anything about how to find a c_I for yes instances. It just says that if we're given it, we can verify (using V) that $I \in \Pi$ (and we can be sure if V says yes, it's correct).

From the definition, we can see $P \subseteq NP$ (the polynomial-time algorithm for a problem in P fits the definition of V by ignoring any hint it's given). Of course, one of the most famous open questions in computer science is whether $P = NP$. Most people expect $P \neq NP$ because many apparently hard problems are in a subset of NP called *NP-complete*. A problem Π is NP-complete if:

- $\Pi \in NP$; and
- for any other problem $\Pi' \in NP$, there is a polynomial-time algorithm $R_{\Pi' \to \Pi}$ (called a *reduction*) that transforms an instance of Π' into an instance of Π such that

$$I' \in \Pi' \iff R_{\Pi' \to \Pi}(I') \in \Pi. \tag{34.2}$$

In other words, if we can solve the NP-complete problem Π, then we can solve any problem in NP by using its reduction to Π. The NP-complete problems are therefore the most powerful and hardest problems in NP. Although from the definition it's not clear whether any NP-complete problems exist, in fact, many problems have been shown to be NP-complete (starting famously with the theorems of Cook [1971] and Levin [1973]).

The key point is that if we reduce a known NP-complete problem Π' to a problem Π we're interested in, then this will show that Π is also NP-complete.

$$
\begin{array}{c}
\Pi_1 \\
\Pi_2 \\
\Pi_3 \longrightarrow \Pi' \longrightarrow \Pi \\
\Pi_4 \\
\vdots
\end{array}
\tag{34.3}
$$

That's our task for the next few sections for some string problems.

Of course, we have only scratched the surface of the theory of NP-completeness, hopefully just enough to orient you; you can find much more detail in many textbooks [e.g., Garey and Johnson, 1979; Papadimitriou, 1994].

34.2 Shortest superstring is NP-complete

Recall the SHORTEST SUPERSTRING problem from Chapter 33: we're given a set of strings S, and we ask whether we can create a superstring S of length $\leq k$ such that every string $s_i \in S$ is a substring of S. This problem is clearly in NP: if someone gives us the superstring (as our hint c_I), then we can check if its length is $\leq k$. We provide a proof (from Gallant et al. [1980]) that this problem is NP-complete.

Gallant et al. [1980] shows this via a reduction from a slightly special case of the HAMILTONIAN PATH problem:

Problem 34.1. *The* RESTRICTED HAMILTONIAN PATH *problem is defined as follows. Let* $G = (V, E)$ *be a directed graph that has two distinguished vertices s and t and where every node except t has out-degree at least 2. The problem is to determine whether G has an s-t path that visits every vertex exactly once.* ◆

This problem is known to be NP-complete (via a straightforward reduction from the standard HAMILTONIAN CYCLE problem [Gallant et al., 1980]). We will reduce it to the SHORTEST SUPERSTRING problem, showing that if we could solve SHORTEST SUPERSTRING in polynomial time, we could use that algorithm to solve Problem 34.1.

Theorem 34.2. SHORTEST SUPERSTRING *is NP-complete.*

To prove Theorem 34.2, across the next two sections we will first describe the reduction, and then show that the reduction is correct.

34.2.1 The reduction

The reduction is from Problem 34.1, so let G be an instance of that problem. We must give an algorithm (the reduction) that transforms G into an instance of SHORTEST SUPERSTRING. That means we have to create a set of strings S and an integer k such that:

$$G \text{ has a Hamiltonian path } \iff$$

$$S \text{ has a superstring of length } \leq k. \qquad (34.4)$$

To construct S, we will construct a set of strings that consists of strings for each node and for each out-edge of every node (and a few other strings).

For each node v except s and t, create a string $v\#\bar{v}$, using two symbols, v and $\bar{v}$, that are unique to this node, and where $\#$ is a special separator character.

For every node v except t, order the outgoing neighbors of v arbitrarily and name them $u_0, \ldots, u_{out(v)-1}$, where $out(v)$ is the number of edges leaving v. For each edge (v, u_i), create two strings: $\bar{v}u_i\bar{v}$ and $u_i\bar{v}u_{i+1 \bmod out(v)}$. The u_i characters are the u vertex characters for the node u that we created when we assigned node u the string u#ū.

In a picture, this looks like (34.5).

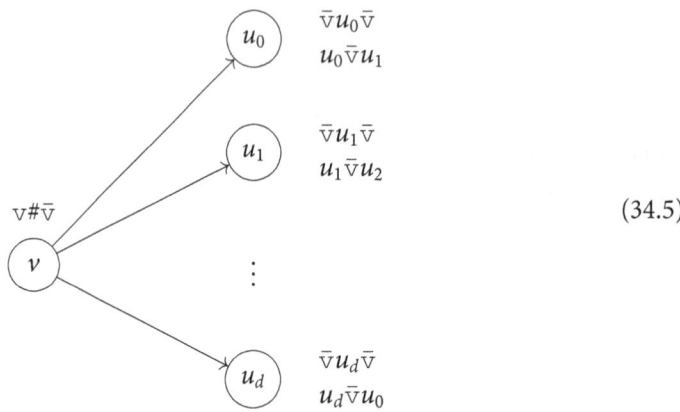

$$(34.5)$$

If we walk down the strings at the right of (34.5), they form a cycle of strings where each successive pair overlaps by two characters.

$$
\begin{array}{cccccc}
\bar{v} & u_0 & \bar{v} & & & \\
 & u_0 & \bar{v} & u_1 & & \\
 & & \bar{v} & u_1 & \bar{v} & \\
 & & & u_1 & \bar{v} & u_2 \\
 & & & \cdots & \cdots & \cdots
\end{array}
\qquad (34.6)
$$

In fact, we can start this walk at any $\bar{v}u_i\bar{v}$ string and complete the cycle back around to the previous $u_{i-1}\bar{v}u_i$. Let $adj(v, u_i)$ be the string spelled by this overlapped concatenation of strings that starts at $\bar{v}u_i\bar{v}$.

Every $adj(v, u_i)$ string has length $2out(v) + 2$ since every u_i adds 2 characters, and the $\bar{v}$ and u_{i-1} at the start and end add two more characters. Each $adj(v, u_i)$ starts with a $\bar{v}$ and ends with an unbarred vertex symbol.

Consider what we must discover to check for a Hamiltonian path: we need a permutation of the nodes $v_1, \ldots, v_n$ where there is an edge from v_i to v_{i+1} for each $i < |V|$. Or, in other words, when we are mid-walk, at a node v_i, we need to know where to go next. The $adj(v, u_i)$ strings can encode that: if v_j follows v_i, then our superstring will have $adj(v_i, v_j)$ as a substring (we have to prove this, though).

We add one more set of strings to $\mathcal{S}$:

$$\{\%\#\bar{s},\, t\#\$\}, \qquad (34.7)$$

where % and $ are new special characters.

To summarize, we have created:

- $|V| - 2$ node strings of the form v#$\bar{v}$;
- $2|E|$ edge strings of the forms $\bar{v}u_i\bar{v}$ and $u_i\bar{v}u_{i+1 \bmod out(v)}$; and
- 2 additional special strings in (34.7);

for a total of $2|E| + |V|$ strings in our SHORTEST SUPERSTRING instance.

This construction of strings gives us a (clearly polynomial-time) reduction $R_{hs} :=$ $R_{\text{HAM PATH}\rightarrow\text{SHORT SUPERSTRING}}$. For any instance G of Problem 34.1, $R_{hs}(G)$ returns the corresponding set of strings as described in this section.

34.2.2 The proof

We need to show that $R_{hs}(G)$ has a short superstring if and only if G has a Hamiltonian path. In particular, we will prove that it has a superstring of length $2|E| + 3|V|$ if and only if G is Hamiltonian. We do this in two parts.

Lemma 34.3. *Suppose that $G = (V, E)$ has a Hamiltonian path $s = v_1, v_2, \ldots, v_{n-1}, t = v_n$. Then $R_{hs}(G)$ has a superstring of length $2|E| + 3|V|$.*

Proof: We can create a superstring of the following form:

$$\boxed{\%\#\bar{s}}\ \boxed{adj(s, v_2)}\ \#\ \boxed{adj(v_2, v_3)}\ \#\ \cdots\ \boxed{t\#\$} \tag{34.8}$$

Diagram (34.8) shows the form of the string, but the actual string would be created by overlapping adjacent strings as much as possible in the order they are given. Every string in $R_{hs}(G)$ is contained in this string. It has length $\sum_{i=0}^{n-1} |adj(v_i, u_{i+1})| + (n-2) + 4$, where the $(n-2)$ term comes from the $\#$ symbols in between the $adj(\cdot, \cdot)$ strings and the $+4$ comes from the starting $\%\#$ and ending $\#\$$. Simplifying:

$$\sum_{i=1}^{n-1} |adj(v_i, v_{i+1})| + (n-2) + 4 \tag{34.9}$$

$$= \sum_{i=1}^{n-1} (2out(v_i) + 2) + (n-2) + 4 \tag{34.10}$$

$$= 2\sum_{i=1}^{n-1} out(v_i) + \sum_{i=1}^{n-1} 2 + n - 2 + 4 \tag{34.11}$$

$$= 2|E| + 2(n-1) + n - 2 + 4 \tag{34.12}$$

$$= 2|E| + 2n - 2 + n - 2 + 4 \tag{34.13}$$

$$= 2|E| + 3|V|. \tag{34.14}$$

So $R_{hs(G)}$ has a superstring of the appropriate length when G has a Hamiltonian path. □

Now, we have to show the other direction: that if S has a superstring of length $2|E| + 3|V|$ then G has a Hamiltonian path. We need first this lemma:

Lemma 34.4. *Let $G = (V, E)$ be an instance of Problem 34.1. No superstring of $R_{hs}(G)$ has length less than $2|E| + 3|V|$.*

Proof: We will argue that if we got the maximum possible overlap for each string, we'd end up with a string of length $2|E| + 3|V|$. There are $3(2|E| + |V|)$ total characters in the strings. If we were able to overlap all of them by 2 characters,

$$(34.15)$$

we would have a superstring of length $3 + (2|E| + |V| - 1) = 2|E| + |V| + 2$. But this is not possible:

- Strings of the form $v\#\bar{v}$ can overlap by at most 1 on each side with any other string. This modifies our estimate by increasing the length by 2 for each such string:

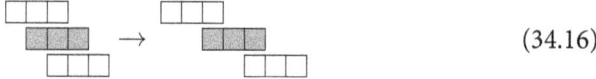

$$(34.16)$$

   so we have to add $2(|V| - 2)$ to our minimum length for each of these strings.
- The string $\%\#\bar{s}$ must be at the start of the string (otherwise what would be before it wouldn't get its maximum overlap). Further, it can only overlap by 1 at its end, so our estimate has to be increased by 1:

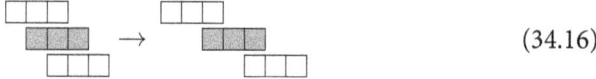

$$(34.17)$$

   so we have to add 1 to our minimum length to account for this.
- The string $t\#\$$ must be at the end of the string (again, because otherwise the following string wouldn't overlap). Further, it can only overlap by 1 at its start:

$$(34.18)$$

   so we have to add 1 to our minimum length to account for this.

Therefore, the best we could do would be $2|E| + |V| + 2 + 2(|V| - 2) + 1 + 1 = 2|E| + 3|V|$.

$\square$

Based on this lemma, we can now prove that if there is a small superstring, there is a Hamiltonian path.

Lemma 34.5. *Let $G = (V, E)$. If $R_{hs}(G)$ has a superstring S of length $2|E| + 3|V|$, then G has a Hamiltonian path.*

Proof: The string x between each successive pair of # in S must be some string $adj(v, u_i)$. To see this: x can only contain edge strings of the form $\bar{v}u_i\bar{v}$ or $u_i\bar{v}u_{i+1 \bmod out(v)}$. Since S has the minimum possible size (by Lemma 34.4), each of the strings within x must overlap by 2, so they must form one of the $adj(v, u_i)$ cycles.

Hence, S must be of the form:

$$\boxed{\%\#\bar{s}}\;\boxed{adj(s, v_j)}\;\#\;\boxed{adj(v_j, v_k)}\;\#\;\cdots\;\boxed{t\#\$}\,. \tag{34.19}$$

Again, the string would be created by "shrinking" the schematic using adjacent overlaps as much as possible; for example, $\boxed{\%\#\bar{s}}\;\boxed{adj(s, v_j)}$ in figure (34.19) should be interpreted as $\%\#\bar{s}v_j\ldots$. The $adj(v_i, v_j)$ strings give successive edges on a path. Since we have to use every string, there must be an $adj(v, \cdot)$ for every v. Hence, this is a Hamiltonian path in G. $\qquad\square$

34.2.3 Additional notes

The alphabet of the strings we put into $\mathcal{S}$ is pretty large, and certainly not of constant size. We have two symbols v and $\bar{v}$ for each vertex, and three special symbols $\$$, $\#$, and $\%$. What this means is that this proof will only show the version of SHORTEST SUPERSTRING that allows such large alphabets is NP-complete. In fact, it's relatively simple to reduce the alphabet in the reduction to use only 3 different characters (Exercise 34.1), and one can show that it remains NP-complete with a binary alphabet [Räihä and Ukkonen, 1981].

34.3 Shortest supersequence is NP-complete

We'll next look at a new problem that is both interesting in its own right and useful to prove the hardness of multiple sequence alignment (Section 34.4).

Problem 34.6. *In the* SHORTEST SUPERSEQUENCE *problem, we are given a set of strings* $\mathcal{S}$, *and an integer n. We ask whether there is a string S, with* $|S| \leq n$ *such that for each* $s = x_1, \ldots, x_\ell \in \mathcal{S}$, *the characters* $x_1, \ldots, x_\ell$ *appear in order in S.* ◆

The SHORTEST SUPERSEQUENCE problem is very similar to the SHORTEST SUPERSTRING, except it removes the restriction that each s appears contiguously in the supersequence—instead we can skip over some characters in S. We prove that this problem is NP-complete using a reduction due to Maier [1978].

Theorem 34.7. SHORTEST SUPERSEQUENCE *is NP-complete.*

Proof: We will reduce from VERTEX COVER, a well-known NP-complete problem where we're given an undirected graph $G = (V, E)$ and an integer k and want to know whether there is a subset U of nodes, with $|U| \leq k$, such that every edge is adjacent to some node in U. That is the nodes of U "cover" the edges of G.

The reduction. Given a vertex cover instance $G = (V, E)$, we create $|V| + 1$ strings. Fix an arbitrary order of the vertices of V. First, we create a template string T of the form:

$$V;\; A;\; E';\; E';\; A;\; V \tag{34.20}$$

where V is the list of the nodes of G, and E' is an arbitrarily ordered list of a symbol for each edge in E duplicated twice, i.e., if e is an edge of G, then E' contains "ee". Let c be $\max\{|V|, |E|\}$. A is a string of $4c$ copies of a special character $\star$. T has length $2|V| + 4|E| + 8c$.

Next, we add a string for each edge $e = \{u, v\}$ in G. Assume that $u < v$ in our arbitrary ordering of V. We then add a string of the form:

$$ee; u; A; v; ee \tag{34.21}$$

for each edge $e = \{u, v\}$. This gives us $|E| + 1$ strings in total. We've used an alphabet that contains a character for every vertex, a character for every edge, the ";" separator, and the special character $\star$. Call the complete set of strings $R(G)$.

Claim. $R(G)$ has a superstring of length $8c + 6|E| + 2|V| + k$ if and only if G has a vertex cover of size k. We will now prove each direction of this claim.

Vertex Cover $\Longrightarrow$ Supersequence. We want to show that if G has a vertex cover C of size k, then $R(G)$ has a superstring of length $8c + 6|E| + 2|V| + k$. Let W be the set of edges $\{u, v\}$ where $u < v$ and u is in C, and let $\bar{W}$ be the edges that are not in W. Construct a string S of the form:

$$W'; \ V; \ A; \ E'; \ C; \ E'; \ A; \ V; \ \bar{W}' \tag{34.22}$$

where W' contains a symbol for each edge in W duplicated twice, and $\bar{W}'$ contains a symbol for each edge in $\bar{W}$ duplicated twice. V, A and E' are defined as in (34.20), and C is a list of the nodes in the vertex cover.

This string S has length $|V| + 4c + 2|E| + k + 2|E| + 4c + |V| + 2|E|$, where the last $2|E|$ term is the total length of W' and $\bar{W}'$—since each edge appears in exactly one of W or $\bar{W}$. Simplifying, S has length $2|V| + 8c + 6|E| + k$, matching the claimed length.

Every string in R is a subsequence of S: clearly T is a subsequence of S. For $s = ee;$ $u; A; v; ee$ we have one of the following cases:

$$
\begin{array}{ccccccccc}
W' & V & A & E' & C & E' & A & V & \bar{W}' \\
ee & u & A & & v & ee & & & \\
& & & ee & v & & A & u & ee
\end{array}
\tag{34.23}
$$

depending on whether $e \in W$ or $e \in \bar{W}$. Hence, we have a supersequence of the claimed length.

Supersequence $\Longrightarrow$ Vertex Cover. Let S be a supersequence of the claimed length. By definition, T is a subsequence, and each $s_e \in \mathcal{S}$ is a subsequence of S, and so each s_e can be aligned to S. Further, we claim that we can assume that the A in s_e can be completely aligned with either the left or right instance of A in T without increasing the length of the superstring (see Exercise 34.2) so we can assume each s_e aligns with T.

We have two options: either the A in s_e aligns to the left or to the right instance of A in T. For $e = \{u, v\}$, we have either:

$$
\begin{array}{ccccccc}
 & V & A & E' & \downarrow & E' & A & V \\
ee & u & A & & v & ee & & \\
& & & ee & u & & A & v & ee.
\end{array}
\tag{34.24}
$$

Form a string by vertically aligning all the strings in this way, matching characters in each region if they are the same. The length of that string is $8c + 6|E| + 2|V|$ plus the length of the $\downarrow$ region since the ee pairs hanging off the ends can all be covered by strings of total length $2|E|$.

By assumption, the length of the superstring is $8c + 6|E| + 2|V| + k$ so the length of the $\downarrow$ region is k. Every edge string has one vertex in the region denoted with $\downarrow$. Hence, the vertices in the $\downarrow$ region form a vertex cover, and therefore there is a vertex cover of size k. $\square$

34.4 MSA with SP-score is NP-complete

Recall the Multiple Sequence Alignment (MSA) problem from Chapter 12: we're given a set of strings S, and we want to construct an alignment $\mathcal{A}$ between them. The alignment is an $|S| \times m$ matrix (for some m no smaller than the length of each of the strings in S), where each row contains the characters of one $s \in S$ with gap characters "$-$" interspersed.

With the SP-score, the quality of the alignment A is measured by adding up all the pairwise scores between characters in a column, and summing that over the columns.

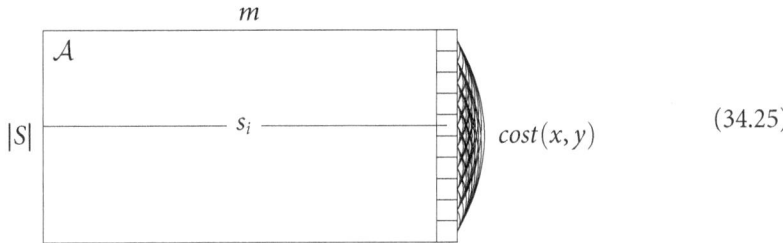

$$(34.25)$$

We show that the problem to determine whether there is an alignment of SP-score less than a given cost c is NP-complete, following the proof in Wang and Jiang [1994].

Theorem 34.8. *MSA with SP-Score is NP-complete.*

Proof: We're going to reduce from Shortest Supersequence (Problem 34.6), which is NP-complete by Theorem 34.7. Let S and m be an instance of Problem 34.6. We will assume that S contains strings over alphabet $\{0, 1\}$, which is still NP-complete [Middendorf, 1994; Räihä and Ukkonen, 1981] (though our proof of the previous section did not show that). Let $\|S\|$ be the total length of the strings in S.

The MSA instance that we will create will use the following pairwise cost function to compute the SP-Score.

	0	1	$\circ$	$\bullet$	$-$
0	2	2	1	2	1
1	2	2	2	1	1
$\circ$	1	2	0	2	1
$\bullet$	2	1	2	0	1
$-$	1	1	1	1	0

$$(34.26)$$

Here, ∘ and • are two new symbols, so the alphabet for our MSA problem is $\{0, 1, \circ, \bullet, -\}$. An interesting thing about this cost function is that any MSA of $|\mathcal{S}|$ binary strings will have cost $(|\mathcal{S}| - 1)\|\mathcal{S}\|$ no matter where we put the gaps.

$$\text{vs.} \tag{34.27}$$

The left alignment has score 12 and so does the right alignment. Exercise 34.4 asks you to write a formal proof of this.

Our reduction from SHORTEST SUPERSEQUENCE to MSA adds two more sequences to align:

$$\mathcal{S} \cup \left\{ \circ^i, \bullet^j \right\}, \tag{34.28}$$

where the notation $\circ^i$ means repeat character $\circ$ i times.

What should we choose for i and j? Well, where we will end up is that the locations of the ∘ and • characters will give us the supersequence, where ∘ corresponds to locations of 0s and • corresponds to locations of 1s in the supersequence. Since we don't know how many 0s and 1s will end up in the supersequence, we take the strategy of creating many MSA problems, one for each possible choice of i and j.

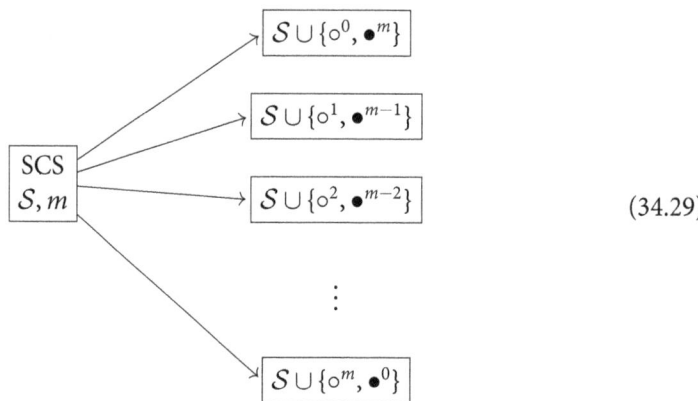

$$\tag{34.29}$$

That is, we're going to create an MSA instance for every $i + j = m$. If any one of these has a particular score, we'll know there is a short supersequence.

Suppose there is a supersequence S of length m with i 0s and j 1s. Then align each $s_i \in \mathcal{S}$ to S (which is possible since S is a supersequence of $\mathcal{S}$). Pick the instance of MSA with $\circ^i$ and $\bullet^j$, and align each ∘ to a 0 in S and each • to a 1 in S. We end up with an alignment that looks like (34.30).

$$\tag{34.30}$$

Focus on the pairs of the SP-score that have at least one character of the pair in the $\circ^i$ or $\bullet^j$ string. For each 0 in $\mathcal{S}$, the cost to align to the $\circ$ and $\bullet$ strings is 2: 1 for the $0 \leftrightarrow \circ$ pair and 1 for the $1 \leftrightarrow -$ pair. The same is true for each 1 in $\mathcal{S}$. For each $-$ added to the sequences in $\mathcal{S}$, we have cost 1: 1 for the $- \leftrightarrow \circ$ or $\bullet$, and 0 for each $- \leftrightarrow -$ pair. Additionally, each $\circ \leftrightarrow -$ and $- \leftrightarrow \bullet$ pair between the last two sequences adds 1 to the cost.

Summing these all up, we have cost

- m for pairs between the $\circ^i$ and $\bullet^j$ strings,
- $+2\|\mathcal{S}\|$ for the alignment of each 0/1 to the $\circ/\bullet$ strings, and
- $+(|\mathcal{S}|m - \|\mathcal{S}\|)$ for the gap characters in $\mathcal{S}$: there are this many gap characters in the $\mathcal{S}$ strings, and each costs 1 to align to one of the $\circ/\bullet$.

This equals $(|\mathcal{S}| + 1)m + \|\mathcal{S}\|$.

Since the contribution of the alignment excluding the $\circ$ and $\bullet$ sequences is $(|\mathcal{S}| - 1)\|\mathcal{S}\|$, the total cost of this alignment is $|\mathcal{S}|\|\mathcal{S}\| + (|\mathcal{S}| + 1)m$. Let this value be c.

Conversely, suppose that for some i and j (with $i + j = m$) there is an alignment with cost c. Since $c = (|\mathcal{S}| + 1)m + \|\mathcal{S}\| + (|\mathcal{S}| - 1)\|\mathcal{S}\|$, and the contribution to the score of the alignment of the sequences in $\mathcal{S}$ is $(|\mathcal{S}| - 1)\|\mathcal{S}\|$, the alignment pairs with the $\circ$ and $\bullet$ sequences must cost $(|\mathcal{S}| + 1)m + \|\mathcal{S}\|$.

No matter the alignment, each character of $\mathcal{S}$ and the $\circ/\bullet$ sequences adds at least 2 to the cost from the pairs with $\circ/\bullet$. Each gap in adds at least 1 from the pairs with $\circ/\bullet$.

If the width of the alignment is m, then there are $|\mathcal{S}|m$ characters in the alignment for strings in $\mathcal{S}$; whether those characters are gaps or 0/1, they will cost at least 1, hence accounting for the $|\mathcal{S}|m$ term. Of the remaining $m + \|\mathcal{S}\|$ cost, the only way to achieve it is to have only one $\circ$ or $\bullet$ in each column (cost m), and most importantly, to achieve that each 0 is aligned to a $\circ$ and each 1 is aligned to a $\bullet$ (total additional cost $\|\mathcal{S}\|$).

Because of this configuration, we can create a sequence S where $S[i] = 0$ if column i contains a $\circ$ and $S[i] = 1$ if column i contains a $\bullet$. By construction (and the argument in the previous paragraph), S is a supersequence of length m.

Therefore, we have shown that if one of the problems in (34.29) has MSA of cost c then the SHORTEST SUPERSEQUENCE problem has a solution of length m and vice versa.

If we had a polynomial-time algorithm for MSA, we could use it to solve each of the problems in (34.29), and therefore provide a polynomial-time algorithm for SHORTEST SUPERSEQUENCE. □

The cost function (34.26) isn't very realistic for any application. The proof doesn't show that MSA is hard for every possible cost function, just for the settings where either (a) we use that particular cost function, or (b) the cost function is part of the input to the MSA instance. Later work [Just, 2001; Bonizzoni and Vedova, 2001] showed that SP-score-based MSA problems with more reasonable cost functions are also NP-complete.

Further, the proof of Theorem 34.8 doesn't exactly follow the approach of Section 34.1. The technique of giving an algorithm for a problem assuming we have an algorithm we can repeatedly call for another problem is a *Turing reduction*, while Section 34.1 describes the more standard *Karp reductions*. Nevertheless, this shows that if we had a fast algorithm for MSA, we'd have one for SHORTEST SUPERSEQUENCE.

34.5 Summary and notes

That SHORTEST SUPERSTRING is NP-complete was shown by Gallant et al. [1980], who also show other restricted variants are also NP-complete. The proof that SHORTEST SUPER-SEQUENCE is NP-complete is from Maier [1978], who also dealt with the version with a restricted alphabet. The proof that MSA is NP-complete is from Wang and Jiang [1994], and subsequent work [e.g., Bonizzoni and Vedova, 2001; Just, 2001; Caucchiolo and Cicalese, 2023] has broadened the classes of MSA problems that are NP-complete.

There are, of course, many other known NP-complete string problems; hopefully this chapter has given you a flavor of some of those types of proofs.

> Presentation Notes
>
> Our presentation of these hardness results follows the original proofs given for each of them.

34.6 Exercises

34.1   Show how to modify the proof that SHORTEST SUPERSTRING is NP-complete so that it only uses characters from the alphabet $\{0, 1, !\}$. Hint: encode the symbols in binary and use ! as a separator.

34.2   Show that we can assume that the A in each edge string is aligned completely to either the right or left instance of A in T in the proof of Theorem 34.7.

34.3   Prove that the cost in (34.26) satisfies the triangle inequality.

34.4   Give a formal proof that any MSA of $|S|$ binary strings will have cost $(|S| - 1)\|S\|$ using the cost function in (34.26).

CHAPTER 35

Conclusion

We've covered a lot of ground in string algorithms. If you understand the topics in this book you are well positioned to begin research in string algorithms and related areas as well as to understand the primary literature related to these and other string algorithm topics.

Of course, there are many more topics that we didn't cover here. An important area that we haven't discussed is parallel string algorithms, though more info can be found, for example, in Kulla and Sanders [2007]; Galil [1992, 1995]; Breslauer and Galil [1990]; Galil [1984, 1986]; Czumaj et al. [1995]; Breslauer and Galil [1991]; Apostolico and Breslauer [1996] among other publications. Online string algorithms is another wide area that we haven't covered—these algorithms must make decisions before seeing the full input. Example papers in this area include Faro and Lecroq [2013]; Galil [1976]; Breslauer and Galil [2014]; Galil [1981]. Quantum string algorithms are explored in, e.g., Akmal and Jin [2022]; Boroujeni et al. [2021]; Ramesh and Vinay [2003]; Wang and Ying [2020].

Though each string problem often requires some reasoning that is unique to that problem, we have seen a few general algorithmic design techniques that are useful in multiple string algorithms and also more broadly. That is one of the nice aspects of focusing on string algorithms: within that relatively narrow context, a number of fundamental algorithmic ideas can be explored.

The very general technique of **dynamic programming** arose in several contexts, most fundamentally in the edit distance (Part II), but also in algorithms like Shift-And (Section 5.2), and those related to hidden Markov models (Chapter 27). Dynamic programming is one of the most useful algorithm design techniques beyond string algorithms, and any algorithm textbook is likely to have a significant discussion of the topic.

Automata such as deterministic finite-state automata were useful ways to think about a problem or an algorithm. We saw this, for example, in the KMP algorithm, and also when discussing regular languages (Chapter 26), hidden Markov models (Chapter 27), and even (viewed broadly enough) genome graphs (Chapter 29). Aho-Corasick (Chapter 6) was also a DFA in disguise.

The idea of **precomputation** was key to several of the algorithms. Most of the standard exact matching algorithms performed some kind of precomputation on the pattern, by for example, computing the Z values (Chapter 2), or the arrays used in Boyer-Moore (Chapter 3). Building a suffix tree, suffix array, or wavelet tree was also often a precomputation step that we undertook before observing the complete input problem. More sophisticated uses of precomputation were used in the Four Russians speed up (Chapter 9) for faster dynamic programming and in the RRR data structure (Chapter 19) to store its tables.

We also used the idea of **chunking** (not a standard term): breaking the input into smaller chunks (often of logarithmic size) and dealing with each of these chunks separately. The RRR data structure, the FM-index, and the Four Russians' speed up were examples of this idea. The related idea of *encoding* groups of characters (as in the suffix array construction algorithm) was also useful.

Hashing also played a role in several algorithms. Obviously, Part IV relied in many places on hash functions, as did Rabin-Karp fingerprinting from Chapter 5.

We also used **amortized analysis** in a number of the proofs of running time. Amortized analysis is another very general and important algorithmic technique with use far afield of string algorithms. Relatedly, and more specific to string algorithms, we often analyzed algorithms by counting different events separately, for example, counting match events and mismatch events separately.

Bibliography

M. I. Abouelhoda, S. Kurtz, and E. Ohlebusch. Replacing suffix trees with enhanced suffix arrays. *Journal of Discrete Algorithms*, 2(1): 53–86, 2004.

M. Agrawal, N. Kayal, and N. Saxena. PRIMES is in P. *Annals of Mathematics*, 160(2):781–793, 2004.

A. V. Aho and M. J. Corasick. Efficient string matching: An aid to bibliographic search. *Commun. ACM*, 18(6):333–340, 1975.

S. Akmal and C. Jin. Near-optimal quantum algorithms for string problems. In *Proceedings of the 2022 Annual ACM-SIAM Symposium on Discrete Algorithms (SODA)*, pages 2791–2832, 2022.

J. Alanko and T. Norri. Greedy shortest common superstring approximation in compact space. In G. Fici, M. Sciortino, and R. Venturini, editors, *String Processing and Information Retrieval*, pages 1–13, Cham, 2017. Springer International Publishing.

F. Almodaresi, P. Pandey, and R. Patro. Rainbowfish: a succinct colored de Bruijn graph representation. In *17th International Workshop on Algorithms in Bioinformatics (WABI 2017)*. Schloss Dagstuhl-Leibniz-Zentrum fuer Informatik, 2017.

F. Almodaresi, P. Pandey, M. Ferdman, R. Johnson, and R. Patro. An efficient, scalable, and exact representation of high-dimensional color information enabled using de Bruijn graph search. *Journal of Computational Biology*, 27(4):485–499, 2020.

S. Altschul, W. Gish, W. Miller, E. Myers, and D. Lipman. Basic local alignment search tool. *J. Mol. Biol.*, pages 403–410, 1990.

S. Altschul, T. Madden, A. Schäffer, J. Zhang, Z. Zhang, W. Miller, and D. Lipman. Gapped BLAST and PSI-BLAST: a new generation of protein database search programs. *Nucleic Acids Res.*, 25:3389–3402, 1997.

A. Apostolico and D. Breslauer. An optimal $O(\log \log n)$-time parallel algorithm for detecting all squares in a string. *SIAM J. Comput.*, 25(6):1318–1331, 1996.

L. Arlazarov, E. A. Dinic, M. A. Kronrod, and I. A. Faradzev. On economic construction of the transitive closure of a directed graph. *Dokl. Akad, Nauk SSSR*, 194:487–488, 1970. [in Russian]. English translation, Soviet Math. Dokl. 11(5):1209–1210 (1970).

D. Arroyuelo, S. González, and M. Oyarzún. Compressed self-indices supporting conjunctive queries on document collections. In *Proc. 17th SPIRE*, pages 43–54, 2010.

M. A. Babenko, P. Gawrychowski, T. Kociumaka, I. I. Kolesnichenko, and T. Starikovskaya. Computing minimal and maximal suffixes of a substring. *Theor. Comput. Sci.*, 638:112–121, 2016.

A. Backurs and P. Indyk. Edit distance cannot be computed in strongly subquadratic time (unless SETH is false). In *Proceedings of the Forty-Seventh Annual ACM Symposium on Theory of Computing*, STOC '15, pages 51–58, New York, NY, USA, 2015. Association for Computing Machinery.

R. A. Baeza-Yates and G. H. Gonnet. A new approach to text searching. *Communication of ACM*, 35(10):74–82, 1992.

Z. Bar-Yossef, T. Jayram, R. Krauthgamer, and R. Kumar. Approximating edit distance efficiently. In *45th Annual IEEE Symposium on Foundations of Computer Science*, pages 550–559, 2004.

L. E. Baum. An inequality and associated maximization technique in statistical estimation for probabilistic functions of Markov processes. In *Inequalities III: Proceedings of the Third Symposium on Inequalities.*, pages 1–8, 1972.

J. L. Bentley, D. D. Sleator, R. E. Tarjan, and V. K. Wei. A locally adaptive data compression scheme. *Communications of the ACM*, 29(4), 1986.

P. Bille, R. Fagerberg, and I. L. Gørtz. Improved approximate string matching and regular expression matching on Ziv-Lempel compressed texts. *ACM Trans. Algorithms*, 6(1):3:1–14, 2010.

P. Bille, P. Gawrychowski, I. L. Gørtz, G. M. Landau, and O. Weimann. Longest common extensions in trees. In *Proceedings of the 26th Annual Symposium on Combinatorial Pattern Matching (CPM 2015)*, pages 52–64, 2015.

O. Birenzwige, S. Golan, and E. Porat. Locally consistent parsing for text indexing in small space. In *Proceedings of the Thirty-First Annual ACM-SIAM Symposium on Discrete Algorithms*, SODA '20, page 607–626, USA, 2020. Society for Industrial and Applied Mathematics.

B. H. Bloom. Space/Time trade-offs in hash coding with allowable errors. *Communications of the ACM*, 13(7):422–426, 1970.

A. Blum, T. Jiang, M. Li, J. Tromp, and M. Yannakakis. Linear approximation of shortest superstrings. *Journal of the ACM (JACM)*, 41(4):630–647, 1994.

H.-J. Boehm, R. Atkinson, and M. Plass. Ropes: an alternative to strings. *Software—Practice and Experience*, 25(12):1315–1330, 1995.

P. Bonizzoni and G. D. Vedova. The complexity of multiple sequence alignment with SP-score that is a metric. *Theoretical Computer Science*, 259(1):63–79, 2001.

K. S. Booth. Lexicographically least circular substrings. *Inf. Process. Lett.*, 10(4/5):240–242, 1980.

A. Borodin and R. El-Yaniv. *Online Computation and Competitive Analysis*. Cambridge University Press, USA, 1998.

M. Boroujeni, S. Ehsani, M. Ghodsi, M. Hajiaghayi, and S. Seddighin. Approximating edit distance in truly subquadratic time: Quantum and MapReduce. *J. ACM*, 68(3):19:1–41, 2021.

C. Boucher, T. Gagie, A. Kuhnle, et al. Prefix-free parsing for building big BWTs. *Algorithms Mol. Biol.*, 14:13, 2019.

A. Bowe. RRR: a succinct Rank/Select index for bit vectors, 2011. URL https://www.alexbowe.com/rrr/.

R. S. Boyer and J. S. Moore. A fast string searching algorithm. *Commun. ACM*, 20(10):762–772, 1977.

D. Breslauer and Z. Galil. An optimal $o(\log n)$ time parallel string matching algorithm. *SIAM J. Comput.*, 19(6):1051–1058, 1990.

D. Breslauer and Z. Galil. A lower bound for parallel string matching. In *Proceedings of the Twenty-Third Annual ACM Symposium on Theory of Computing*, STOC '91, pages 439–443, New York, NY, USA, 1991. Association for Computing Machinery.

D. Breslauer and Z. Galil. Real-time streaming string-matching. *ACM Trans. Algorithms*, 10(4):22, 2014.

A. Broder and M. Mitzenmacher. Network applications of Bloom filters: A survey. *Internet Mathematics*, 1(4):485–509, 2005.

A. Z. Broder. On the resemblance and containment of documents. In *Proceedings of Compression and Complexity of Sequences*, pages 21–29, 1997.

C. Burge and S. Karlin. Prediction of complete gene structures in human genomic DNA. *J. Mol. Biol.*, 268(1):78–94, 1997.

M. Burrows and D. J. Wheeler. A block sorting lossless data compression algorithm. Technical Report 124, Digital Equipment Corporation, 1994.

A. Carlson, J. Betteridge, B. Kisiel, B. Settles, E. R. H. Jr., and T. M. Mitchell. Toward an architecture for never-ending language learning. In *Proceedings of the Twenty-Fourth Conference on Artificial Intelligence (AAAI 2010)*, pages 1306–1313, 2010.

H. Carrillo and D. Lipman. The multiple sequence alignment in biology. *SIAM J. Appl. Math.*, 48:1073–1082, 1988.

A. Caucchiolo and F. Cicalese. Hardness and approximation of multiple sequence alignment with column score. *Theoretical Computer Science*, 946:113683, 2023.

I. W. Chang and E. L. Lawler. Sublinear approximate string matching and biological applications. *Algorithmica*, 12:327–344, 1994.

M. Charikar. Similarity estimation techniques from rounding algorithms. In *STOC '02: Proceedings of the thiry-fourth annual ACM symposium on Theory of computing*, pages 380–388, 2002.

B. Chazelle, J. Kilian, R. Rubinfeld, and A. Tal. The Bloomier filter: An efficient data structure for static support lookup tables. In *Proceedings of the Fifteenth Annual ACM-SIAM Symposium on Discrete Algorithms*, SODA '04, pages 30–39. Society for Industrial and Applied Mathematics, 2004.

N. Cherniavsky and R. Ladner. Grammar-based compression of DNA sequences. Technical Report 2007-05-02, UW CSE, 2004.

N. Chomsky. Three models for the description of language. *IRE Transactions on Information Theory*, 2(3):113–124, 1956.

A. Cicherski and N. Dojer. From de Bruijn graphs to variation graphs—relationships between pangenome models. In F. M. Nardini et al., editors, *SPIRE 2023, LNCS 14240*, pages 114–128, 2023.

F. Claude and G. Navarro. Practical rank/select queries over arbitrary sequences. In *Proc. 15th SPIRE, LNCS 5280*, pages 176–187, 2008.

F. Claude, P. Nicholson, and D. Seco. Space efficient wavelet tree construction. In *Proc. 18th SPIRE*, pages 185–196, 2011.

J. Cocke and J. T. Schwartz. Programming languages and their compilers: Preliminary notes. Technical report, Courant Institute of Mathematical Sciences, New York University, 1970.

R. Cole. Tight bounds on the complexity of the Boyer-Moore string matching algorithm. In *Proceedings of the Second Annual ACM-SIAM Symposium on Discrete Algorithms*, SODA '91, pages 224–233. Society for Industrial and Applied Mathematics, 1991.

Computational Pan-Genomics Consortium. Computational pangenomics: status, promises and challenges. *Briefings in Bioinformatics*, 19(1):118–135, 2018.

M. E. Consens, C. Dufault, M. Wainberg, D. Forster, M. Karimzadeh, H. Goodarzi, F. J. Theis, A. Moses, and B. Wang. To transformers and beyond: Large language models for the genome. *ArXiv*, page 2311.07621, 2023.

S. Cook. The complexity of theorem proving procedures. In *Proceedings of the Third Annual ACM Symposium on Theory of Computing*, pages 151–158, 1971.

T. H. Cormen, C. E. Leiserson, R. L. Rivest, and C. Stein. *Introduction to Algorithms*. MIT Press, 2009.

M. Crochemore and W. Rytter. *Jewels of Stringology*. World Scientific Publishing Co. Pte. Ltd., Singapore, 2003.

M. Crochemore, C. Hancart, and T. Lecroq. *Algorithms on Strings*. Cambridge University Press, New York, 2007.

C. Crowley. Data structures for text sequences. Technical report, University of New Mexico, 1998.

H. Cui, C. Wang, H. Maan, et al. scGPT: toward building a foundation model for single-cell multi-omics using generative AI. *Nat Methods*, 2024.

A. Czumaj, Z. Galil, L. Gąsieniec, K. Park, and W. Plandowski. Work-time-optimal parallel algorithms for string problems. In *Proceedings of the Twenty-Seventh Annual ACM Symposium on Theory of Computing*, STOC '95, pages 713–722, New York, NY, USA, 1995. Association for Computing Machinery.

N. de Bruijn. A combinatorial problem. *Proceedings of the Section of Sciences of the Koninklijke Nederlandse Akademie Van Wetenschappen Te Amsterdam*, 49(7):758–764, 1946.

D. DeBlasio, F. Gbosibo, C. Kingsford, and G. Marçais. Practical universal k-mer sets for minimizer schemes. In *Proceedings of the 10th ACM International Conference on Bioinformatics, Computational Biology and Health Informatics*, pages 167–176, 2019.

A. Delcher, D. Harmon, S. Kasif, O. White, and S. Salzberg. Improved microbial gene identification with GLIMMER. *Nucleic Acids Research*, 27(23):4636–4641, 1999.

A. Delcher, K. Bratke, E. Powers, and S. Salzberg. Identifying bacterial genes and endosymbiont DNA with Glimmer. *Bioinformatics*, 23(6):673–679, 2007.

S. Deorowicz, M. Kokot, S. Grabowski, et al. KMC 2: Fast and resource-frugal k-mer counting. *Bioinformatics*, 31(10):1569–1576, 2015.

J. Devlin, M.-W. Chang, K. Lee, and K. Toutanova. BERT: pre-training of deep bidirectional transformers for language understanding. arXiv:1810.04805v2, 2018.

R. Durbin, S. Eddy, A. Krogh, and G. Mitchison. *Biological Sequence Analysis*. Cambridge University Press, 1998.

A. Dutta, D. Pellow, and R. Shamir. Parameterized Syncmer schemes improve long-read mapping. *PLoS Comput. Biol.*, 18(10): e1010638, 2022.

A. Ebrahimpour Boroojeny, A. Shrestha, A. Sharifi-Zarchi, S. R. Gallagher, S. C. Sahinalp, and H. Chitsaz. Graph traversal edit distance and extensions. *Journal of Computational Biology*, 27(3):317–329, 2020.

S. R. Eddy and R. Durbin. RNA sequence analysis using covariance models. *Nuc. Acids Res.*, 22(11):2079–2088, 1994.

R. Edgar. Syncmers are more sensitive than minimizers for selecting conserved k-mers in biological sequences. *PeerJ*, 9:e10805, 2021.

B. Ekim, B. Berger, and Y. Orenstein. A randomized parallel algorithm for efficiently finding near-optimal universal hitting sets. In R. Schwartz et al., editors, *International Conference on Research in Computational Molecular Biology*, pages 37–53. Springer, 2020.

M. Equi, T. Norri, J. Alanko, et al. Algorithms and complexity on indexing founder graphs. *Algorithmica*, 85:1586–1623, 2023.

L. Fan, P. Cao, J. Almeida, and A. Broder. Summary cache: a scalable wide-area web cache sharing protocol. *IEEE / ACM Transactions on Networking*, 8:281–293, 2000.

M. Farach. Optimal suffix tree construction with large alphabets. In *Proceedings of the 38th Annual Symposium on Foundations of Computer Science (FOCS 1997)*, pages 137–143, 1997.

M. Farach and M. Thorup. String matching in Lempel-Ziv compressed strings. In *Proceedings of the Twenty-Seventh Annual ACM Symposium on Theory of Computing*, STOC '95, pages 703–712, New York, NY, USA, 1995. Association for Computing Machinery.

M. Farach and M. Thorup. String matching in Lempel-Ziv compressed strings. *Algorithmica*, 20:388–404, 1998.

S. Faro and T. Lecroq. The exact online string matching problem: A review of the most recent results. *ACM Comput. Surv.*, 45(2), 2013.

H. Ferrada, T. Gagie, S. Gog, and S. J. Puglisi. Relative Lempel-Ziv with constant-time random access. In *International Symposium on String Processing and Information Retrieval*, pages 13–17. Springer, 2014.

P. Ferragina and G. Manzini. Opportunistic data structures with applications. In *Proceedings of the 41st Annual Symposium on Foundations of Computer Science*, pages 390–398, 2000.

P. Ferragina and G. Manzini. Indexing compressed texts. *J. ACM*, 52(4):552–581, 2005.

P. Ferragina, G. Manzini, V. Mäkinen, and G. Navarro. An alphabet-friendly FM-index. In *Proc. 11th SPIRE, LNCS 3246*, pages 150–160, 2004.

P. Ferragina, R. Giancarlo, and G. Manzini. The myriad virtues of Wavelet Trees. *Information and Computation*, 207(8):849–866, 2009a.

P. Ferragina, F. Luccio, G. Manzini, and S. Muthukrishnan. Compressing and indexing labeled trees, with applications. *J. ACM*, 57(1), 2009b.

D. Ferrucci, E. Brown, J. Chu-Carroll, J. Fan, D. Gondek, A. A. Kalyanpur, A. Lally, J. W. Murdock, E. Nyberg, J. Prager, N. Schlaefer, and C. Welty. Building Watson: An overview of the DeepQA project. *AI Magazine*, 2010.

C. A. Finseth. *The Craft of Text Editing*. 1999. http://www.finseth.com/craft.

C. Flye Sainte-Marie. Solution to question nr. 48. *L'Intermédiaire Des Mathématiciens*, 1:107–110, 1894.

G. D. Forney Jr. The Viterbi algorithm. *Proc. IEEE*, 61:268–278, 1973.

H. N. Gabow and R. E. Tarjan. A linear-time algorithm for a special case of disjoint set union. In *Proceedings of the 15th ACM Symposium on Theory of Computing (STOC)*, pages 246–251, 1983.

T. Gagie, G. Navarro, and S. J. Puglisi. New algorithms on wavelet trees and applications to information retrieval. *Theoretical Computer Science*, 426-427:25–41, 2012.

T. Gagie, S. J. Puglisi, and D. Valenzuela. Analyzing relative Lempel-Ziv reference construction. In *International Symposium on String Processing and Information Retrieval*, pages 160–165. Springer, 2016.

T. Gagie, G. Manzini, and J. Sirén. Wheeler graphs: A framework for BWT-based data structures. *Theor. Comput. Sci.*, 698:67–78, 2017.

T. Gagie, G. Navarro, and N. Prezza. Fully functional suffix trees and optimal text searching in BWT-runs bounded space. *J. ACM*, 67:2:1–2:54, 2020.

Z. Galil. Real-time algorithms for string-matching and palindrome recognition. In *Proceedings of the Eighth Annual ACM Symposium on Theory of Computing*, STOC '76, pages 161–173, New York, NY, USA, 1976. Association for Computing Machinery.

Z. Galil. On improving the worst case running time of the Boyer-Moore string matching algorithm. *Commun. ACM*, 22(9): 505–508, 1979.

Z. Galil. String matching in real time. *J. ACM*, 28(1):134–149, 1981.

Z. Galil. Optimal parallel algorithms for string matching. In *Proceedings of the Sixteenth Annual ACM Symposium on Theory of Computing*, STOC '84, pages 240–248, New York, NY, USA, 1984. Association for Computing Machinery.

Z. Galil. Optimal parallel algorithms for string matching. *Inf. Control*, 67(1–3):144–157, 1986.

Z. Galil. A constant-time optimal parallel string-matching algorithm. In *Proceedings of the Twenty-Fourth Annual ACM Symposium on Theory of Computing*, STOC '92, pages 69–76, New York, NY, USA, 1992. Association for Computing Machinery.

Z. Galil. A constant-time optimal parallel string-matching algorithm. *J. ACM*, 42(4):908–918, 1995.

J. Gallant, D. Maier, and J. Astorer. On finding minimal length superstrings. *Journal of Computer and System Sciences*, 20(1):50–58, 1980.

M. R. Garey and D. S. Johnson. *Computers and Intractability*. W.H. Freeman and Company, New York, 1979.

E. Garrison, J. Sirén, A. M. Novak, G. Hickey, J. M. Eizenga, E. T. Dawson, W. Jones, S. Garg, C. Markello, M. F. Lin, B. Paten, and R. Durbin. Variation graph toolkit improves read mapping by representing genetic variation in the reference. *Nature Biotechnology*, 36(9):875–879, 2018.

P. Gawrychowski. Optimal pattern matching in LZW compressed strings. In *Proceedings of the Twenty-Second Annual ACM-SIAM Symposium on Discrete Algorithms*, SODA '11, pages 362–372. Society for Industrial and Applied Mathematics, 2011a.

P. Gawrychowski. Pattern matching in Lempel-Ziv compressed strings: Fast, simple, and deterministic. In C. Demetrescu and H. M. M., editors, *Algorithms – ESA 2011*, pages 421–432, Berlin, Heidelberg, 2011b. Springer Berlin Heidelberg.

P. Gawrychowski. Optimal pattern matching in LZW compressed strings. *ACM Trans. Algorithms*, 9(3), 2013.

O. Gotoh. An improved algorithm for matching biological sequences. *Journal of Molecular Biology*, 162(3):705–708, 1982.

S. Grabowski and M. Raniszewski. Sampling the suffix array with minimizers. In C. Iliopoulos, S. Puglisi, and E. Yilmaz, editors, *String Processing and Information Retrieval*, number 9309. In Lecture Notes in Computer Science, pages 287–298. Springer International Publishing, 2013.

R. Grossi, A. Gupta, and J. S. Vitter. High-order entropy-compressed text indexes. In *Proceedings of the 14th Annual SIAM/ACM Symposium on Discrete Algorithms (SODA)*, pages 841–850, 2003.

J. Guo, D. Hermelin, and C. Komusiewicz. Local search for string problems: Brute-force is essentially optimal. *Theor. Comput. Sci.*, 525:30–41, 2014.

D. Gusfield. Efficient methods for multiple sequence alignment with guaranteed error bounds. *Bull. Math. Biol.*, 55:141–154, 1993.

D. Gusfield. *Algorithms on Strings, Trees, and Sequences: Computer Science and Computational Biology*. Cambridge Press, 1997.

D. S. Hirschberg. A linear space algorithm for computing maximal common subsequences. *Communications of the ACM*, 18(6):341–343, 1975.

M. Hoang, H. Zheng, and C. Kingsford. Differentiable learning of sequence-specific minimizer schemes with DeepMinimizer. *J. Comput. Biol.*, 29(12):1288–1304, 2022.

G. Holley and P. Melsted. Bifrost: highly parallel construction and indexing of colored and compacted de Bruijn graphs. *Genome Biology*, 21(1):249–269, 2020.

J. Hopcroft. An $n \log n$ algorithm for minimizing states in a finite automaton. In *Proceedings of the International Symposium on the Theory of Machines and Computations*, pages 189–196, 1971.

J. E. Hopcroft, W. J. Paul, and L. G. Valiant. On time versus space and other related problems. In *Proceedings of the 16th Annual Symposium on Foundations of Computer Science*, pages 57–64, 1975.

D. A. Huffman. A method for the construction of minimum-redundancy codes. *Proceedings of the IRE*, pages 1098–1101, 1952.

C. S. Iliopoulos and W. F. Smyth. Optimal algorithms for computing the canonical form of a circular string. *Theor. Comput. Sci.*, 92(1):87–105, 1992.

P. Indyk. A small approximately min-wise independent family of hash functions. *J. Algorithms*, 38(1):84–90, 2001.

P. Indyk and R. Motwani. Approximate nearest neighbors: Towards removing the curse of dimensionality. In *Proceedings of the Thirtieth Annual ACM Symposium on Theory of Computing*, STOC '98, pages 604–613, New York, NY, USA, 1998. Association for Computing Machinery.

Z. Iqbal, M. Caccamo, I. Turner, P. Flicek, and G. McVean. De novo assembly and genotyping of variants using colored de Bruijn graphs. *Nat. Genet.*, 44(2):226–232, 2012.

P. Ivanov, B. Bichsel, H. Mustafa, A. Kahles, G. Rätsch, and M. Vechev. AStarix: Fast and optimal sequence-to-graph alignment. In R. Schwartz, editor, *Research in Computational Molecular Biology*, pages 104–119. Springer International Publishing, 2020.

C. Jain, H. Zhang, Y. Gao, and S. Aluru. On the complexity of sequence-to-graph alignment. *J. Comput. Biol.*, 27(4):640–654, 2020.

Y. Ji, Z. Zhou, H. Liu, and R. V. Davuluri. DNABERT: pre-trained bidirectional encoder representations from transformers model for DNA-language in genome. *Bioinformatics*, 37(15):2112–2120, 2021.

N. C. Jones and P. A. Pevzner. *An Introduction to Bioinformatics Algorithms*. MIT Press, 2004.

W. Just. Computational complexity of multiple sequence alignment with SP-score. *J. Comput. Biol.*, 8(6):615–623, 2001.

J. Kärkkäinen. Fast BWT in small space by blockwise suffix sorting. *Theoretical Computer Science*, 387(3):249–257, 2007.

J. Kärkkäinen and P. Sanders. Simple linear work suffix array construction. In *Automata, Languages and Programming. ICALP 2003, Lecture Notes in Computer Science, vol 2719*, 2003.

J. Kärkkäinen, G. Navarro, and E. Ukkonen. Approximate string matching on Ziv-Lempel compressed text. *Journal of Discrete Algorithms*, 1(3):313–338, 2003.

R. M. Karp and M. O. Rabin. Efficient randomized pattern-matching algorithms. *IBM Journal of Research and Development*, 31(2):249–260, 1987.

R. M. Karp, R. E. Miller, and A. L. Rosenberg. Rapid identification of repeated patterns in strings, trees and arrays. In *Proceedings of the Fourth Annual ACM Symposium on Theory of Computing*, STOC '72, pages 125–136, New York, NY, USA, 1972. Association for Computing Machinery.

T. Kasai, G. Lee, H. Arimura, S. Arikawa, and K. Park. Linear-time longest-common-prefix computation in suffix arrays and its applications. In A. Amir, editor, *Combinatorial Pattern Matching*, pages 181–192, Berlin, Heidelberg, 2001. Springer Berlin Heidelberg.

T. Kasami. An efficient recognition and syntax-analysis algorithm for context-free languages. Technical Report R-257, Coordinated Science Laboratory, University of Illinois—Urbana, Illinois, 1966.

Z. Khan, J. S. Bloom, L. Kruglyak, and M. Singh. A practical algorithm for finding maximal exact matches in large sequence datasets using sparse suffix arrays. *Bioinformatics (Oxford, England)*, 25(13):1609–1616, 2009.

B. Kille, R. Groot Koerkamp, D. McAdams, A. Liu, and T. J. Treangen. A near-tight lower bound on the density of forward sampling schemes. *Bioinformatics*, 41(1):btae736, 2025.

J. Kleinberg and E. Tardos. *Algorithm Design*. Pearson Education, 2006.

D. Knuth, J. H. Morris, and V. Pratt. Fast pattern matching in strings. *SIAM Journal on Computing*, 6(2):323–350, 1977.

A. Kuhnle, T. Mun, C. Boucher, et al. Efficient construction of a complete index for pan-genomics read alignment. *J. Comput. Biol.*, 27:500–513, 2020.

F. Kulla and P. Sanders. Scalable parallel suffix array construction. *Parallel Computing*, 33(9):605–612, 2007.

S. Kurtz. Reducing the space requirement of suffix trees. *Software: Practice and Experience*, 29(13):1149–1171, 1999.

S. Kuruppu, S. J. Puglisi, and J. Zobel. Optimized relative Lempel-Ziv compression of genomes. In *Proceedings of the Thirty-Fourth Australasian Computer Science Conference-Volume 113*, pages 91–98. Australian Computer Society, Inc., 2011.

G. M. Landau and U. Vishkin. Introducing efficient parallelism into approximate string matching and a new serial algorithm. In *Proceedings of the Eighteenth Annual ACM Symposium on Theory of Computing*, STOC '86, pages 220–230, New York, NY, USA, 1986. Association for Computing Machinery.

B. Langmead. Introduction to the Burrows-Wheeler Transform and FM Index, 2013. https://www.cs.jhu.edu/~langmea/resources/bwt_fm.pdf.

B. Langmead. Lecture notes, 2024. https://www.langmead-lab.org/teaching.html.

B. Langmead and S. Salzberg. Fast gapped-read alignment with Bowtie 2. *Nature Methods*, 9:357–359, 2012.

B. Langmead, C. Trapnell, M. Pop, and S. Salzberg. Ultrafast and memory-efficient alignment of short DNA sequences to the human genome. *Genome Biol.*, 10:R25, 2009.

K. Lari and S. Young. The estimation of stochastic context-free grammars using the Inside-Outside algorithm. *Computer Speech & Language*, 4(1):35–56, 1990.

N. Larsson and A. Moffat. Off-line dictionary-based compression. *Proceedings of the IEEE*, 88(11):1722–1732, 2000.

C. Lee, C. Grasso, and M. F. Sharlow. Multiple sequence alignment using partial order graphs. *Bioinformatics*, 18(3):452–464, 2002.

V. I. Levenshtein. Binary codes capable of correcting deletions, insertions, and reversals. *Soviet Physics Doklady*, 10(8):707–710, 1966.

L. Levin. Universal search problems. *Problems of Information Transmission (in Russian)*, 9(3):115–116, 1973. Translated into English by B. A. Trakhtenbrot (1984). "A survey of Russian approaches to perebor (brute-force searches) algorithms." Annals of the History of Computing. 6(4):384–400.

H. Li. Aligning sequence reads, clone sequences and assembly contigs with BWA-MEM. arXiv:1303.3997v2 [q-bio.GN], 2013.

H. Li. Minimap and miniasm: Fast mapping and de novo assembly for noisy long sequences. *Bioinformatics*, 32(14):2103–2110, 2016.

H. Li and I. Birol. Minimap2: Pairwise alignment for nucleotide sequences. *Bioinformatics*, 34(18):3094–3100, 2018.

H. Li and R. Durbin. Fast and accurate short read alignment with Burrows-Wheeler transform. *Bioinformatics*, 25:1754–1760, 2009.

H. Li and R. Durbin. Fast and accurate long-read alignment with Burrows-Wheeler transform. *Bioinformatics*, 26:589–595, 2010.

H. Li, X. Feng, and C. Chu. The design and construction of reference pangenome graphs with minigraph. *Genome Biology*, 21(1):265–283, 2020.

S. Li et al. CodonBERT: Large language models for mRNA design and optimization. *BioRxiv*, page 2023.09.09.556981, 2023.

Y.-L. Lin, C. Ward, B. Jain, and S. Skiena. Constructing orthogonal de Bruijn sequences. In *Proceedings of the 12th International Conference on Algorithms and Data Structures*, WADS'11, pages 595–606, Berlin, Heidelberg, 2011. Springer-Verlag.

J. Liu, M. Yang, Y. Yu, H. Xu, K. Li, and X. Zhou. Large language models in bioinformatics: applications and perspectives. *ArXiv*, 2401.04155v1, 2024.

C. Ma, H. Zheng, and C. Kingsford. Exact transcript quantification over splice graphs. In C. Kingsford and N. Pisanti, editors, *20th International Workshop on Algorithms in Bioinformatics (WABI 2020)*, volume 172 of *Leibniz International Proceedings in Informatics (LIPIcs)*, pages 12:1–12:18, Dagstuhl, Germany, 2020. Schloss Dagstuhl–Leibniz-Zentrum für Informatik.

D. Maier. The complexity of some problems on subsequences and supersequences. *J. ACM*, 25(2):322–336, 1978.

V. Mäkinen and G. Navarro. Rank and select revisited and extended. *Theor. Comput. Sci.*, 387:332–347, 2007a.

V. Mäkinen and G. Navarro. Implicit compression boosting with applications to self-indexing. In *SPIRE*, pages 214–226, 2007b.

V. Mäkinen, D. Belazzougui, F. Cunial, and A. I. Tomescu. *Genome-scale Algorithm Design*. Cambridge University Press, Cambridge, UK, 2015.

V. Mäkinen, B. Cazaux, M. Equi, T. Norri, and A. I. Tomescu. Linear time construction of indexable founder block graphs. In C. Kingsford and N. Pisanti, editors, *20th International Workshop on Algorithms in Bioinformatics (WABI 2020)*, volume 172 of *Leibniz International Proceedings in Informatics (LIPIcs)*, pages 7:1–7:18, Dagstuhl, Germany, 2020. Schloss Dagstuhl–Leibniz-Zentrum für Informatik.

C. Makris. Wavelet trees: a survey. *Comp. Sci. Inf. Sys.*, 9(2):585–625, 2012.

U. Manber and G. Myers. Suffix arrays: a new method for on-line string searches. In *First Annual ACM-SIAM Symposium on Discrete Algorithms*, pages 319–327, 1990.

U. Manber and G. Myers. Suffix arrays: a new method for on-line string searches. *SIAM Journal on Computing*, 22(5):935–948, 1993.

U. Manber and S. Wu. Fast text searching allowing errors. *Communications of the ACM*, 35(10):83–91, 1992.

G. Manzini. An analysis of the Burrows-Wheeler transform. *J. ACM*, 48(3):407–430, 2001.

G. Marçais, D. Pellow, D. Bork, Y. Orenstein, R. Shamir, and C. Kingsford. Improving the performance of minimizers and winnowing schemes. *Proceedings of ISMB and Bioinformatics*, 33(14):i110–117, 2017.

G. Marçais, D. DeBlasio, and C. Kingsford. Asymptotically optimal minimizers schemes. *Bioinformatics (ISMB)*, 34(13):i13–i22, 2018.

G. Marçais, D. DeBlasio, P. Pandey, and C. Kingsford. Locality sensitive hashing for the edit distance. *Bioinformatics, Proceedings of ISMB*, 35(14):i127–i135, 2019a.

G. Marçais, B. Solomon, R. Patro, and C. Kingsford. Sketching and sublinear data structures in genomics. *Annual Review of Biomedical Data Science*, 2(1):93–118, 2019b.

W. J. Masek and M. S. Paterson. A faster algorithm for computing string edit distances. *J. Comput. System Sci.*, 20(1):18–31, 1980.

E. M. McCreight. A space-economical suffix tree construction algorithm. *Journal of the ACM*, 23(2):262–272, 1976.

M. Middendorf. More on the complexity of common superstring and supersequence problems. *Theoretical Computer Science*, 125:205–228, 1994.

T. Mitchell, W. Cohen, E. Hruschka, P. Talukdar, J. Betteridge, A. Carlson, B. Dalvi, M. Gardner, B. Kisiel, J. Krishnamurthy, N. Lao, K. Mazaitis, T. Mohamed, N. Nakashole, E. Platanios, A. Ritter, M. Samadi, B. Settles, R. Wang, D. Wijaya, A. Gupta, X. Chen, A. Saparov, M. Greaves, and J. Welling. Never-ending learning.

In *Proceedings of the Twenty-Ninth AAAI Conference on Artificial Intelligence (AAAI-15)*, pages 2302–2310, 2015.

J. Moore, R. Boyer, and D. Davies. The 77-Editor. Technical Report 62, Department of Computational Logic, University of Edinburgh, 1973.

M. Muggli, A. Bowe, N. Noyes, P. Morley, K. Belk, R. Raymond, T. Gagie, S. Puglisi, and C. Boucher. Succinct colored de Bruijn graphs. *Bioinformatics*, 33(20):3181–3187, 2017.

M. D. Muggli, B. Alipanahi, and C. Boucher. Building large updatable colored de Bruijn graphs via merging. *Bioinformatics*, 35(14):i51–i60, 2019.

T. Mun, A. Kuhnle, C. Boucher, et al. Matching reads to many genomes with the r-index. *J. Comput. Biol.*, 27:514–518, 2020.

E. Myers. An $O(ND)$ difference algorithm and its variations. *Algorithmica*, 1:251–266, 1986.

E. W. Myers. The fragment assembly string graph. *Bioinformatics*, 21(suppl_2):ii79–ii85, 2005.

E. W. Myers and W. Miller. Optimal alignments in linear space. *Bioinformatics*, 4(1):11–17, 1988.

G. Myers. A fast bit-vector algorithm for approximate string matching based on dynamic programming. *J. ACM*, 46(3):395–415, 1999.

J. Mykkeltveit. A proof of Golomb's conjecture for the de Bruijn graph. *J. Comb. Theory Series B*, 13(1):40–45, 1972.

G. Navarro. Wavelet trees for all. *Journal of Discrete Algorithms*, 25:2–20, 2014.

G. Navarro and J. Tarhio. Boyer-Moore string matching over Ziv-Lempel compressed text. In R. Giancarlo and D. Sankoff, editors, *Combinatorial Pattern Matching*, pages 166–180, Berlin, Heidelberg, 2000. Springer Berlin Heidelberg.

S. B. Needleman and C. D. Wunsch. A general method applicable to the search for similarities in the amino acid sequence of two proteins. *Journal of Molecular Biology*, 48(3):443–453, 1970.

C. Nevill-Manning and I. Witten. Compression and explanation using hierarchical grammars. *The Computer Journal*, 40:103–116, 1997.

A. M. Novak, G. Hickey, E. Garrison, S. Blum, A. Connelly, A. Dilthey, J. Eizenga, M. S. Elmohamed, S. Guthrie, A. Kahles, et al. Genome graphs. *BioRxiv*, 101378, 2017.

R. Nussinov and A. B. Jacobson. Fast algorithm for predicting the secondary structure of single-stranded RNA. *Proceedings of the National Academy of Sciences of the United States of America*, 77(11):6309–6313, 1980.

Y. Orenstein, D. Pellow, G. Marçais, R. Shamir, and C. Kingsford. Designing small universal k-mer hitting sets for improved analysis of high-throughput sequencing. *PLoS Computational Biology*, 13(10):e1005777, 2017.

R. Ostrovsky and Y. Rabani. Low distortion embeddings for edit distance. *Journal of the ACM (JACM)*, 54(5):23–es, 2007.

A. Pagh, R. Pagh, and S. S. Rao. An optimal Bloom filter replacement. In *Proceedings of the Sixteenth Annual ACM-SIAM Symposium on Discrete Algorithms*, SODA '05, pages 823–829. Society for Industrial and Applied Mathematics, 2005.

R. Pagh. Low redundancy in dictionaries with $O(1)$ worst case lookup time. In *Proceedings of ICALP (LNCS)*, volume 1644, pages 595–604, 1999.

P. Pandey, Y. Gao, and C. Kingsford. VariantStore: a large-scale genomic variant search index. *Genome Biology*, 22:231, 2021.

C. H. Papadimitriou. *Computational Complexity*. Addison-Wesley, Reading, MA, 1994.

B. Paten, M. Diekhans, D. Earl, J. S. John, J. Ma, B. Suh, and D. Haussler. Cactus graphs for genome comparisons. *Journal of Computational Biology*, 18(3):469–481, 2011.

B. Paten, A. M. Novak, J. M. Eizenga, and E. Garrison. Genome graphs and the evolution of genome inference. *Genome Research*, 27(5):665–676, 2017.

P. Pevzner. *Computational Molecular Biology: an Algorithmic Approach*. MIT Press, 2000.

P. A. Pevzner, H. Tang, and M. S. Waterman. An Eulerian path approach to DNA fragment assembly. *Proceedings of the National Academy of Sciences of the United States of America*, 98(17), 2001.

S. J. Puglisi, W. F. Smyth, and A. H. Turpin. A taxonomy of suffix array construction algorithms. *ACM Computing Surveys*, 39(2):4, 2007.

Y. Qiu and C. Kingsford. Constructing small genome graphs via string compression. In *Proceedings of ISMB 2021, Bioinformatics*, volume 37, pages i205–i213, 2021.

Y. Qiu, Y. Shen, and C. Kingsford. Revisiting the complexity of and algorithms for the graph traversal edit distance and its variants. *Algorithms Mol. Biol.*, 19:17, 2024.

A. Radford, K. Narasimhan, T. Salimans, and I. Sutskever. Improving language understanding by generative pre-training. Technical report, OpenAI, 2018.

K.-J. Räihä and E. Ukkonen. The shortest common supersequence problem over binary alphabet is NP-complete. *Theoretical Computer Science*, 16(2):187–198, 1981.

R. Raman, V. Raman, and S. R. Satti. Succinct indexable dictionaries with applications to encoding k-ary trees, prefix sums and multisets. *ACM Transactions on Algorithms*, 3(4):43, 2007.

H. Ramesh and V. Vinay. String matching in $\tilde{O}(\sqrt{n} + \sqrt{m})$ quantum time. *J. Discrete Algorithms*, 1(1):103–110, 2003.

M. Roberts, W. Hayes, B. R. Hunt, S. M. Mount, and J. A. Yorke. Reducing storage requirements for biological sequence comparison. *Bioinformatics*, 20(18):3363–3369, 2004.

M. Rossi, M. Oliva, B. Langmead, T. Gagie, and C. Boucher. MONI: a pangenomic index for finding maximal exact matches. *J. Comput. Biol.*, 29(2):169–187, 2022.

I. Sakai. Syntax in universal translation. In *1961 International Conference on Machine Translation of Languages and Applied Language Analysis*, volume II, pages 593–608, 1962.

Y. Sakakibara, M. Brown, R. Hughey, I. S. Mian, K. Sjolander, R. C. Underwood, and D. Haussler. Stochastic context-free grammars for tRNA modeling. *Nuc. Acids Res.*, 22(23): 5112–5120, 1994.

S. Salzberg, A. Delcher, S. Kasif, and O. White. Microbial gene identification using interpolated Markov models. *Nucleic Acids Research*, 26(2):544–548, 1998.

S. Schleimer, D. Wilkerson, and A. Aiken. Winnowing: Local algorithms for document fingerprinting. In *Proceedings of the 2003 ACM SIGMOD International Conference on Management of Data*, pages 76–85, New York, NY, 2003. ACM.

R. Schwartz. *Biological Modeling and Simulation*. MIT Press, Cambridge, MA, 2008.

R. M. Sherman and S. L. Salzberg. Pan-genomics in the human genome era. *Nature Reviews Genetics*, 21(4):243–254, 2020.

R. M. Sherman, J. Forman, V. Antonescu, D. Puiu, M. Daya, N. Rafaels, M. P. Boorgula, S. Chavan, C. Vergara, V. E. Ortega, et al. Assembly of a pan-genome from deep sequencing of 910 humans of African descent. *Nature Genetics*, 51(1):30–35, 2019.

Y. Shiloach. Fast canonization of circular strings. *J. Algorithms*, 2(2):107–121, 1981.

M. Sipser. *Introduction to the Theory of Computation*. Cengage Learning, 3rd edition, 2012.

J. Sirén, N. Välimäki, and V. Mäkinen. Indexing finite language representation of population genotypes. In T. M. Przytycka and M.-F. Sagot, editors, *Algorithms in Bioinformatics*, pages 270–281, Berlin, Heidelberg, 2011. Springer Berlin Heidelberg.

J. Sirén, N. Välimäki, and V. Mäkinen. Indexing graphs for path queries with applications in genome research. *IEEE/ACM Transactions on Computational Biology and Bioinformatics*, 11(2):375–388, 2014.

T. F. Smith and M. S. Waterman. Identification of common molecular subsequences. *Journal of Molecular Biology*, 147(1):195–197, 1981.

B. Solomon and C. Kingsford. Fast search of thousands of short-read sequencing experiments. *Nat. Biotechnol.*, 34(3):300–302, 2016.

B. Solomon and C. Kingsford. Improved search of large transcriptomic sequencing databases using split sequence bloom trees. In *Proceedings of RECOMB 2017: Research in Computational Molecular Biology*, pages 257–271, 2017.

J. A. Storer and T. G. Szymanski. Data compression via textual substitution. *Journal of the ACM (JACM)*, 29(4):928–951, 1982.

C. Sun, R. S. Harris, R. Chikhi, and P. Medvedev. AllSome sequence bloom trees. *Journal of Computational Biology*, 25:467–479, 2018.

R. E. Tarjan and A. C.-C. Yao. Storing a sparse table. *Communications of the ACM*, 22:606–611, 1979.

The Unicode Consortium. The unicode standard, version 6.0.0, 2011. URL http://www.unicode.org/versions/Unicode6.0.0/.

C. Theodoris, L. Xiao, A. Chopra, et al. Transfer learning enables predictions in network biology. *Nature*, 618:616–624, 2023.

I. Tomohiro, J. Kärkkäinen, and D. Kempa. Faster sparse suffix sorting. In E. W. Mayr and N. Portier, editors, *31st International*

Symposium on Theoretical Aspects of Computer Science (STACS 2014), volume 25 of *Leibniz International Proceedings in Informatics (LIPIcs)*, pages 386–396, Dagstuhl, Germany, 2014. Schloss Dagstuhl – Leibniz-Zentrum für Informatik.

J. S. Turner. Approximation algorithms for the shortest common superstring problem. *Information and Computation*, 83(1):1–20, 1989.

E. Ukkonen. On-line construction of suffix trees. *Algorithmica*, 14(3):249–260, 1995.

J. Van Leeuwen. On the construction of Huffman trees. In *ICALP*, pages 382–410, 1976.

A. Vaswani, N. Shazeer, N. Parmar, J. Uszkoreit, L. Jones, A. N. Gomez, L. Kaiser, and I. Polosukhin. Attention is all you need. In *Proceedings of the 31st International Conference on Neural Information Processing Systems*, NIPS'17, page 6000–6010, 2017.

A. Viterbi. A personal history of the Viterbi algorithm. *IEEE Signal Processing Magazine*, 23(4):120–142, 2006.

A. J. Viterbi. Error bounds for convolutional codes and an asymptotically optimum decoding algorithm. *IEEE Trans. Inform. Theory*, IT-13:260–269, 1967.

VS Code, 2018. https://code.visualstudio.com/blogs/2018/03/23/text -buffer-reimplementation.

R. A. Wagner and M. J. Fischer. The string-to-string correction problem. *J. ACM*, 21(1):168–173, 1974.

L. Wang and T. Jiang. On the complexity of multiple sequence alignment. *J. Comp. Biol.*, 1(4):337–348, 1994.

Q. Wang and M. Ying. Quantum algorithm for lexicographically minimal string rotation. *CoRR*, page abs/2012.09376, 2020.

P. Weiner. Linear pattern matching algorithms. In *14th Annual IEEE Symposium on Switching and Automata Theory*, pages 1–11, 1973.

D. Wood and S. Salzberg. Kraken: Ultrafast metagenomic sequence classification using exact alignments. *Genome Biol.*, 15(3):R46, 2014.

D. Wood, J. Lu, and B. Langmead. Improved metagenomic analysis with Kraken 2. *Genome Biol.*, 20(1):257, 2019.

D. H. Younger. Recognition and parsing of context-free languages in time n^3. *Information and Control*, 10(2):189–208, 1967.

H. Zheng, C. Kingsford, and G. Marçais. Improved design and analysis of practical minimizers. *Bioinformatics*, 36(Suppl 1):i119–i127, 2020.

H. Zheng, C. Kingsford, and G. Marçais. Sequence-specific minimizers via polar sets. *Bioinformatics*, 37(Suppl 1):i187–i195, 2021a.

H. Zheng, C. Kingsford, and G. Marçais. Lower density selection schemes via small universal hitting sets with short remaining path length. *J. Comput. Biol.*, 28(4):395–409, 2021b.

H. Zheng, C. Kingsford, and G. Marçais. Sequence-specific minimizers via polar sets. *Bioinformatics*, 37(Supplement_1):i187–i195, 2021c.

H. Zheng, G. Marçais, and C. Kingsford. Creating and using minimizer sketches in computational genomics. *Journal of Computational Biology*, 30:1–26, 2023.

J. Ziv and A. Lempel. A universal algorithm for sequential data compression. *IEEE Transactions on Information Theory*, 23:337–343, 1977.

List of Algorithms

Index

GPSR Authorized Representative: Easy Access System Europe - Mustamäe tee
50, 10621 Tallinn, Estonia, gpsr.requests@easproject.com